International Conference and Exhibition in Advanced Materials and Microscopy (ICEAMM 2017)

IOP Conference Series: Materials Science and Engineering
Volume 343

Kuta, Indonesia
25 - 26 October 2017

Editors:

Ir. Haftirman
Haris Wahyudi

ISBN: 978-1-5108-6119-0
ISSN: 1757-8981

Printed from e-media with permission by:

Curran Associates, Inc.
57 Morehouse Lane
Red Hook, NY 12571

Some format issues inherent in the e-media version may also appear in this print version.

Printed by Curran Associates, Inc. (2018)

For permission requests, please contact the Institute of Physics
at the address below.

Institute of Physics
Dirac House, Temple Back
Bristol BS1 6BE UK

Phone: 44 1 17 929 7481
Fax: 44 1 17 920 0979

techtracking@iop.org

Additional copies of this publication are available from:

Curran Associates, Inc.
57 Morehouse Lane
Red Hook, NY 12571 USA
Phone: 845-758-0400
Fax: 845-758-2633
Email: curran@proceedings.com
Web: www.proceedings.com

Table of contents

Volume 343

The International Conference and Exhibition in Advanced Materials and Microscopy 2017 (ICEAMM 2017)

25–26 October 2017, Bali, Indonesia

Accepted papers received: 19 March 2018
Published online: 4 April 2018

Preface

Papers

International Conference & Exhibition in Advanced Materials and Microscopy 2017

ICEAMM 2017

25-26 October 2017, Kuta-Bali, Indonesia

Organized by

Department of Mechanical Engineering, Faculty of Engineering, Mercu Buana University

In cooperation with

Indonesian Society for Microscopy (InaSM) and Committee of Asia Pacific Societies for Electron Microscopy (CAPSEM)

IOP Conf. Series: Materials Science and Engineering **343** (2018) 011001 doi:10.1088/1757-899X/343/1/011001

Editorial Board

Dr. Ir. Haftirman, M. Eng
Department of Mechanical Engineering, Universita Mercu Buana
E-mail: haftirman@gmail.com

Ir. Haris Wahyudi, M.Sc
Department of Mechanical Engineering, Universita Mercu Buana
E-mail: haris.wahyudi@mercubuana.ac.id

Organizing Committee

Advisory Committee
- Assoc. Prof. Dr.-Ing. Yupiter HP Manurung (UiTM Malaysia)
- Prof. Dr. Hing Han Lian (CAPSEM)
- Assoc. Prof. Dr. Edy Herianto Majlan (UKM Malaysia)
- Prof. Dr.-Ing. Bambang Suharno (Universitas Indonesia)
- Assoc. Prof. Dr. Muhaimin Ismoen (ITB Brunei)
- Prof. Dr. Sulaiman Hasan (Malaysia)
- Dato' Prof. Dr. Musa Ahmad (USIM Malaysia)
- Assoc. Prof. Dr. Gunawan Witjaksono (UTP, Malaysia)
- Prof. Dr. Seong Cho Yu (CBNU Korea)
- Prof. Dr. Dong Seok Yang (CBNU Korea)
- Prof. Dr. Erry Yulian Triblas Adesta (UIA Malaysia)
- Prof. Dr. Sreekantha Reddy Dugasani (Sunkyunk Kwan University Korea)
- Assoc. Prof. Dr. Phan The Long. (HUFS Korea)

Executive Committee (Patron)
- Dr. Arissetyanto Nugroho, MM (Rector UMB)
- Ir. Nasri Sebayang (Director PT. PLN Sumatera-Bali)
- Dr.-Ing. Pudji Untoro (Surya University Indonesia, Committee Asia Pacific Microscopy)
- Dr.-Ing. Arbi Dimyati (Indonesian Society of Microscopy, InaSM)
- Poempida Hidayatullah, PhD (BPJS)

General Chairs
- Prof. Dana Santoso Saroso, MEng. Sc. Ph.D
- Prof. Ngadino Surip
- Dr. Purwanto
- Dr. Danto Sukmajati
- Dr. Sagir Alva

Conference Chair
- Prof. (Em). Dr.-Ing. Ir. Darwin Sebayang

Conference Secretaries
- Dr. Ir. Haftirman, M.Eng
- Dr. Kontan Tarigan

Technical Program Chairs
- Ir. Haris Wahyudi, M.Sc
- Dr. Ir. Ignatius Agung Wibowo, M.Sc
- Dr. Rita Sundari, M.Sc

Publication and Publicity Committee
- Nur Indah, S.ST., MT
- Swandya Eka Pratiwi, ST. M.Sc
- Alief Avicenna Luthfie, ST., M.Eng

Treasurer
Feby Febriyanti

IOP Conf. Series: Materials Science and Engineering **343** (2018) 011001 doi:10.1088/1757-899X/343/1/011001

Preface

ICEAMM 2017 is the International Conference & Exhibition in Advanced Materials and Microscopy 2017 (ICEAMM 2017) held on 25th-26th October 2017 at Kuta, Bali, Indonesia. ICEAMM 2017 was organized by Department of Mechanical Engineering, Faculty of Engineering, Mercu Buana University in cooperation with Indonesian Society for Microscopy (InaSM) and Committee of Asia Pacific Societies for Electron Microscopy (CAPSEM). At the same time, the conference was conducted to commemorate the National Technology of Indonesia (HAKTEKNAS).

The conference aims to provide a high level national and international research for scholars, scientists, researchers, and engineers to present their results in the field of materials science and engineering mainly in the advanced materials, microscopy and mechanical engineering. The event was also intended to have collaboration as well as build networks among participants and committees.

The issue covers all the aspects of materials and mechanical engineering, from characterization, synthesize of materials to applications and technologies for materials and mechanical engineering. All papers have been reviewed by appropriate expert relevant to their topic disciplines.

The editors hope that this issue will provide the reader a broad overview of the recent field of materials science and engineering, and that they will be a valuable references source for further research. The editors would like to express their sincere appreciations and thanks to all the committee members of the ICEAMM 2017 for their excellent efforts. Thanks also to IOP Conference Series for producing the issue with full articles.

Conference Chair
Prof. (Em). Dr.-Ing. Ir. Darwin Sebayang

IOP Conf. Series: Materials Science and Engineering **343** (2018) 011001 doi:10.1088/1757-899X/343/1/011001

IOP Conf. Series: Materials Science and Engineering **343** (2018) 011001 doi:10.1088/1757-899X/343/1/011001

IOP Conf. Series: Materials Science and Engineering **343** (2018) 011001 doi:10.1088/1757-899X/343/1/011001

Peer review statement

All papers published in this volume of *IOP Conference Series: Materials Science and Engineering* have been peer reviewed through processes administered by the proceedings Editors. Reviews were conducted by expert referees to the professional and scientific standards expected of a proceedings journal published by IOP Publishing.

IOP Conf. Series: Materials Science and Engineering **343** (2018) 012001 doi:10.1088/1757-899X/343/1/012001

Magnetic properties near the ferromagnetic-paramagnetic transformation in the austenite phase of $Ni_{43}Mn_{44}X_2Sn_{11}$ (X = Fe and Co) Heusler alloys

W Z Nan[1], T D Thanh[1], T S You[2], H G Piao[3], S C Yu[1]

[1]Department of Physics, Chungbuk National University, Cheongju 28644, South Korea
[2]Department of Chemistry, Chungbuk National University, Cheongju 28644, South Korea
[3]College of Science, China Three Gorges University, Yichang 443002, China

scyu@chungbuk.ac.kr

Abstract. In this work, we present a detail study on the magnetic properties in the austenitic phase (A phase) $Ni_{43}Mn_{44}X_2Sn_{11}$ alloy with X = Fe and Co, which were prepared by an arc-melting method in an argon atmosphere. The *M*(*T*) curves of two samples exhibits a single magnetic phase transition at the Curie temperature of the ferromagnetic (FM) austenitic phase with T_C^A = 298 K and 334k for (X = Fe and Co) respectively. Based on the Landau theory and *M*(*H*) data measured at different temperatures, we found that the FM-PM phase transitions around T_C^A in both samples were the second-order phase transition. Under an applied field change of 30 kOe, around T_C^A, the magnetic entropy changes were found to be 0.66 J Kg^{-1} K^{-1} and 1.62 J Kg^{-1} K^{-1} for (X = Fe and Co) respectively.

1.Introduction

Heusler alloys have attracted lots of attention since the first discovery of the ferromagnetic (FM) Cu-Mn-based alloys by Heusler et. al. [1] in 1903. It was also found that the composition, crystal structure, and sample-preparation conditions are important parameters for determining the magnetic properties of these alloy systems. In recent years, Heusler alloys have been of intensive attention due to their physical properties related to martensitic distortion, ferromagnetism and various exotic phenomena [2-4], which are suitable for high-density magnetic recording technology and magneto-optic applications. Besides, the metamagnetic shape memory effect at room temperature was also found [2], together with many interesting effects derive from the unique transformation of these Heusler alloys [5–7]. Among these, the magnetocaloric effect (MCE) of the Ni-Mn-Z (Z = In, Sn, Sb) Heusler alloys is one of the most interesting properties. It has been widely studied and can be referenced in a large number of publications [8,9]. It also reported that materials undergoing a first-order phase transition (FOPT) show giant MCE with a large ΔS_m value. However, this MCE only occurs in a narrow temperature range. In contrast, materials with a second-order phase transiton (SOPT) offer moderate ΔS_m values, but these ΔS_m values are stable over a wide temperature range so their refrigerant capacity (RC) values can be larger than those of FOPT.

The structures and physical properties of Heusler alloys are thus very sensitive to their compositions

[10,11,12]. Krenke et al. [10] have found a close relationship between structural and magnetic transformations of the Ni-Mn-Sn systems and their valence electron concentrations per atom ratio (*e/a*), which is the concentration-weighted sum of *s*, *d*, and *p* valence electrons of Ni ($3d^8 4s^2$), Mn ($3d^5 4s^2$), and Sn ($5s^2 5p^2$) atoms. According to Tan et al. [13], the anti-FM interactions in the M phase appear mainly due to the hybridization between the Ni and Mn atoms and obviously, the Ni-Mn distance affect strongly on the strength of this hybridization. Several previous reports pointed out that the magnetic properties in the vicinity the FM-PM transition in the austenitic phase (A phase) of Ni-Mn-Sn systems is also very sensitive to the composition and the value of *e/a*. The short-range FM order in the A phase of $Ni_{50}Mn_{37}Sn_{13}$ will be modified to the long-range FM order by a small change in sn-content [12] or by a partial replacement of Ni by Cu [14], Gd [15]. In this work, we present the magnetic properties and MCE in the austenitic phase (A phase) of $Ni_{43}Mn_{44}X_2Sn_{11}$ alloys with (X = Fe and Co).

2.Experimental detail

Two alloy ingot samples of $Ni_{43}Mn_{44}X_2Sn_{11}$ alloys with (X = Fe and Co) were prepared by arc-melting under an Ar atmosphere using Ni, Mn, Sn, Fe, and Co metals (99.9% purity) and then, annealed at 1323 K for 48h. The compositions of the produces were check by the energy dispersive X-ray (EDX) spectroscopy. Crystal structures of the title compounds were checked at room temperature by a powder X-ray diffractometer using a Cu-Ka radiation source. The magnetization measurements vs. temperature as much as magnetic field were performed on a vibrating sample magnetometer (Versa-Lab, Quantum design) using a warming mode. Here, the temperature interval between magnetic field isotherms was 2 K near the phase transition temperature.

3. Results and discussion

Figure 1 show the EDX spectra of the samples recorded at room temperature attached with the scanning electron microscopy (SEM) images. The composition of the samples has been analyzed on the large areas and showed in Table 1. This confirms that the values of the atomic percentage determined based on EDX spectra are quite close to the prescribed ratio, within experimental error. Based on EDX results, we calculated the valence electron concentrations per atom *e/a* for the samples, thus *e/a* = 7.95 and 7.97 for X = Fe and Co, respectively. These values are very close to those obtained from the nominal compositions (*e/a* = 7.98 and 8.00 for X = Fe and Co, respectively). X-ray diffraction patterns (not show here) obtained at room temperature indicate that the samples are single phase products adopting the cubic $L2_1$ structure type (space group Fm3m) belonging to the A phase. Corresponding to the valence electron concentrations per atom *e/a* = 7.98–8.00. These values are consistent with those reported by Krenke et al. [10] that the structure of Ni-Mn-Sn Heusler alloys at room temperature is cubic $L2_1$ if their *e/a* in the range of 7.723–8.041.

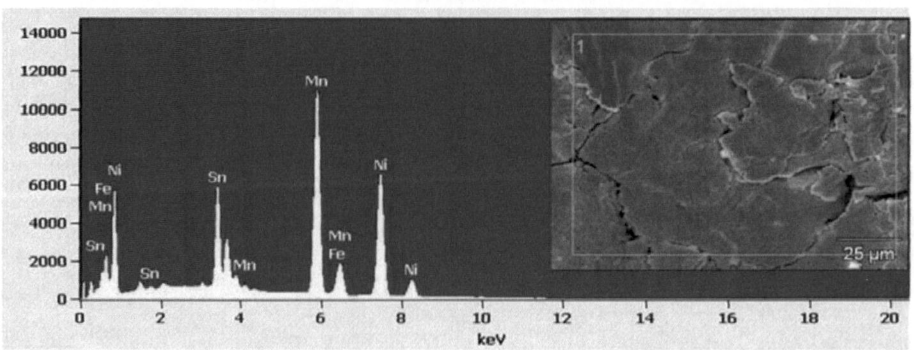

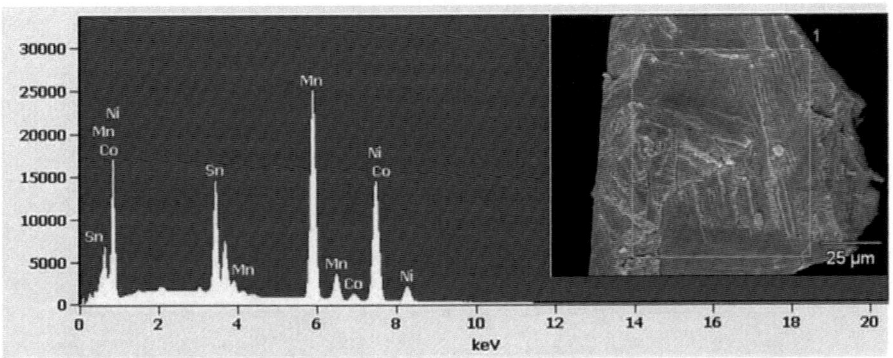

Figure 1. EDX spectra attached with the SEM images for (a) Fe and (b) Co sample.

Table 1. Atomic percentage of the chemical species and the valence electron concentrations per atom *e/a* for $Ni_{43}Mn_{44}X_2Sn_{11}$ alloys deduced from the EDX analyses and their nominal compositions.

Sample	EDX analysis results					Normal	
	Ni (%)	Mn (%)	Sn (%)	X (%)	*e/a*	Compositon	*e/a*
Fe	42.18	44.65	11.16	2.01	7.95	$Ni_{43}Mn_{44}Fe_2Sn_{11}$	7.98
Co	42.41	44.23	11.29	2.07	7.97	$Ni_{43}Mn_{44}Co_2Sn_{11}$	8.00

Following the structure analyses, we have investigated the magnetic properties of the $Ni_{43}Mn_{44}X_2Sn_{11}$ alloys with (X = Fe and Co). As shown in Figure 2 temperature dependences of zero-field-cooled (ZFC) magnetization *M* (*T*) curve, for $Ni_{43}Mn_{44}X_2Sn_{11}$ alloys with X = Fe and Co alloys under an applied field H= 100 Oe. Clearly there only a FM-PM phase transition of the austenitic phase (A phase) taking place at T_C^A=298 K and 334 K for (X = Fe and Co) respectively. It can be related to the strength of the hybridization between the Ni and Mn atoms may be reduced as the Fe and Co substitution occurs, which can be contributed to the suppression of the anti-FM interaction in the alloy [11]. A similar behavior was also observed by Ag substitution, or increasing Sn concertration up to 18% [10] in Ni-Mn-based Heusler alloys. The nature of FM-PM phase transition, and how does the FM interacts in the A phase of these alloys should be clarified. For this reason, we measured the isothermal magnetization *M*(*H*) curves at different temperatures around their T_C^A.

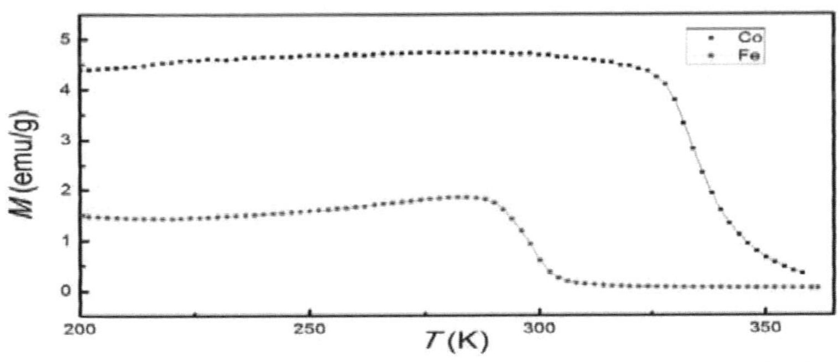

Figure 2. *M* (*T*) curves of the $Ni_{43}Mn_{44}X_2Sn_{11}$ alloys for (a) Fe and (b) Co sample.

Figure 3 All the isothermal *M*(*H*) curves show a progressive decrease of the magnetization values, and the nonlinear *M*(*H*) curves become linear as increasing temperature, which is a signature of the FM-PM transition in the vicinity of T_C^A. Following contents, we shall use T_C instead of using T_C^A to mention the Curie temperature of the A phase in the samples. To understand the nature of the magnetic phase transition in the A phase of $Ni_{43}Mn_{44}X_2Sn_{11}$ alloys, we have plotted M^2 *vs.* *H/M* curves [16] based on the isothermal *M(H)* curves measured at different temperatures, see Figure 4 We can see that M^2 *vs.* *H/M* curves at high field are quasi linear and nearly parallel. However, at low field region, M^2 *vs.* *H/M* curves are nonlinear and they are driven towards two opposite directions revealing to the FM and PM states.

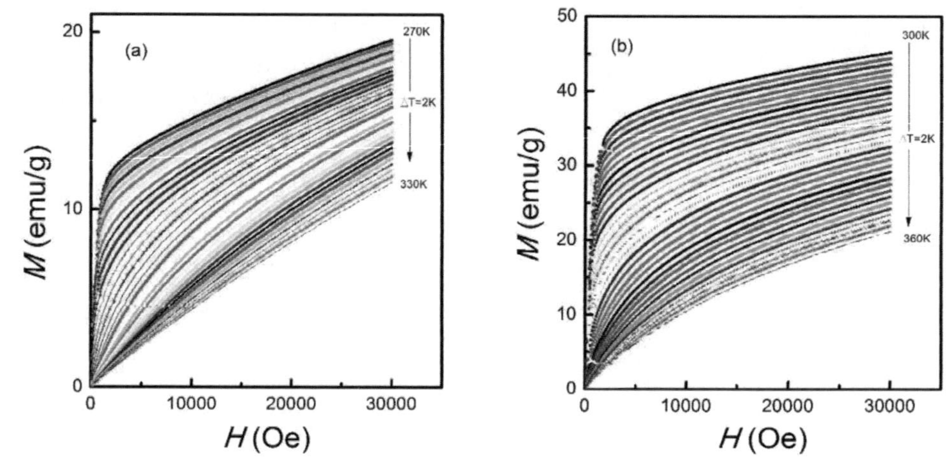

Figure 3. *M*(*H*) curves of the $Ni_{43}Mn_{44}X_2Sn_{11}$ alloys for (a) Fe and (b) Co sample at selected temperatures.

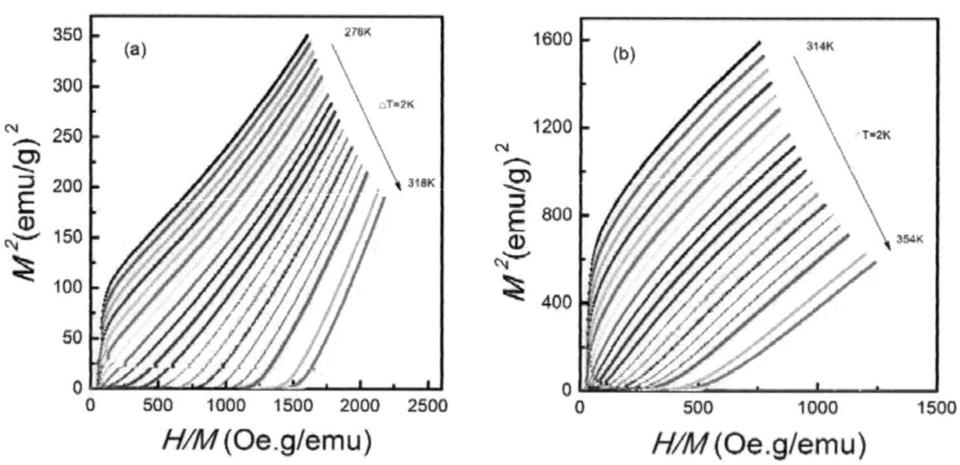

Figure 4. *H/M* versus M^2 of the $Ni_{43}Mn_{44}X_2Sn_{11}$ alloys for (a) Fe and (b) Co sample at selected temperatures.

Basing on the *M*(*H*) data above, the MCE can be measured directly or it can be indirectly from the measured magnetization. Magnetization measured experimentally as a function of temperature has been rightfully suggested as a useful technique. The magnetic entropy change caused by the variation of the external magnetic field. On the basis of the thermodynamic theory, the magnetic entropy change caused by the variation of the external magnetic field from 0 to H_{max} is given by

$$\Delta S_M = \int_0^{H_{\max}} (\frac{\partial S}{\partial H})_T dH \quad (1)$$

From the Maxwell's thermodynamic relationship

$$(\frac{\partial S}{\partial H})_T = (\frac{\partial M}{\partial T})_H \quad (2)$$

Equation (1) can be rewritten as follows:

$$\Delta S_M = \int_0^{H_{\max}} (\frac{\partial M}{\partial T})_H dH \quad (3)$$

Numerical evaluation of the magnetic entropy change was carried out from formula (3) using isothermal magnetization measurements at small discrete field and temperature intervals, can be computed approximately from (3) by

$$|\Delta S_M| = \sum_i \frac{M_i - M_{i+1}}{T_{i+1} - T_i} \Delta H \quad (4)$$

where M_i and M_{I+1} are the magnetization values measured at temperature T_i and T_{i+1} in a field H, respectively. Thus, the magnetic entropy changes associated with applied field variations can be calculated from Eq. (4). Figure 5 shows ΔS_m (T) curves of the $Ni_{43}Mn_{44}X_2Sn_{11}$ alloys with X = Co for the applied fields ranging from 5 to 30 kOe. Magnetic entropy changes were found to be 0.66 J Kg^{-1} K^{-1} and 1.62 J Kg^{-1} K^{-1} for X = Fe and Co respectively. An increase of H enhances absolute values of ΔS_m Around the temperatures T_C^A, ΔS_m reaches to the maxima. The normal MC effect around the austenitic FM–PM transition is considered to be more helpful for magnetic refrigeration [2,14,15]. Materials with a second-order phase transiton (SOPT) offer moderate ΔS_m values, but these ΔS_m values are stable over a wide temperature range. The above results prove that, among prepared $Ni_{43}Mn_{44}X_2Sn_{11}$ alloys, X = Co sample gives more optimal features helpful for magnetic refrigeration applications around room temperature compared with the X = Fe sample.

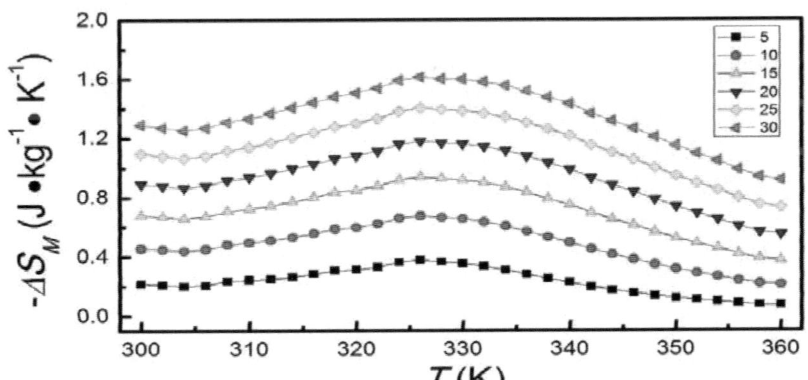

Figure 5. ΔS_m (T) curves with different applied fields (H = 5–30 kOe) for the $Ni_{43}Mn_{44}X_2Sn_{11}$alloy with X = Co.

4. Conclusion

We prepared $Ni_{43}Mn_{44}X_2Sn_{11}$ alloys (X = Fe and Co) ingots by arc-melting, and then studied their magnetic properties and MCE around the FM-PM phase transition of the A phase. We pointed out that

Fe and Co substitution contributed to the suppression the anti-FM interaction. Even the magnetic entropy changes not so high, the Curie temperature both around room temperature. There seems no magnetic contribution of the A phase in the sample X = Fe compared with X = Co. Such the features make it (the alloy with X = Co) become a promising candidate for refrigeration applications around room temperature.

5. Acknowledgment

This work was supported by the National Natural Science Foundation of China (Grant Nos. 11474183 and 51371105).

6. References

[1] Heusler F, Starck W and Haupt E 1903 *Verh. Phys. Ges.* **5** 220
[2] Kainuma R, Imano Y, Ito W, Sutou Y, Morito H, Okamoto S, Kitakami O, Oikawa K, Fujita A, Kanomata T and Ishida K 2006 *Nature* **439** 957
[3] Liu J, Gottschall T, Skokov K P, Moore J D and Gutfleisch O 2012 *Nat. Mater.* **11** 620
[4] Taulats E S, Castillo-Villa P O, Manosa L, Frontera C, Pramanick S, Majumdar S and Planes A 2014 *J. Appl. Phys.* **115** 173907
[5] Krenke T, Duman E, Acet M, Wassermann E F, Moya X, Manosa L and Planes A 2005 *Nat. Mater.* **4** 450
[6] Buchelnikov V D and Sokolovskiy V V 2011 *Phys. Met. Metallogr.* **112** 633
[7] Koyama K, Okada H, Watanabe K, Kanomata T, Kainuma R, Ito W, Oikawa K and Ishida K 2006 *Appl. Phys. Lett.* **89** 182510
[8] Zhang A, Zhang X X, Yu S Y, Chen J L, Cao Z X and Wu G H 2007 *Appl. Phys. Lett.* **91** 012510
[9] Biswas A, Phan T L, Dan N H, Zhang P, Yu S C, Srikanth H and Phan M H 2013 *Appl. Phys. Lett.* **103** 162410
[10] Krenke T, Acet M, Wassermann E F, Moya X, Manosa L and Planes A 2005 *Phys. Rev. B* **72** 014412
[11] Khan M, Dubenko I, Stadler S, Jung J, Stoyko S S, Mar A, Quetz A, Samanta T, Ali N and Chow K H 2013 *Appl. Phys. Lett.* **102** 112402
[12] Phan T L, Zhang P, Dan N H, Yen N H, Thanh P T, Thanh T D, Phan M H and Yu S C 2012 *Appl. Phys. Lett.* **101** 212403
[13] Tan C L, Huang Y W, Tian X H, Jiang J X and Cai W 2012 *Appl. Phys. Lett.* **100** 132402
[14] Huu D T, Yen N H, Thanh P T, Mai N T, Thanh T D, Phan T L, Yu S C and Dan N H 2015 *J. Alloys Compd.* **622** 535
[15] Zhang P, Phan T L, Dan N H, Thanh T D and Yu S C 2014 *J. Alloys Compd.* **615** S335
[16] Arrott A and Noakes J E 1967 *Phys. Rev. Lett* **19** 786

Sol gel method for synthesis of semiconducting ferrite and the study of FTIR, DTA, SEM and CV

Sagir Alva[1], Tang Ing Hua[2], Umar Kalmar Nizar[3], Haris Wahyudi[1], and RitaSundari[1*]

[1] Faculty of Engineering, Universitas Mercu Buana, 11650 Jakarta, Indonesia
[2] Department of Chemistry, Faculty of Science, Universiti of Technologi Malaysia, 81310 Skudai, Johor, Malaysia
[3]Department of Chemistry, Universitas Negeri Padang, 25131 Padang, Indonesia

ritsun2003@yahoo.com

Abstract. In this study, a sol gel method using citric acid as anionic surfactant is used for synthesis of magnesium ferrite. Calcinations of magnesium ferrite at temperature (300°C, 600°C and 800°C) have been conducted after sol gel process. Characterization study of the prepared magnesium ferrite related to calcinations using Fourier transform infrared spectrometry (FTIR), Differential thermogravic analysis (DTA), and Scanning electron microscope (SEM) has been discussed. The study of Cyclic voltammetry (CV) of the prepared magnesium ferrite has been examined to assay the semiconducting behavior of magnesium ferrite in relation to its electrochemical behavior.

1. Introduction

Previous publications reported that ferrites were used for many applications such as composites, heterogeneous catalysts, magnetic materials, sensors and semi conductors [1 – 5]. The electrical and magnetic properties of ferrites depend on their size, shape, crystalline structure and cationic distributions, which can be controlled during fabrication process [3]. Furthermore, ferrites with their general formula, MFe_2O_4 (M= Co, Zn, Mg, etc.), possess important electrical and magnetic properties, where they form close packed face-centered cubic (fcc) crystals while the M^{2+} and Fe^{3+} ions can form either tetrahedral or octahedral structures placed inside the fcc unit cell [3,6].

The fabrication of ferrites has drawn much greater interest because ferrites show remarkable properties in many applications. A number of methods have been developed to fabricate ferrites including electrospinning [2,7], co-precipitation [8,9], high energy ball milling [10], solid state reaction [11], and oxidative thermal decomposition [12]. To the best of our knowledge, the sol–gel method yields more promising results in the synthesis of ferrites because this method gives many advantages with respect to good stoichiometric control and particle size distribution at fairly low temperature [6,9]. The study of characterizations of ferrites would provide a meaningful correlation of their structures and effective changes in the environment of ions related to different calcinations temperatures [3]. Therefore, we wish to expose the application of sol gel method using citric acid as anionic surfactant for the synthesis of ferrite materials with respect to magnesium ferrite as an object of this study. The study of characterizations using FTIR spectroscopy to examine functional groups, DTA analysis to investigate thermal behavior of magnesium ferrite, and SEM mapping to study surface morphology are included in this study. To assay the electrical properties of the prepared magnesium ferrites, a cyclic voltammetry (CV) examination will be used to highlight the electrochemical behavior of the fabricated magnesium ferrites.

2. Experimental

2.1.Materials

All chemicals were purchased and used without further purification. Deionized distilled water was used throughout the experiments. Magnesium nitrate hydrate (HmbG, 99.5%), iron nitrate hydrate (OReC, AR grade), citric acid monohydrate (OReC, AR grade), and sodium hydrogen phosphate dihydrate (GCE, AR grade) were used in this study.

2.2.Methods

2.2.1. Synthesis of Mg-ferrite. The magnesium ferrite was prepared through the method of Hua and Sundari [3]. Iron nitrate hydrate, magnesium nitrate hydrate and citric acid monohydrate were mixed and dissolved in deionized distilled water and made up to 100 mL in order to obtain concentration ratio of 0.5 M : 0.25 M : 0.5 M. The solution mixture was stirred and heated at 60°C for 3h and followed by heating up to 80°C until the mixture changed to gel form. Then the gel was dried in an oven at 100°C for 24h and followed by calcinations in a muffle furnace at 300, 600 and 800°C for 2h, respectively.

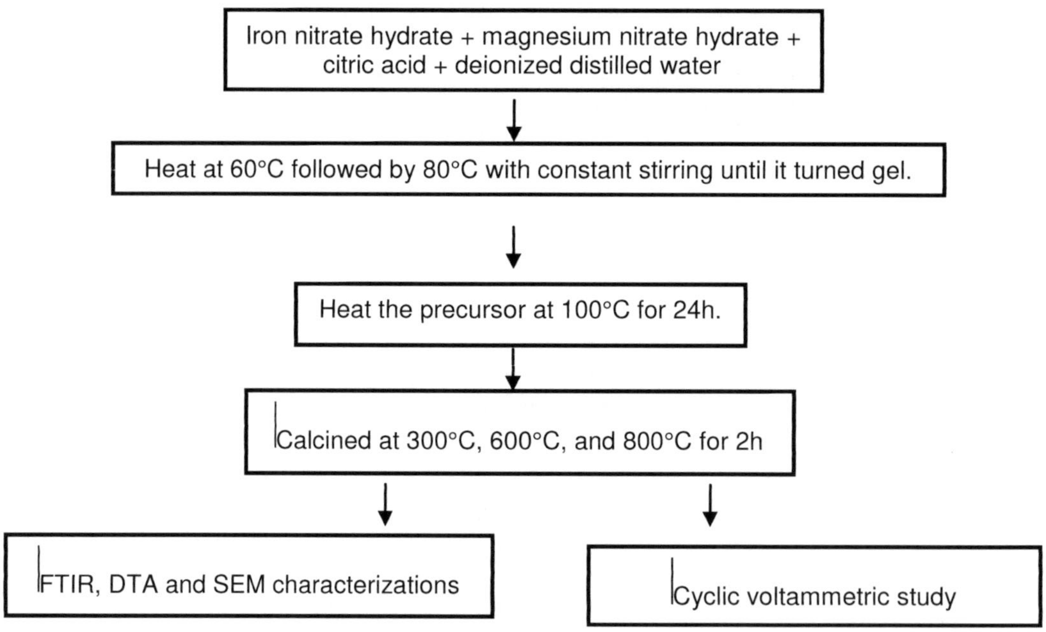

Figure 1. The sol gel process for synthesis of magnesium ferrite using citric acid followed by calcinations.

2.2.2. Characterizations. The prepared magnesium ferrite was characterized using the following instruments. The functional group of the magnesium ferrite was determined by Fourier transform infrared spectrometer (FTIR, Perkin Elmer) using KBr pelletizer as reference for running samples of 20 scans per run. The exothermic and endothermic decomposition of the magnesium ferrite was examined by Differential thermogravic analyzer (DTA, Mettler Toledo TGA/SDTA 851) with nitrogen flow rate at 10 scale units per minute. The surface morphology of the magnesium ferrite was examined by Scanning electron microscope mapping (SEM, Philips XL) using an accelerating voltage of 25.0 kV integrated with Energy dispersive X-ray (EDX) to determine atomic distribution on material surface.

2.2.3 Semi conducting test. The semiconducting test was used to assay the electrical properties of the prepared magnesium ferrite. The test was based on the cyclic voltammetry (CV) performance executed by a system of three electrode configuration transduction, i.e. (i) the Ag/AgCl/KCl reference electrode, (ii) the glassy carbon counter electrode, and (iii) the screen printed carbon (SPC) working electrode attached by the prepared magnesium ferrite. All the electrodes immersed in a 5 mL beaker glass filled by a phosphate buffer solution. An autolab PGSTAT 30 using a GPES software was applied to perform cyclic voltammograms (CVs) recorded between -0.5V and +0.5V with a scan rate of 0.02 V/s.

3. Results and discussion

3.1 Functional group of magnesium ferrite

The functional group of the prepared magnesium ferrites were determined by FTIR spectroscopy. As shown in Figure 2, the prepared magnesium ferrite (100°C) showed the presence of strong stretching vibration of O-H group at 3421 cm^{-1}, whereas the calcinations temperature increased the stretching vibration of O-H group gradually disappeared. A strong absorption peak at 1400 cm^{-1} (spectrum of 100°C) attributed to trace CO_2 adsorption related to oxidation of citric acid during sol gel process. When magnesium ferrite was calcined at 300°C, some functional groups were eliminated attributed by the absence of several absorption peaks except for O-H and NO_3- groups in the range of 1000 – 400 cm^{-1}, which were retained in the tetrahedral or octahedral structure due to stronger chemical bonding. This phenomenon is in agreement with previous investigations [3 – 4, 6, 13].

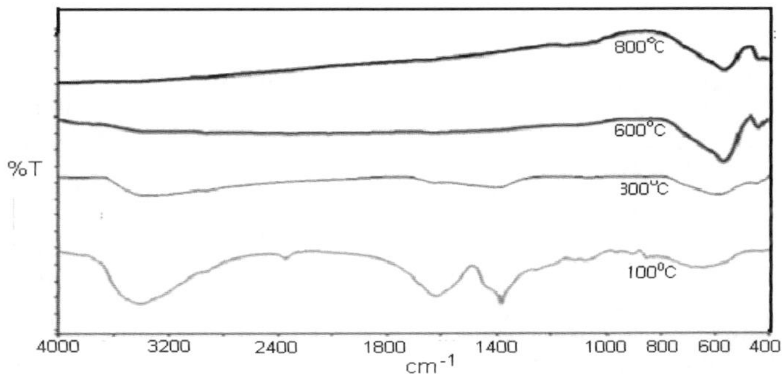

Figure 2. FTIR spectra of magnesium-ferrites. Sol gel method using citric acid at 100°C followed by calcinations at 300°C, 600°C and 800°C.

3.2Thermal properties of magnesium ferrite

The exothermic or endothermic decomposition of the prepared magnesium ferrite was investigated by DTA examination. As shown in Figure 3, the sol gel process at 80 – 100°C showed an endothermic dehydration attributed by H_2O and CO_2 removal. The DTA profile showed a flat line when the calcination of magnesium ferrite increased to 800°C due to exothermic decomposition of C-N bond and removal of NO and NO_2 groups. A similar phenomenon is also found in the study of Lick and Soria [12] in the synthesis of manganese ferrite, as well as in the study of ferrite using electro spinning method [2].

IOP Conf. Series: Materials Science and Engineering **343** (2018) 012002 doi:10.1088/1757-899X/343/1/012002

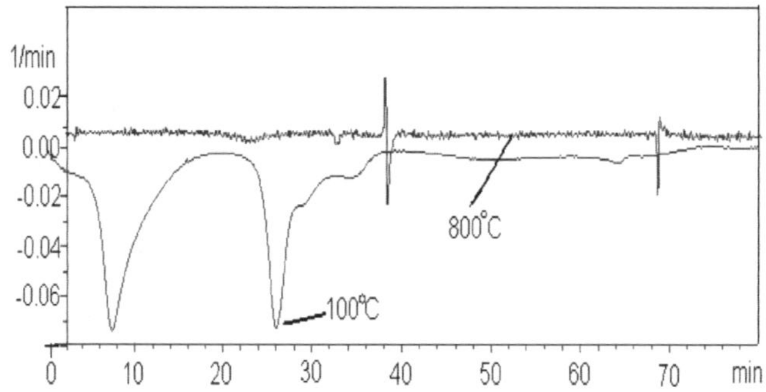

Figure 3. The DTA profiles of magnesium-ferrite at 100°C and 800°C. Sol gel method using citric acid followed by calcinations.

3.3 Surface morphology of magnesium ferrite

The surface analysis of the prepared magnesium ferrite was examined by SEM mapping. Figure 4 showed the SEM mapping of magnesium ferrite after sol gel process at 80-100°C (Fig. 3a). When magnesium ferrite was calcined until 600°C, the density of iron distribution on material surface increased from 26.2% to 50.6% as recorded by the EDX data (Fig. 3b). The white dots in Figure 3 indicate iron particles scattering over the material surface. High temperature calcination (600°C) yielded stronger thermal agitation causing higher iron distribution on the material surface [3].

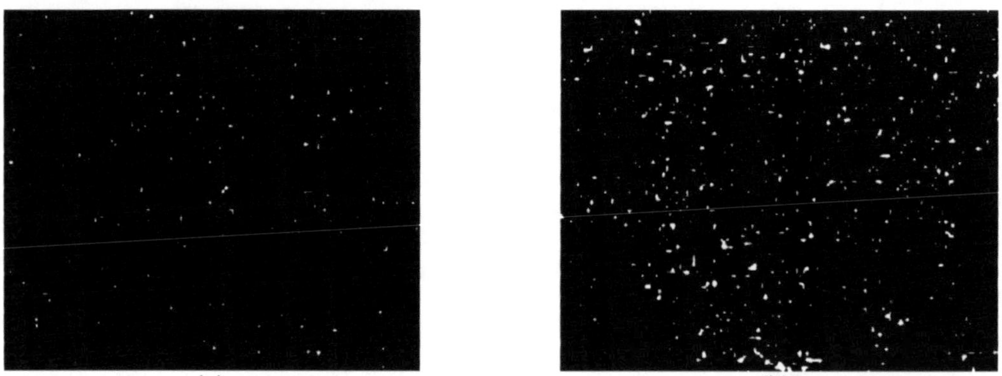

(a) (b)

Figure 4. SEM mapping of magnesium-ferrite for iron distribution at 100°C (a) and 600°C (b). Sol gel method using citric acid followed by calcinations. White dots indicated iron particles scattering.

3.4Semi conducting test of magnesium ferrite

The electrical properties of prepared magnesium ferrite were demonstrated by CV profiles related to electrochemical behavior of magnesium ferrite. Figure 5 showed CV profiles of magnesium ferrite calcined at 300°C and 600°C, respectively. Both reduction and oxidation peaks of the CV profile of 600°C – magnesium ferrite was shown stronger than that of 300°C – magnesium ferrite caused by higher calcination temperature induced higher electrical conductivity of the ferrite to some extent. This phenomenon is similar to that one of earlier investigation using magnesium ferrite doping by nickel and cobalt elements [5].

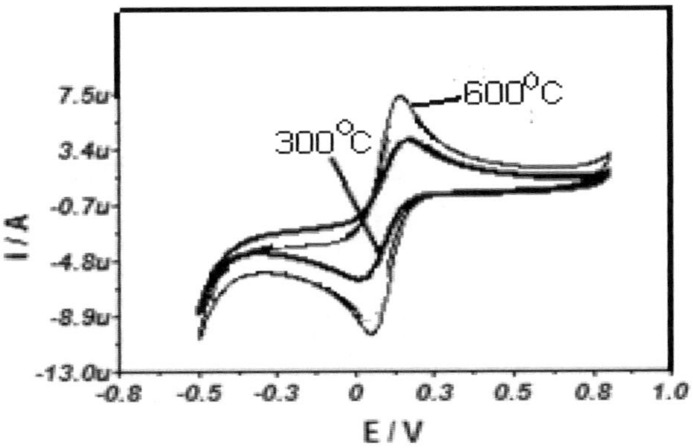

Figure 5. Cyclic voltammograms (CV) of magnesium-ferrite. Sol gel method using citric acid followed by calcinations.

4. Conclusions

The CV profiles related to electrochemical behavior of the prepared magnesium ferrite show that the sol gel method has been successful in the synthesis of semi conducting magnesium ferrite. The sol gel method followed by calcinations process at higher temperatures may enhance the electrical properties of the prepared magnesium ferrite to some extent. The characterization study showed meaningful correlation between structures and effective changes related to different calcination temperatures.

Acknowledgement

The research was financially supported by GUP tier 2 through RMC-UTM project funding, which is gratefully acknowledged.

Reference

[1] Willey R J, Noirclerc P and Busca, G 1993 *Chem. Eng. Commun.*123
[2] Maensiri S, Sangmanee M and Wiengmoon A 2009 *Nanoscale Res. Lett.* **4** 221
[3] Hua T I and Sundari R 2012 *Malaysian J. Fundamental Appl. Sci.* **8** 149
[4] Sundari R, Hua T I, Aziz M and Nizar U K 2014 *Malaysian J. Anal. Sci.* **18** 485
[5] Erdawati, Yusmaniar and Sundari R 2015 *Asian J Chem.* **27** 4693
[6] Pradeep A, Priyadharsini P and Chandrasekaran G 2008 *J Magnetism & Magnetic Mater.* **320** 2774
[7] Liu C P, Li M W, Zhong C, Huang J R, Tian Y L, Tong L and Mi W B 2007 *J Mater. Sci.* **42** 6133
[8] Siemons M, Weirich Th, Mayer J and Simon U 2004 *J Inorg. General Chem.* **630** 2083
[9] Pradeep A and Chandrasekaran G 2005 *Mater. Lett.* **60** 371
[10] Pradhan S K, BidB S, Gateshki M and Petkov V 2005 *J Mater. Chem. Phys.* **93** 224
[11] Mishra S, Kundu T K, Barick K C, Bahadur D and Chakravorty D 2006 *J Magnetism & Magnetic Mater.* **307** 222
[12] Lick I D and Soria D B 2009 *J. Argentine Chem. Soc.* **97** 102
[13]Tolstoy V P, Chernyshova I V and Skryshecsky V A 2003 *Handbook of Infrared Spectroscopy of Ultrathin Films, John Wiley & Sons, USA, Hoboken, New Jersey* 670

Characterization of fabricated MnO_2-amberlite photocatalyst by FTIR, XRD and SEM for alizarin removal

Rita Sundari[1], Sagir Alva[1], Darwin Sebayang[1], Haris Wahyudi[1], Soleha Jonit[2], Atiqah Kamaruddin[2]

[1]Faculty of Engineering, Universitas Mercu Buana, 11650 Jakarta, Indonesia
[2]Department of Chemistry, Faculty of Science, Universiti Teknologi Malaysia, 81310 UTM Johor Bahru, Malaysia

ritsun2003@yahoo.com

Abstract. This study investigates the characterizations of fabricated MnO_2-amberlite using for alizarin photocatalytic degradation. Mixtures of fabricated MnO_2-amberlite were magnetically stirred for 6-, 12-, 18-, 24-, and 30h, respectively. Some functional groups of amberlite like C-O, C-C and N-H were eliminated after doping with MnO_2 as shown by FTIR (Fourier Transform Infrared) spectra in the range of 1600 – 900 cm^{-1}, change in crystalline structures due to linkage of MnO_2 with amberlite was observed at 2θ of 25° in the XRD (X-ray diffraction) analysis, and change in surface morphology due to MnO_2 doping in amberlite was investigated in SEM (Scanning electrone microscope) examination. Since alizarin is a harmful material causing environmental problem, this preliminary study found that the 18h-stirred fabricated MnO_2-amberlite showed the best photocatalytic properties, which enabled to degrade 23.4% alizarin under 1h UV exposure without heating.

1. Introduction

Nowadays the development of catalysts has taken greater interest especially in purification of wastewater for human health reason. Among the environmental catalysts, MnO_2 photocatalyst is the most promising one due to thermocatalytic activity for the decomposition of organic pollutant in wastewater [1]. Several publications reported MnO_2 as photocatalyst for dyestuff degradation [2 – 5]. As we have known advance oxidation process is associated with free radicals generated by irradiation of UV and/or visible light on semiconducting material such as MnO_2, which are responsible for the decomposition of organic compound and changed it into harmless products such as CO_2 and H_2O. The utilization of photocatalyst has attracted great interest due to its simple, low cost, and friendly environmental technology.

Amberlite or resin has largely used as solid phase adsorbent for removing contaminants in aqueous solution. Previous investigations reported amberlite used for extracting selected heavy metals (copper, cadmium and lead) from dairy cow's milk, and fractionized amberlite for removing boron from water, as well as amberlite XAD 7 for extracting bismuth (III) from hydrochloric acid solution [6-8]. In addition, several sensors have used modified amberlite as immobilized matrix [9-11]. The utilization of amberlite as strong acid cationic exchanger has attracted considerable interest due to its stability and low cost. As resin produced by many plant particularly woody plants made resin in greater utility. Industrial effluents contain various chemicals including synthetic dyes, which are carcinogenic in nature. Several dyes decompose aerobically and anaerobically yielding the formation of carcinogenic compounds. Alizarin or mordant red is known as an organic compound, which is largely used as dye in textile industries yielding environmental problems. Roopaei et al. [12] used hematite (Fe_2O_3)/NiS photocatalyst to degrade alizarin in aqueous solution on account of chemical stability and environmental friendliness, however, this study used MnO_2 immobilized in amberlite, which is more stable and has broader band gap for alizarin photocatalytic degradation. Different stirring times (6h, 12h, 18h, 24h and 30h) of MnO_2 and amberlite solution mixture followed by calcinations was used for the synthesis of MnO_2-amberlite. The mixture of MnO_2-amberlite and alizarin solution was exposed to UV light at 265 nm to degrade alizarin. Characterization study of MnO_2-amberlite photocatalyst by FTIR, XRD and SEM was investigated.

2. Experimental

2.1 Materials

All chemicals were purchased and used without further purification. Deion distilled water was used throughout the experiments. MnO_2 (Merck OHG, 99.5 %) for the synthesis of photocatalyst, amberlite IRA 400 (Sigma Aldrich GmbH, $\geq$ 95%) for immobilized matrix, and alizarin for assayed dyestuff (Fluka Chemie, 99%) were used in this study.

2.2 Methods

2.2.1 Synthesis of MnO_2-amberlite. About 15g amberlite was grinded to fine powder and dried in an oven at 98-99°C and then 0.500g dried amberlite was dissolved in 15 ml deion distilled water and stirred for 5h. 0.200g MnO_2 was added to amberlite solution and stirred vigorously with magnetic stirrer at different time (6-, 12-, 18-, 24-, and 30h), respectively. All MnO_2-amberlite solutions with respect to different stirring times were centrifuged at 3700 rpm for 15min. to obtain suspended solid, respectively. The MnO_2-amberlite products were dried in an oven at 98-99°C followed by calcination at 250°C for 5h, respectively.

2.2.2 Characterizations. The fabricated MnO_2-amberlite was characterized using the following instruments. The functional group of the MnO_2-amberlite was determined by FTIR (Perkin Elmer) with potassium bromide as the reference sample. The structural property of the fabricated MnO_2-amberlite was investigated by XRD (Siemens D5000) using CuKα radiation. The surface morphology of the fabricated MnO_2-amberlite was examined by SEM (Philips XL 40) using gold sputter (Polaron Division SEM coating system) to coat the material powder.

2.2.3 Photocatalytic degradation test. The catalytic degradation of alizarin by the fabricated MnO_2-amberlite was investigated by UV-Vis spectrometer (Perkin Elmer Lambda 25). Standard solutions of alizarin from 10 ppm to 50 ppm were prepared for calibration at 265 nm. A mixture of fabricated MnO_2-amberlite and alizarin solution of 50 ppm (part per million) placed in a bottle covered by aluminium foil was exposed to UV light at 265 nm for 1h stirring. The aluminium foil protects the sample from other contaminants.

3. Results and discussion

3.1 Functional group of the MnO_2-amberlite

The functional groups of the fabricated MnO_2-amberlite was determined by FTIR spectroscopy. As shown in Figure 1 (c), the peaks presence at 3421 cm^{-1}, 1614 cm^{-1}, 1488 $cm^{-1}$and 1475cm^{-1} are corresponding with benzene sulphonic groups in cross-linked amberlite polymer. The pure amberlite shows the aromatic rings at 3022 cm^{-1} and C-H stretching at 2921 cm^{-1} (Figure 1c), while the N-H stretching at 3421 cm^{-1} (Figure 1c) was getting weaker and showed at 3426 cm^{-1} in Figure 1b due to MnO_2 doping in amberlite. Furthermore, the FTIR spectrum range of 1600 – 900 cm^{-1} of pure amberlite shows vibration of C-O and C-C stretching, and N-H bending in the amberlite polymer. When pure amberlite mixed and magnetically stirred with MnO_2, almost all of the functional groups initially present were eliminated due to mechanical effect and the inclusion of MnO_2 in the cross-linked polymer. The presence of remarkable peaks at 588 cm-1 (Figure 1b) and 589 cm-1 (Figure 1a) was due to the formation of Mn-O linkage in relation to tetrahedral and octahedral structure [13 – 15].

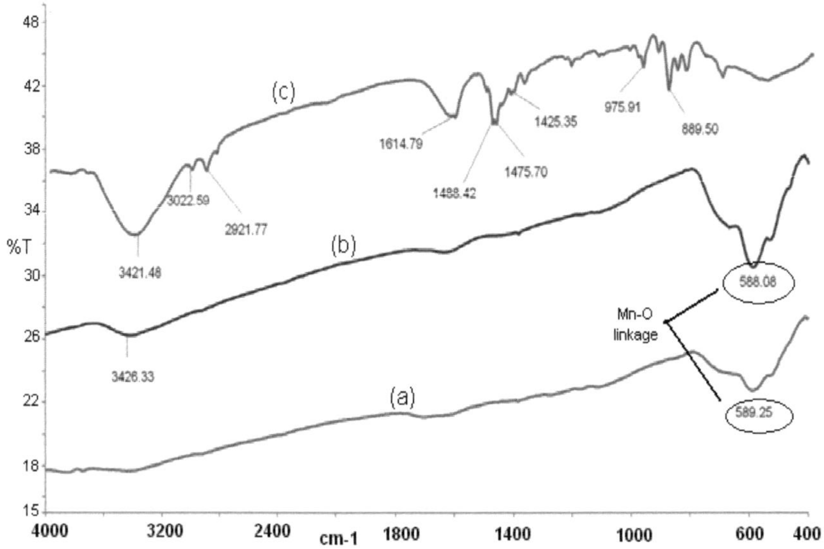

Figure 1. FTIR spectra of (a) 6h-stirred MnO_2-amberlite, (b) 18h-stirred MnO_2-amberlite, and (c) pure amberlite. KBr reference sample.

3.2 Structure of the MnO_2-amberlite

The structural property of the MnO_2-amberlite was investigated by the XRD. Figure 2 shows a peak cluster at 2θ of 25°, which forms unstable structure. It is apparent that the stirring action on the MnO_2-amberlite did not give meaningful effect on the crystalline structure of pure MnO_2. Previous work reported that the diffraction angles of 25°, 37° and 57° are corresponding with (211), (400) and (440) planes, respectively [15].

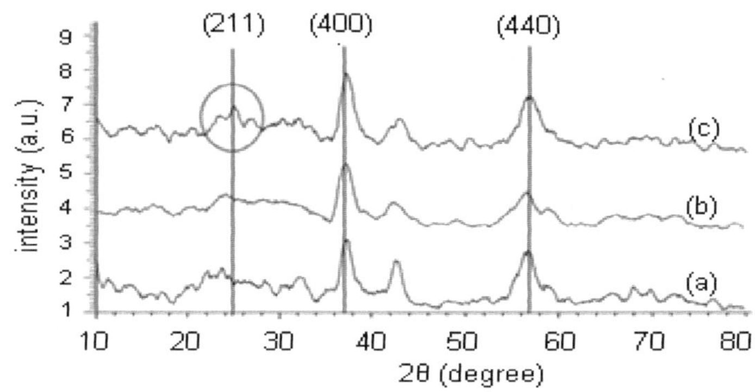

Figure 2. XRD spectra of (a) pure MnO_2, (b) 6h-stirred MnO_2-amberlite, and (c) 18h-stirred MnO_2-amberlite (50 nA, 40 kV, and angular range of 20.80°).

3.3 Surface morphology of MnO_2-amberlite

The surface structure of the fabricated MnO_2-amberlite was exmined by SEM. Theoretically, the longer stirring time yields finer particles, however, agglomeration of particles may occur during stirring process due to polarization and induction effects. Previous work reported manganese distribution on the ferrite surface were varied due to thermal effect [15-16], however, this study shows the particles distribution on the material surface due to mechanical effect. Figure 3 shows the SEM imaging of pure amberlite and fabricated MnO_2-amberlite at different stirring times.

3.4 Photocatalytic degradation of alizarin

The utilization of fabricated MnO_2-amberlite for photocatalytic degradation of alizarin was conducted by using UV-Vis spectrometer. Based on the surface adsorption test of BET (Brunauer Emmett Teller) (not shown), the 18h-stirred MnO_2-amberlite showed the largest surface area (65.11m^2 g^{-1} of 0.032g sample), which assumed to yield the best catalytic property in this study and therefore, this sample was used in the examination of degradation of alizarin. Based on the calibration curve of UV-Vis absorption at 265 nm the fabricated MnO_2-amberlite might degrade 23.4% of alizarin in aqueous solution under non heat condition and 6.8% of alizarin under heat condition.

(a) Pure amberlite

(b) 6h-stirred MnO_2-amberlite

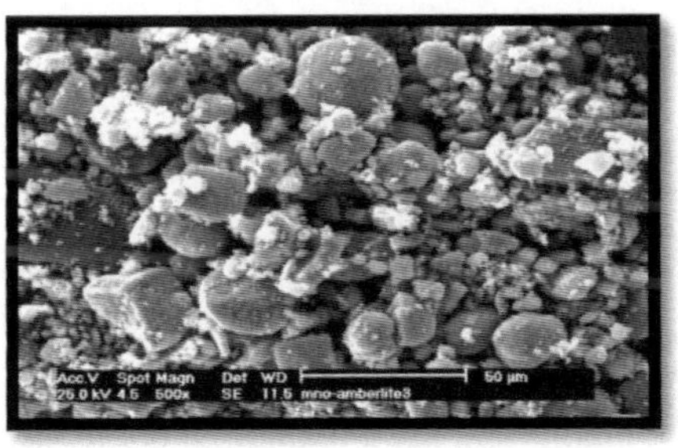

(c) 18h-stirred MnO_2-amberlite

(d) 30h-stirred MnO_2-amberlite

Figure 3. SEM imaging of pure amberlite and fabricated MnO_2-amberlite. 50 μm working distance. 500 x magnification. 25.0 kV.

4. Conclusion

The study presented important features related to the characterization study of FTIR, XRD and SEM of fabricated MnO_2-amberlite, which gives meaningful information for photocatalytic degradation of alizarin. Some important functional groups were removed and replaced by remarkable peaks at 588 - 589 cm^{-1} due to MnO_2 inclusion as shown in the study of FTIR spectra, however, the XRD characterization at 2θ of 25° showed that amberlite matrix and stirring process did not give significant effect on the crystalline structure of pure MnO_2. The SEM imaging showed finer particle distribution on the material surface after MnO_2 mixed with amberlite under stirring process.

Acknowledgement

This study was sponsored and financially supported by the Research Management Center (RMC), Universiti of Technologi Malaysia through FAV Grant 77545, which is gratefully acknowledged.

References

[1] Qiliang J, Hiroshi A, Musashi F and Hiroaki T 2013 *Catalysts* **3** 444
[2] Li S J, Ma Z, Wang L and Liu J Z 2008 *Sci. China Series B: Chem.* **51** 179
[3] Lifang Z, Daiping H and Ping J 2009 *Catalysis Commun.* **10** 1414
[4] Liu Y, Luo C, Sun J, Li H, Sun Z and Yan S 2015 *J. Mater. Chem.* A **3** 5674
[5] Xue M, Huang L, Wang J O, Wang Y, Gao L, Zhu J and Zhou Z G 2008 *Nanotechnol.* **19** 5604
[6] Hoque M I, Didarul A C, Holze R, Nakib C and Azam M S 2015 *J.Environ.Chem.Engineer.* **3** 831
[7] Nawaf D, Victor K and Nidal H 2015 *Desalination* **370** 1
[8] Navarro R, Ruiz P, Saucedo I and Guibal E 2014 *Separation. Purifi. Tech.* **135** 268
[9] Moghaddam M K, Hellmann J and Lang W 2016 *Procedia Engineer.* **168** 1110
[10] Bliznyuk V N, Duval C E, Apul O G, Seliman A F and DeVol T A 2015 *Polymer* **56** 271
[11] Bairi A, Nithyadevi N, Bairi I, Garin A M and Garcia J A M 2017 *Energy Build.* **141** 218
[12] Roopaei H, Zohdi A R, Abbasi Z and Bazrafkan M 2014 *Indian J Sci. Technol.* **7** 1882
[13] Pradeep A and Chandrasekaran G 2005 *Mater. Lett.* **60** 371
[14] Pradeep A, Priyadharsini P and Chandrasekaran G 2008 *J Magnetism & Magnetic Mater.* **320** 2774
[15] Hua T I and Sundari R 2012 *Malaysian J. Fundamental Appl. Sci.* **8** 149
[16] Sundari R, Hua T I, Aziz M and Nizar U K 2014 *Malaysian J. Anal. Sci.* **18** 485

IOP Conf. Series: Materials Science and Engineering **343** (2018) 012004 doi:10.1088/1757-899X/343/1/012004

The Cross-Sectional Investigation of Oxide Scale FeCr Alloys and Commercial Ferritic Steel Implanted with Lanthanum and Titanium Dopants after Oxidation Test at 900°C

Hendi Saryanto[1], Darwin Sebayang[2], Pudji Untoro[3], Tjipto Sujitno[4]

[1]Gajah Tunggal Polytechnic, Kel. Pasir Jaya, Tangerang-Banten, Indonesia
[2]Mercu Buana University, Jln. Meruya Selatan, Jakarta Barat, Indonesia
[3]Surya University, Jln. Boulevard Gading Serpong, Banten-Indonesia
[4]Centre For Accelerator and Material Process Technology. BATAN JL. Babarsari, D.I Yogayakarta, 55281, Indonesia

hsaryanto@gmail.com

Abstract. The cross-sectional examinations of oxide scales formed by oxidation on the surface of FeCr alloys and Ferritic Steel that implanted with lanthanum and titanium dopants were observed and investigated. Scanning electron microscopy (SEM) coupled with energy dispersive X-ray spectroscopy (EDS) has been used to study the cross-sectional oxides produced by specimens after oxidation process. X-ray diffraction (XRD) analysis was used to strengthen the analysis of the oxide scale morphology, oxide phases and oxidation products. Cross-sectional observations show the effectiveness of La implantation for improving thinner and stronger scale/substrate interface during oxidation process. The result shows that the thickness of oxide scales formed on the surface of La implanted FeCr alloy and ferritic steel was found less than 3 µm and 300 µm, respectively. The oxide scale formed on the surface of La implanted specimens consisted roughly of Cr_2O_3 with a small amount of FeO mixture, which indicates that lanthanum implantation can improve the adherence, reduce the growth of the oxide scale as well as reduce the Cr evaporation. On the other side, the oxide scale formed on the surface of FeCr alloys and ferritic steel that implanted with titanium dopant was thicker, indicating that significant increase in oxidation mass gain. It can be noticed that titanium implantation ineffectively promotes Cr rich oxide. At the same time, the amount of Fe increased and diffused outwards, which caused the formation and rapid growth of FeO.

1. Introduction

The main problem with the use of conventional steel alloy is their inadequate high-temperature oxidation resistance [1]. Under SOFC operating conditions, ferritic steel form the oxide scale that is susceptible to spalling during normal fuel cell operation, in particular during thermal cycling encountered at start-up and shut down [2]. The extensive oxide losses from the surface or oxide spalling usually occur, which has become a serious challenge that leads to serious deterioration of the cell performances [3, 15]. Therefore, the alloys stabilization and oxidation performances at the high temperatures are essential to enhanced performance of the cell, chromium evaporation is known to be a major problem in SOFC stacks since volatile chromium is gets deposited on the cathode and poisons

the oxygen electrode [4]. To enable practical application, two classes of technology can be approached to reduce the growth rate of the oxide scale, the first is to develop nanostructured alloys, and the other is to conduct surface treatment of the developed or available alloy with a reactive element [5,15]. FeCr alloys and ferritic stainless steels exhibit excellent properties as interconnect material in SOFCs [6]. The oxidation performance of FeCr used in high-temperature applications (such as Interconnector in SOFC) depends on their potential to form and maintain a Cr-rich protective oxide scale [7-8]. Technical requirements to be used as interconnect in SOFC should fulfill a number of specific requirements, i.e. a high oxidation resistance, a high electrical conductivity of the surface oxide scales, gas tightness, and a coefficient of thermal expansion matched to the electrolyte and the electrodes [9-10]. The role of the interconnect component in SOFC units is to electrically connect individual cells in series to form a multi-cell stack with higher power density. FeCr and Ferritic stainless steels have many desirable properties for use in solid oxide cell stacks and are the most common type of materials used as interconnect. When Cr based alloys and high-Cr ferritic steels are exposed to an oxidizing environment at elevated temperatures, an oxide layer is formed on the surface, acting as a barrier between the metal and the gas. Chromium increases the oxidation resistance of stainless steels by the formation of a chromia (Cr_2O_3) scale on the surface. Other oxides which are often encountered in the layers are iron oxides (wüstite FeO, magnetite Fe_3O_4, or hematite Fe_2O_3) [8,11-12].
Refer to our previous study [8,13-14] the cross-sectional investigation of FeCr alloy with smallest crystallite sizes can significantly reduces the growth of oxide scale and improve the high-temperature oxidation and corrosion resistance of FeCr alloy that oxidized at 900°C, However, the research did not investigate the characterization of oxide scale on a cross-sectional analysis and the growth mechanism of the oxide scale of implanted specimens. Therefore, the objective of this study is to investigate the cross-sectional characterization of oxide scale formed on the surfaces of implanted specimens compare with un-implanted specimens oxidized at 900°C. Oxidation kinetics, and as well as the evolution of oxide scale composition and microstructure are systematically studied. The mechanism of the formation of oxide scale on both of steels is discussed.

2. Specimens and Experimental Procedure

2.1 Specimens
The selected FeCr alloy with smallest crystallite size (38.51 nm) as the best oxidation results in our previous study [8] were selected and implanted with lanthanum and titanium dopants. Implantation of lanthanum and titanium dopants with nominal doses10^{17} ions/cm^2 was undertaken using 100 keV ion beam energy and 200 keV acceleration potential and the beam current density was maintained at 10 μA/cm^2. The depth concentration profiles of lanthanum and titanium dopants in to substrate of both type specimens were collective in Table. 1 [13-15].

Table 1. The depth concentration profiles of lanthanum and titanium dopants in to substrate of steels

Sample	Depth profile (Å)	
	Implanted with lanthanum	Implanted with titanium
FeCr alloys	297	706
Commercial FS	296	639

2.2. Oxidation Test
The samples of oxidation test were cut from the specimens to the size of 4x5x3 mm. The samples were subjected to oxidation tests in laboratory air at atmospheric pressure in a PROTHERM box furnace at 900°C for 100 h in laboratory air at the atmospheric pressure.

2.3. Preparation of cross-section specimens
Cross-section samples are mounted using "cold-mounting" resins. Samples were placed on a commercial sample holder (cylindrical), fixed through a drop of wax. Resin (plus hardener and accelerant) is then poured in the sample. After the sample and resin becoming a hard block, the

samples were polished on the cross-sectional face up to 1500 grade SiC paper and final polishing using 0.05 μm diamond paste.

2.4. *Microstructure Characterization*

The oxidized alloys were investigated by a BRUKER D8 Advance X-ray diffraction (XRD) equipped with CuKα radiation and phase identification as referred to the ICDD PDF-2 database. The surface and cross-sectional morphology analyses of oxidized specimens were carried out with a JEOL JSM-6380 scanning electronic microscope (SEM) equipped with an energy-dispersive X-ray spectroscopy (EDS) analyzer. Furthermore, the best selected of oxidized specimens were analyses more detail using EDS line/mapping analysis.

3. Results and Discussion

3.1. *Previous Results*

The previous results on developed of FeCr Alloys, oxidation test, and parabolic rate were collected in Table 1 and Fig.1 [7, 8, 9, 10 & 11], where the parabolic rate constant (k_p) is determined by the mass gain (ΔW) per unit surface area of a specimen (A) and exposure time (t) [10, 13-15].

$$(\Delta W/A)^2 = k_p t \quad (1)$$

Table 2. Mass gain change and parabolic rate constant (k_p) of FeCr alloys after oxidized at 900°C

Specimens	Crystallite size (nm)	Mass Gain (mg/cm^2)	Parabolic Rate, k_p (mg^2/cm^4)
FeCr "A"	53.33	21.81	6.191
FeCr "B"	38.51	7.72	0.794
FeCr "C"	76.60	35.18	13.58

Table 1 show the mass gain curves of the studied alloys with different crystallite size oxidized at 900°C. Mass gain change (mg/cm^2) and parabolic rate (mg^2/cm^4) indicate that FeCr "B" as the smallest crystallite size was found as the lowest mass gain change and parabolic rate, which indicates as the best oxidation resistance.

3.2. *Results of the current study*

SEM morphologies of the cross-section of the oxide scale on the surface of implanted and the selected of un-implanted (FeCr "B") alloys after oxidized at 900°C for 100 h in air atmosphere are presented in Fig 1 below.

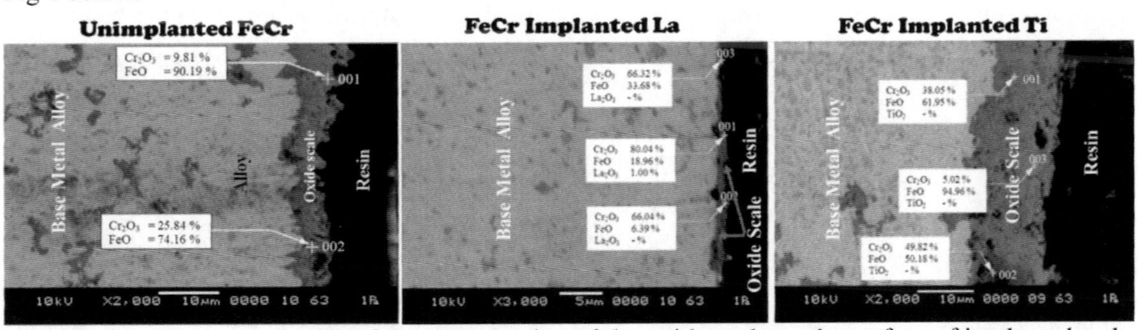

Figure 1. SEM morphologies of the cross-section of the oxide scale on the surface of implanted and un-implanted FeCr alloys oxidized at 900°C

Fig. 1 shows the cross-sectional morphology of oxide scale on the surface of implanted and un-implanted specimens oxidized at 900°C. A very thin oxide scale, less than 3 μm, had formed on the surface of FeCr alloy that implanted with lanthanum. While, the thickness of the oxide scale formed on the surface of un-implanted FeCr alloy and FeCr alloy that implanted with titanium was formed about 8 and 15 μm, respectively. According to EDX analysis, it can be seen that the oxide scale on the surface of FeCr alloy implanted with lanthanum was dominated by the presence of Cr_2O_3 about 60.05-80.04 %, which indicates that lanthanum implantation can protect the Cr_2O_3 layer, reduce the rate of Cr evaporation and reduce the growth of the oxide scale as evidenced by the thin layer of oxide scale. However, the oxide scale formed on the surface FeCr alloy that implanted with titanium dopant indicated covered and distributed evenly with the main composition of FeO approximately 50.18-94.96%. It can be noticed that titanium implantation ineffectively promotes Cr rich oxide. At the same time, the amount of Fe increased and diffused outwards, which caused the formation and rapid growth of FeO. Furthermore, Cr_2O_3 as the protective layer becomes non-protective, due to the formation of volatile Cr_2O_3 upon exposure to oxidizing atmospheres for extended periods.

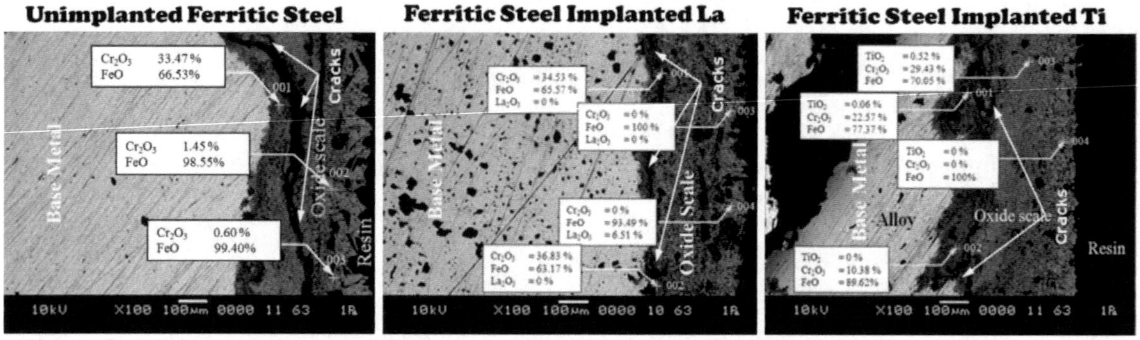

Figure 2. SEM morphologies of the cross-section of the oxide scale on the surface of implanted and un-implanted Ferritic Steel oxidized at 900°C

Fig. 2 shows the cross-sectional of oxide scale formed on the implanted and un-implanted ferritic steels oxidized at 900°C. The oxide scale on the surface of an-implanted ferritic steel is formed up to a scale thickness of 600 μm, where the oxide scales show mainly consisted of Cr_2O_3 about 33.47 and FeO 99.40 at%. However, the adherence between scale and a base metal reduced causes the occurrences of the largest crack along the scale. For the specimen that implanted with lanthanum show better oxide scale indicated by the thickness of the oxide scale was found about 300 μm thick, although the cracking also occurs but the presence of Cr_2O_3 as the protective layers is still significant in the range of 34.53-36.83 %. While, the presence of Cr_2O_3 on the oxide scale of FeCr that implanted with titanium was found in the range 10.38-29.43 at %. This phenomenon indicates that the implantation of lanthanum has influenced the formation stable layer of Cr_2O_3 and plasticity as well as protects it from evaporation and cracks on high temperature oxidation. The results of the cross-sectional analysis as described above have been confirmed by the results of the oxidation kinetics of both type specimens oxidized at 900°C as shown by following picture.

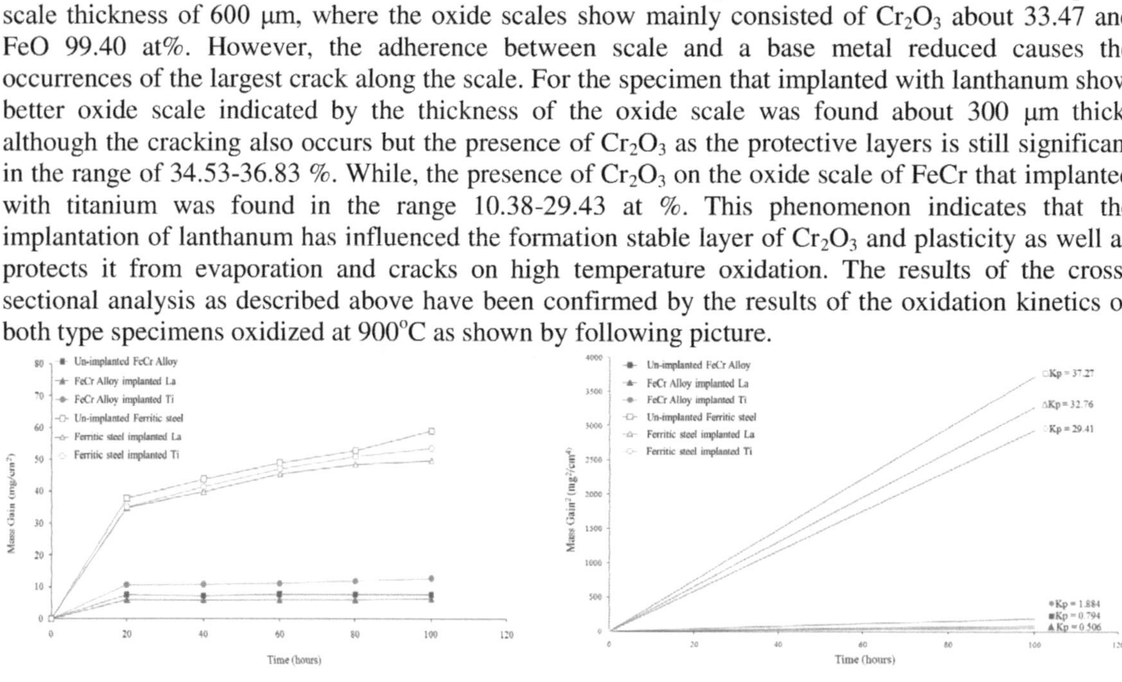

Mass gain curves
a

Parabolic rate curves
b

Figure 3. Mass gain curves and parabolic rate curves of un-implanted and implanted specimens oxidized at 900°C

Figs. 3 show the mass gain curves and parabolic rate curves of the implanted and un-implanted of FeCr alloys and Ferritic steel oxidized at 900°C. Mass gain curves and parabolic rate curves indicates that ferritic steels displayed higher mass gain trend and exhibits poor oxidation resistance. The beneficial effect of lanthanum implantation is clearly visible, the oxidation behaviour being improved. In fact, the mass gain and parabolic rate more lowest, if compared with the un-implanted and Ti implanted specimens. The effect of lanthanum implantation for specimen of FeCr alloy that implanted with lanthanum is significantly to reduce the mass gain change and parabolic rate from 7.73 mg/cm^2 and 0.794 mg^2/cm^4 to 6.29 mg/cm^2 and 0.506 mg^2/cm^4, respectively, in comparing with un-implanted FeCr alloy. Furthermore, The effect of lanthanum implantation for specimen of ferritic steel that implanted with lanthanum is significantly to reduce the mass gain change and parabolic rate from 59 mg/cm^2 and 37.27 mg^2/cm^4 to 49.7 mg/cm^2 and 29.41 mg^2/cm^4, respectively, in comparing with un-implanted ferritic steel. While, the effect of titanium implantation becomes ineffective to reduce the growth of oxide scale which indicates by the highest mass gain change and parabolic rate for both type of specimens. In order to strengthen the analysis result described above, XRD analysis were applied to identified the oxide phase of both type of specimens. The XRD patterns for specimens oxidized at 900°C for100 h are presented in Fig.4.

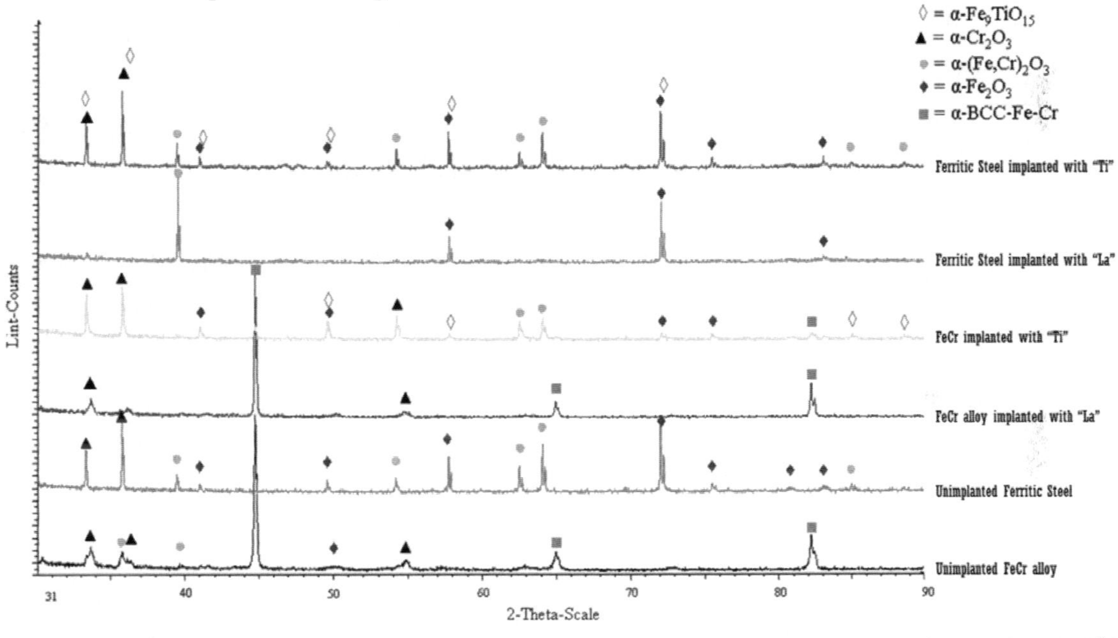

Figure 4. X-rays diffractograms of unimplanted and implanted specimens oxidized for 100 h at 900°C

Fig. 4 above shows the diffraction patterns of FeCr alloys and ferritic steels oxidized at 900°C. For unimplanted FeCr alloy, the signal peaks can be identified as α-(Fe,Cr)$_2$O$_3$, α-Fe$_2$O$_3$, and α-Cr$_2$O$_3$. The FeCr phase is also shown, but only on as the developed specimen which indicates the thin of oxide scale on the substrate surface. For the implanted specimens, the main oxide scale obtained were α-(Fe, Cr)$_2$O$_3$, α-Fe2O3, α-Cr$_2$O$_3$. The presence of the peak of the original alloy indicates that oxide scale was formed as a thin layer which dominated by the presence of Cr-rich oxide and small peaks of solid solution chromium iron oxide (Fe,Cr)$_2$O$_3$ [10, 13-15]. On the contrary, FeCr alloy that implanted with titanium dopant, all strong peaks were dominated by the presence of α-Fe$_9$TiO$_{15}$, hematite Fe$_2$O$_3$ and small peaks of solid solution chromium iron oxide (Fe,Cr)$_2$O$_3$, although the presence of the peaks of the original alloys still found with a small intensity which indicates that oxide scale was growing rapidly with the dominating presence of Fe-rich oxide due to ineffectively of titanium implantation. In order to understand more detail about the mechanism of oxide growth on the surface of specimens, EDX line/mapping analysis was used to verify the results in the formation of the oxide scale of FeCr and ferritic steel that implanted with lanthanum as the best oxidation results as shown in Fig.5 and 6.

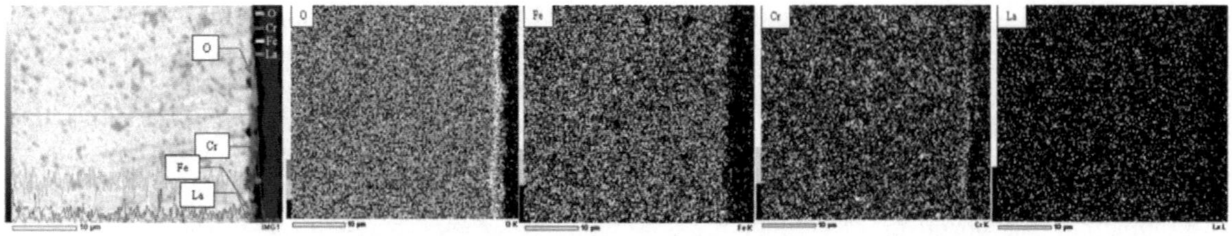

Figure 5. Cross-sectional EDS mapping analysis of FeCr implanted with La oxidized at 900°C

Fig.5 above shows that the scale formed on FeCr alloy that implanted with lanthanum consisted of corundum oxides layer of Cr_2O_3 wich confirmed by EDX mapping analyses which indicated by high presence of Cr rich oxide on the inner of oxide scale, while, the EDX mapping analyses indicates that the presence of Fe was found on the outer of oxide scale with small amount of Fe on the formation of FeO. However, EDX mapping analyses is not clear to detect the presence of Lanthanum on the oxide scale due to low doses of Lanthanum that implanted in to substrate of the alloy. This phenomenon indicates that the lanthanum implantation significantly to reduce the growth of oxide scale by reducing the Cr^{3+} ions diffuses to the surface of the specimens. where the enhancement of Cr diffusion guarantee the stable growth of the Cr-rich in the inner of the oxide scales, which significantly improve oxidation resistance, even a Cr_2O_3 scale may be formed at higher temperatures owing to the higher diffusion rate of Cr.

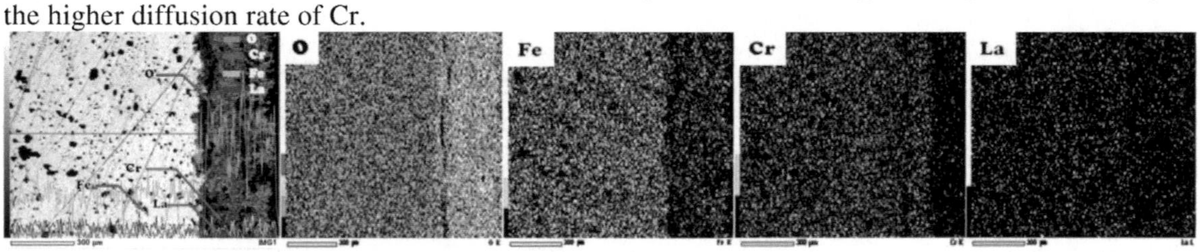

Figure 6. Cross-sectional EDS mapping analysis of ferritic steel implanted with La oxidized at 900°C

Fig.6 above shows that the Cr concentration in ferritic steel is still not sufficiently high to allow the formation of a more protective Cr_2O_3 layer, while the iron ions dissolve in and diffuse rapidly through the Cr_2O_3 scale and eventually an outer layer of fairly pure iron oxides will result. It may be caused by Fe and Cr ion transport is predominant then make the scale growth will proceed outwards, the oxygen transports through the FeO layer to the Cr_2O_3 layer and the Cr ions diffuse to the same layer; both of them react with each other, which make the oxide scale become thicker. Furthermore, due to the loose and incompact scale, thus causing the crack in the surface of the specimen which increase the rate of Cr evaporation. The oxygen may diffuse to the oxidation/alloy interface through the cracks and pores in the oxide scale. This phenomenon was confirmed by XRD results as shown in Fig. 3, where the result shows that the oxide phase that formed on the surface of ferritic steel that implanted with lanthanum is dominated by the presence of Fe_2O_3. It's also reinforcement by mass gain change and parabolic rate constant (k_p) as presented in Fig. 3. It can be noticed that a discontinuous Cr_2O_3 scale was formed, but could not protect the alloy from internal oxidation. At the same time, Fe became enriched and diffused outwards, which caused the formation and rapid growth of FeO, although it has been implanted with lanthanum due to the higher crystallite size effect.

4. Conclusion

Based on the results obtained in the oxidation of implanted and un-implnted FeCr alloys and ferritic steel at 900°C for 100 h the following conclusions can be formulated:

1. The oxidation resistance of FeCr alloy that implanted with Lanthanum showed better quality than that of commercial alloy. The beneficial effect of La implanted sample on the oxidation resistance was clearly visible to improve the adherence of oxide scale. Cross-sectional observations revealed the effectiveness of La implantation for improving thinner and stronger

scale/substrate interface during oxidation process which presents the homogenous structure with good adherence and a uniform inter-diffusion layer between substrate and oxide layer.

2. The specimens that implanted with lanthanum show better promoting the formation of Cr_2O_3 in large amount compositions which indicated that chromium preference diffused outward to form a Cr_2O_3 layer.
3. The oxide scale formed on the surface of FeCr alloys and ferritic steel that implanted with titanium dopant was thicker oxide scales indicating that significant increase in oxidation weight gain. It can be noticed that titanium implantation ineffectively promotes Cr rich oxide. At the same time, the amount of Fe increased and diffused outwards, which caused the formation and rapid growth of FeO. Furthermore, FeO nucleated and grew at the discontinuous areas such as the incompact Cr_2O_3 scale or its micro-defective areas; leading to some bulges and cracks.

5. References

[1] Kulekci, M. K. 2008. Magnesium and its alloys applications in automotive industry. *The International Journal of Advanced Manufacturing Technology*, *39*(9), 851-865.

[2] Cooper, L., Benhaddad, S., Wood, A., & Ivey, D. G. 2008. The effect of surface treatment on the oxidation of ferritic stainless steels used for solid oxide fuel cell interconnects. *Journal of Power Sources*, *184*(1), 220-228.

[3] Korsunsky, A. (Ed.). 2012. *Current Themes in Engineering Science 2010*. American Institute of Physics.

[4] Hui, R. 2013. Materials and Processing for Metal-Supported Solid Oxide Fuel Cells. *Materials for High-Temperature Fuel Cells*, 309-340.

[5] Saryanto, H., Deni, S. K., Untoro, P., Saleh, M. H., & Sebayang, D. 2010. Determination of Nanocrystalline Fe80Cr20 Powder based Alloys using Williamson-Hall Method. In *Advanced materials research* (Vol. 129, pp. 999-1003). Trans Tech Publications.

[6] Sachitanand, R., Sattari, M., Svensson, J. E., & Froitzheim, J. 2013. Evaluation of the oxidation and Cr evaporation properties of selected FeCr alloys used as SOFC interconnects. *international journal of hydrogen energy*, *38*(35), 15328-15334.

[7] Ali-Löytty, H. 2013. Microalloying Mediated Segregation and Interfacial Oxidation of FeCr Alloys for Solid-Oxide Fuel Cell Applications. *Tampereen teknillinen yliopisto. Julkaisu-Tampere University of Technology. Publication; 1114.*

[8] Saryanto, H., Sebayang, D., & Untoro, P.2017, May. Cross-Sectional Investigations of Oxide Scale Nanocrystalline FeCr Alloys after High-Temperature Oxidation Test at 900° C. In *IOP Conference Series: Materials Science and Engineering* (Vol. 196, No. 1, p. 012030). IOP Publishing.

[9] Hilpert, K., Quadakkers, W. J., & Singheiser, L. 2003. Interconnects. *Handbook of Fuel Cells.*

[10] Sebayang, D., Khaerudini, D. S., Saryanto, H., Othman, M. A., Sujitno, T., & Untoro, P. (2011).
Effect of Nanocrystalline and Ti implantation on the oxidation behaviour of Fe80Cr20 alloy and commercial ferritic steel. In *Key Engineering Materials* (Vol. 474, pp. 2134-2139). Trans Tech Publications.

[11] Lessing, P. A. 2007. A review of sealing technologies applicable to solid oxide electrolysis cells. *Journal of Materials Science*, *42*(10), 3465-3476.

[12] Akanda, S. R. 2012. *Mechanical Characterization of Coating-Interconnect Interfaces and Anode-Electrolyte Interfaces for Solid Oxide Fuel Cells.* The Ohio State University.

[13] Sebayang D *et al* Oxidation Resistance of $Fe_{80}Cr_{20}$ Alloys Treated by Rare Earth Element Ion Implantation. Current Themes In Engineering Science 2010: *AIP Conference Proceedings.* 1394. No. 1 pp 90-102.

[14] Hendi, S., Darwin, S., Pudji, U, & Tjipto, S. 2009. Ion implantation process of Lanthanum and Titanium Dopants into a substrate of Fe80 Cr20.

[15] Sebayang, D., Khaerudini, D. S., Saryanto, H., Omar, B., Othman, M. A., Hamid, A., ... &

IOP Conf. Series: Materials Science and Engineering **343** (2018) 012004 doi:10.1088/1757-899X/343/1/012004

Untoro, P. 2011. Oxidation Resistance of Unimplanted and Implanted of Nanocrystalline FeCr Alloy and Commercial Alloy with Lanthanum. *Journal of Advanced Microscopy Research*, *6*(4), 263-277.

IOP Conf. Series: Materials Science and Engineering **343** (2018) 012005 doi:10.1088/1757-899X/343/1/012005

Effect of heat treatment On Microstructure of steel AISI 01 Tools

Melya Dyanasari Sebayang[1], Sesmaro Max Yudo[1], Charlie Silitonga[2]
[1] Mechanical Lecturer University of Christian Indonesia
[2] Mechanical student University of Christian Indonesia

melcan_sebayang@yahoo.co.id

Abstract. This study discusses the influence of quenching, normalizing , and annealing to changes in hardness , tensile , and microstructure of materials tool steel AISI 01 after the material undergo heat treatment process. This heat treatment process includes an initial warming of 600° C and a 5-minute detention time, followed by heating to an austenisation temperature of 850°C. After that a different cooling process, including annealing process, normalizing and quenching oil SAE 40. Tests performed include tensile, hard, and microstructure with shooting using SEM (Scanning Electron Microscope). This is done to see the effect of different heat treatment and cooling process. The result of this research is difference of tensile test value, hard, and micro structure from influence of difference of each process. The quenching process obtains the highest tensile and hard values followed by the normalizing process, annealing, and the lowest is in the starting material, this is because the initial material does not undergo heat treatment process. The resulting microstructure is pearlit and cementite, the difference seen from the shape and size of the grains. The larger the grain size, the greater the hardness.

1. Introduction

The standard tool steel AISI 01 is a metal alloy between Silicon, Chromium, and Mangan. In the industrial world widely used dies, punch, and others. Use of this type of metal in the industry because it has good machine capable properties. In materials capable of machining, often require further processing to improve the mechanical properties and materials.

Factors affecting the manufacturing process of the dies are the strength and hardness of the metal as well as changes in the microstructure that occurs during the heat treatment process. By looking for a comparison between the Quenching, Normalizing, and Annealing processes through tensile testing, hard testing, and micro testing, it can be known which process experienced changes in the value of hardness and microstructure in this material.

This research is to determine the effect of heat treatment experienced on carbon steel AISI 01 at austenisation temperature conditions 850°C and the process of quenching, normalizing and annealing to changes in tensile values, hardness and microstructure.

2. Materials and Experimental Procedure

In this research selected materials, AISI 01 tool steel in the world of industry is widely used as a material punch maker and dies. Here is the chemical composition of AISI 01 tool steel. From Table 1

Table 1. Chemical composition of AISI 01

NO	Composisi	Persentase (%)
1	Carbon (C)	0.9 – 1.05
2	Crom (Cr)	0.5 – 0.7
3	Vanadium (Vn)	0.05 – 0.15
4	Mangan (Mn)	1 – 1.2
5	Silikon (Si)	0.15 – 0.35

it shows that AISI 01 contains high carbon content, carbon content has an effect on hardness value of steel. The alloying elements are useful for improving the properties of steel.

The manufacture of specimens is done by various processes, starting from the process of cutting the object from the shape of the rod (shaft) with a diameter of 16 mm and the length of each 250 mm. Then the test object is latched into 12mm inner diameter.

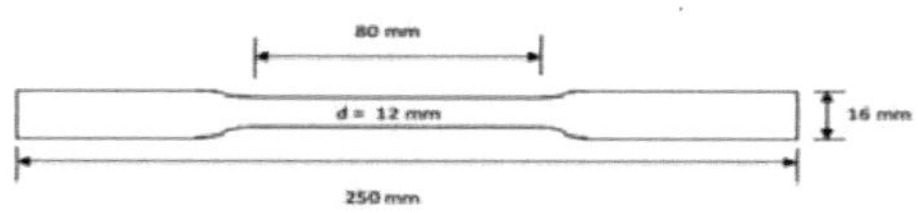

Figure 1. Specimen dimension

Before entering the test, the specimen is classified according to the group of each process. The grouping of test objects is:

- 1st group: 4 specimens for the annealing process
- 2nd group: 4 specimens for the normalizing process
- 3rd group: 4 specimens for the quenching process
- 4rd group: 4 specimens for comparative specimens which are not given heat treatment, but only hardness, tensile, and microstructure testing are performed.

After the test object is ready then start the process of heating or heat treatment. Then other tests, such as tensile tests, hard tests, and microstructures.

The heat treatment process is carried out by heating the steel in the heating kitchen by first warming up at 600 ° C with 5-minute hold time. The purpose of the initial heating is to produce a uniform heating throughout the specimen's surface to produce evenly homogeneous structures. Then proceed with the heating process at austenisation temperature with a temperature of 850 ° C - 900 ° C and held for 60 minutes. In the cooling process (immersion) is done after the steel experienced austenisation process. In this process cooling is done at a certain speed, in order to produce tensile, hard, and different microstructures. In this cooling process using methods of cooling annealing, normalizing, and quenching oil.

In testing to be performed, the dye used is oil (SAE 40), because it can reduce the occurrence of stress, while water or salt can be used for simple forms. The use of SAE 40 oil was chosen because it has a viscosity of 40 at 100 ° C. The use of SAE 40 oil as a cooling medium will cause the occurrence of the carbon membrane in the specimen depending on the oil viscosity. for the normalizing cooling process is carried out by air cooling. After the heating process, the material is removed from the kitchen heater and placed in free air. For the annealing process is done by slow cooling in the kitchen heater (furnace). The furnace temperature is lowered by turning off the furnace so that the temperature drops slowly until it reaches room temperature.

After through the process of heat treatment and cooling the test is done based on the classification of each specimen that has been through a different cooling process. In this test, three tests are tested, tensile test, hard test and microstructure. The tensile test was conducted using a tensile test machine with a maximum capacity of 30,000 kgf. This hard test is performed to determine the effect of heat treatment and its cooling variation on the hardness value of each specimen tested. The

IOP Conf. Series: Materials Science and Engineering **343** (2018) 012005 doi:10.1088/1757-899X/343/1/012005

test method uses the Rockwell B hard test, as it is easier and does not damage the test specimen on the surface of each specimen. Testing of microstructure on each material must be tested by metallographic.

Table 2. Hard Test

NO	AISI 01	Annealing	Normalizing	Quench Oli
1	98,95	91,82	99,57	99,55
2	79,5	95,67	99,6	99,5
3	92	97,82	99,67	99,5
4	94	96,50	99	99,55
HRB	**91,11**	**95,45**	**99,46**	**99,52**

3. Results and Discussion

Tensile Strength Test Analysis

Based on tests that have been performed such as tensile, hard, and microstructure tests conducted on specimens under conditions following and before heat treatment with different cooling methods. After experiencing the heat treatment process each material or specimen tested tensile strength has different tens compared to the original specimens of steel AISI 01.

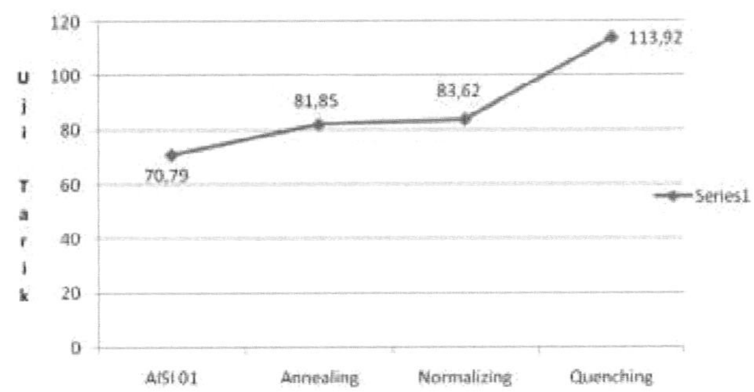

Figure 2. Tensile test graph

From the picture above diagram shows that the quenching process material has the highest tensile strength of 113.92 kgf / mm^2, compared with the normalizing process that has a tensile strength of 83.62 kgf / mm^2. The difference in tensile strength is 28.3 kgf / mm^2. This is because the cooling rate process is different. In the annealing process material, it is seen that the value of tensile strength has a tensile strength of 81.85 kgf / mm^2, slightly lower than the normalizing process which has at least a difference in tensile strength value of 1.77 kgf / mm^2. While the initial material (AISI 01) has a tensile strength value of 70.79 kgf / mm^2 lowest compared to other processes. This is because the material does not conduct heat treatment process so that there is no change of structure and phase in the material. In the tensile test chart, it can be concluded that different tensile strength values are caused by different cooling processes. The more the cooling process or the rapid cooling rate, the higher the tensile value of the material. This is also because in the process of heat treatment and different cooling rates produce different structures and phases.

Hardness Test Analysis

The heat treatment in AISI 01 steel affects the hardness value of this steel, either decreasing or increasing depending on the cooling process after heat treatment.

In Figure 3, there is a difference in the value of violence.

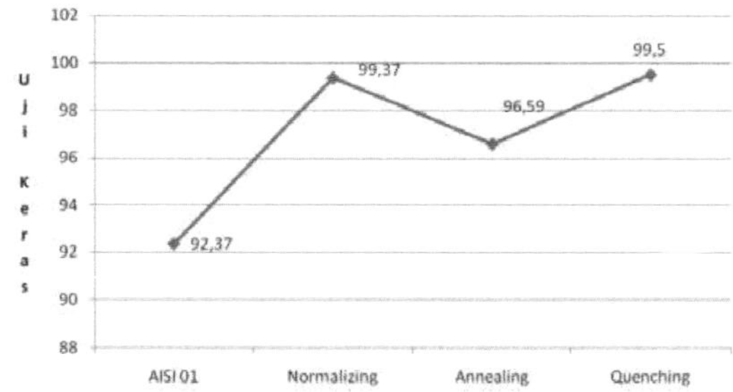

Figure 3. Hardness test graph

From the picture it can be seen that the initial material that does not experience the heat treatment hardness value is the lowest compared to the material or other specimens. This is because no structure changes. Seen on the process of annealing that experienced cooling in the furnace (furnace) has a hardness value of 95.45 HRB. The slow cooling process in the furnace results in a more dominantly soft mechanical property than the violent carbon. When compared to the material that is normalizing the process of cooling the air, obtained a hardness of 99.49 HRB. At quenching process can be seen hard value of 99.52 HRB. Quenching process difference with air cooling is 0.03 HRB. The cooling process of each specimen is different or it can be concluded the specimen that experienced cooling process faster, he will change the structure and the nature becomes harder. This is evident in the hard test graph above where the AISI 01 steel specimen after the heat treatment process is then cooled with an SAE 40 oil quenching medium which includes rapid cooling is harder than any other cooling process.

Elongation Analysis

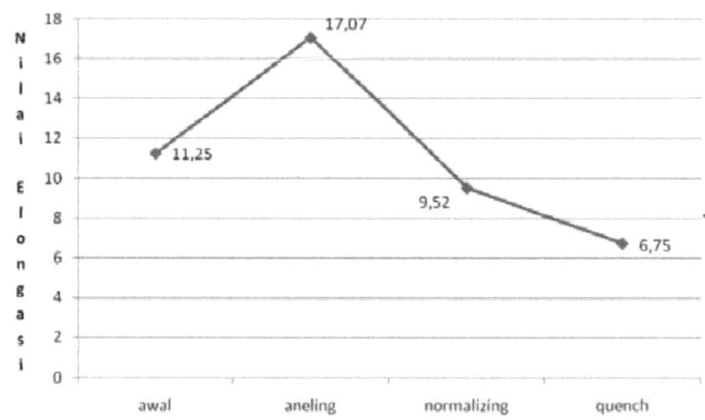

Figure 4. Elongation graph

For elongation values, in accordance with the properties produced, the specimen with rapid cooling leads to a decrease in the value of elongation. In the specimen without heat treatment the elongation value was 11.25% and the annealing specimens of 17.07% increased by 5.82%. This is because the specimen with annealing process is softer compared to the initial specimen. For normalizing specimens of 9.52% compared to the specimen quenching process which has a value of elongation of 6.75% there was a decrease of 2.77%.

So the greater the elongation value, the material or specimen will be more soft and vice versa if the elongation value is smaller then the specimen is getting harder.

Micro Structures Analysis

In the non-heat treatment specimens, the visible microstructures were pearlite and cementite visible microstructures as Fig 5 until Fig.8.

Figure 5. Microstructure of 2000 x enlarged annealing specimen

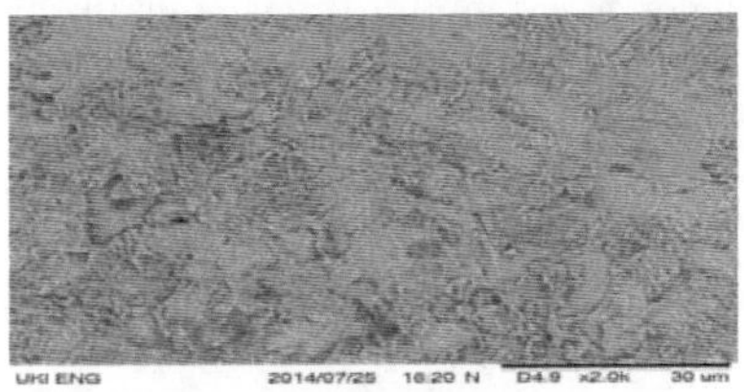

Figure 6. Microstructure of 2000 x enlarged normalizing specimen

Figure 7. Microstructure specimen AISI 01 magnification 2000x

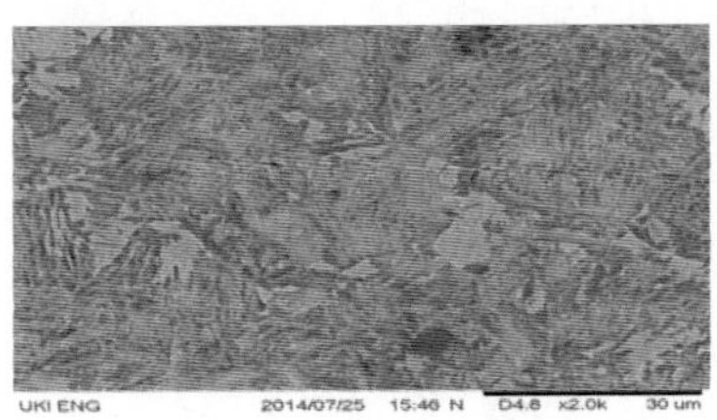

Figure 8. Microstructure of 2000 x enlarged quenching specimen

For the structure that forms pearlite has a rather dark colour while darker than pearlite is cementite. Then on the micro structure of the annealing process will form a fine pearlite structure because it can be seen from the size of the grain, because the cooling process is quite long. In the normalizing process will form pearlite and cementite with a larger size of the microstructure annealing, because the cooling process is fairly fast. In the process of quenching oil to form pearlite and cementite structure with the largest grain size due to the quenching process of oil is the fastest cooling process among other processes. Thus, the structures formed are pearlite and cementite, but have differences in grain size. When viewed from the observation of micro structures with varying magnification using the Scanning Electron Microscope (SEM) the structure or phase formed is pearlite and cementite, which can distinguish the hard and tensile values are the size and shape of the grain. The larger the grain size, the greater the value of tensile and hardness. Seen in the process of quenching oil has the highest tensile and hard value compared with other processes.

4. Conclusion

Based on research and observations made on AISI 01 tool steel material, it can be concluded as follows:

1. Differences in the cooling process may alter the mechanical properties of the material. This is evident in the SAE 40 oil quenching process which has the highest tensile and hardness value followed by the normalizing, annealing and the smallest of tensile and hard values found in the starting material or heatless specimen.
2. For elongation values in the specimen without heat treatment the elongation value was 11.25% and the annealing specimens of 17.07% increased by 5.82%. This is because the specimen with annealing process is softer compared to the initial specimen.
3. For normalizing specimens of 9.52% compared to the specimen quenching process which has a value of elongation of 6.75% there was a decrease of 2.77%.

5. References

[1]. Adnyanā, D.N.dr. Ir., 1989, *Logam dan Panduan*, Jakarta;
[2]. Amanto, Hai, 1999, *Ilmu bahan*, Bumi Aksara, Jakarta;
[3]. Van Vlack, L.H., 1992, *Ilmu dan Teknologi bahan*, terjemahan Sriati Djaprie, Erlangga, Jakarta
[4]. Darmawan, A.S. 2001, *Bahan Kuliah Pemilihan Bahan dan proses*, Surakarta
[5]. American Sociaty for Metals, 1995, *Metals Handbook*, desk Edition,
[6]. Ir. Erikson Samosir MT, 2003, *Diktat Mata kuliah Material Teknik*, UKI,
[7]. Surdia T, 1995, *Pengetahuan Bahan Teknik*, Pradya Paramitha, Jakarta
[8]. Sumanto, 1996, *Pengetahuan Bahan*, Andi offset, Yogyakarta,
[9]. Alamsyah, Aumran, 1993, *Pemilihan Baja Perkakas dan Perlakuan Panas*, Medan,

Evaluation of band gap energy and surface roughness for tin indium zinc oxide thin films by atomic force microscopy and electron spectroscopy

Yus Rama Denny[1,*)], Teguh Firmansyah[2)], Isnaeni[3)], Sovian Aritonang [4)], Kartina AM[5)]

[1)]Department of Physics Education, University Sultan Ageng Tirtayasa, Banten, 42435, Indonesia

[2)]Department of Electrical Engineering, University Sultan Ageng Tirtayasa, Banten, 42435, Indonesia

[3)]Research Center for Physics, Indonesian Institute of Science, Building 442 Puspiptek Serpong, South Tangerang, Banten, Indonesia 15314

[4)]Department of Defense Industry, Indonesian Defense University, IPSC, Sentul Bogor, Indonesia

[5)]Department of Agroekoteknologi, University Sultan Ageng Tirtayasa, Banten, 42435, Indonesia

yusramadenny@untirta.ac.id

Abstract. The SnInZnO thin film was prepared at room temperature by RF magnetron sputtering. The band gap of this film was investigated by REELS and UV-Spectrometer analysis. The surface morphology and the root mean square (RMS) roughness of SnInZnO thin films measured using atomic force microscopy (AFM). By using REELS with low energy (500 eV) and high energy (1500 eV) electron beams, the surface band gap as well as and the bulk energy band gap were measured. In addition, the surface roughness was evaluated by atomic force microscopy. The result shows that the band gaps are affected by the primary electron energies. The transmittance value of 80% in the visible light region has been achieved. The optical band gap was calculated by using Tauc's relation from UV-Spectrometer analysis and compared to that of the band gap obtained by the REELS spectra.

Keywords: transparent conducting oxide; REELS; UV-Spectrometer; band gap energy.

1. Introduction

ZnO-based films have applications as active channel for transparent conducting oxides (TCO) materials in thin film transistors (TFTs). This is due to their higher carrier mobility and transparency in the visible light region compared to the conventional TFTs, which are based on hydrogenated amorphous silicon (a-Si:H) having a carrier mobility of less than 1 $cm^2V^{-1}s^{-1}$ [1]. Tin indium zinc tin oxide (SnInZnO) films, particularly, can be one of the possible alternatives for transparent electrodes of various optoelectronic devices such as flat-panel displays, solar cells, and organic light-emitting diodes because of their good electrical conductivity, chemical stability, and low deposition temperature [2, 3]. There are several deposition techniques to deposit SnInZnO thin films such as electron beam evaporation, laser deposition, spray pyrolisis, organic chemical vapor deposition, and sputtering. However, the sputtering method takes advantages of relatively high deposition rate and a high uniformity so that this method has been used in large-area and high quality depositions. In this paper, the SnInZnO thin film was deposited with RF magnetron sputtering. We used both REELS spectra and UV-Spectrometer to determine the band gap of the SnInZnO thin films and atomic force microscopy to evaluate the surface roughness of the film.

2. Experiments

SnInZnO thin films were deposited on glass substrates at room temperature in argon mixed with oxygen gas (Ar: O_2 = 15: 85) by RF magnetron sputtering with the RF power of 200 W. The target composition ratios of In: Zn: Sn in IZTO thin films is 13: 60.2: 26.8. Thin film was annealed at 350 °C for an hour in air. The band gap of SnInZnO thin films will be examined with two different methods such as REELS spectra and UV-Spectrometer. REELS spectra were obtained by using a VG ESCALAB 210 equipment. The REELS spectra were measured with the primary electron energy of 500 and 1500 eV for excitation and with the constant analyzer pass energy of 20 eV. The full width at half maximum (FWHM) of the elastic peak was 0.8 eV. The transmittance and absorbance spectra of the IZTO thin films were measured by Genesys 6 model from Thermo Electron Corporation in the wavelength range of 300 to 1000 nm at RT at increments of 0.1 nm. The surface morphology including the roughness calculation of the films was performed by atomic force microscopy (AFM) from Nanofocus Inc. system with a scanning area of 2x 2 mm^2.

3. Results and discussion

The basic principle of REELS spectra can be explained in Fig. 1. The REELS spectra consist in bombarding the surface of a sample with a beam of monoenergetic of electrons and detecting the energy distribution of the backscattered electrons [4]. The REELS spectra contains of two area, the elastic peak is due to electron that have lost no energy (elastic backscattering electrons) and at lower kinetic energies refer to electrons that have lost part of their energy through electronic excitations or phonons excitations within the solid as being shown in the Schematic of Fig. 1. The losses produced by the interactions with phonons can be observed only if the experimental set-up is high enough. Hence, this study were performed an experimental resolution of typically 0.8 eV, which can be estimated from full width at half maximum (FWHM) of the elastic peak from REELS experiment.

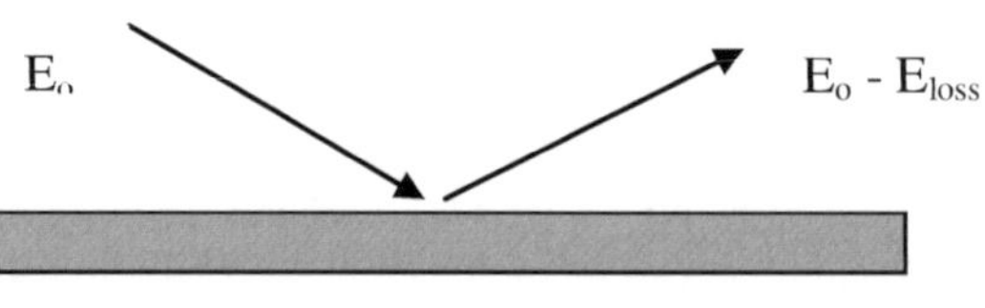

Figure 1. The basic principle of REELS spectra.

The response of a solid to an external field can be characterized by the dielectric response function $\varepsilon(\bar{q}, \omega)$ of the solid. The energy loss probability that an electron loses energy and transfers momentum $\hbar\bar{q}$ in a single collision in the bulk of the solid is proportional to $1/(q^2)\text{Im}(-1/\ \varepsilon(\bar{q}, \omega))$. Despite the fact that in a REELS experiment non-zero momentum transfers occurred this technique has been known however shown to be a powerful tool to obtain information on the gap energy of materials oxide. The shape of the loss spectrum corresponding to single losses is in this case identical to that of the optical loss function $\text{Im}(-1/\ \varepsilon(\bar{q}, \omega))$ and then optical absorption defect related bands can be compared to defect-related losses. The flat part of the loss spectrum near the elastic peak of the REELS spectra is related to the energy gap [5].

REELS spectra of SnInZnO thin films on the glass substrate can be shown in Fig. 2. It shows that REELS spectra contain both elastically backscattered electrons and inelastically backscattered electrons. The kinetic energy corresponds to the energy of electrons that coming out from the surface of thin films. For the elastic peak case, the value of kinetic energy is similar as the primary energy. Thus, there is no loss energy for the elastic peak (zero energy loss). The loss energy can be calculated by subtracting the kinetic energy with the primary energy. In addition, the REELS also consist of the Plasmon peak (known as plasma oscillation) [6]. The Plasmon peak can be explained as follows: the valence electron in a solid can be thought of as a set of coupled oscillators which interact with each other and with reflected or transmitted electron via electrostatic forces. In the simplest situation, the valence electrons behave essentially as free particles and constitute a "free-electron gas" also known as "Fermi sea" or "jellium". The collective excitation of valence electrons is called plasmon. The longe-range Coulomb interactions then make possible collective oscillation of large number of electrons. The plasmon loss signals are observed with a rather broad spectrum shifted toward the lower kinetic energy side of the primary elastic peak by plasmon energy of ~15 eV or more as can be seen from Fig. 2. The plasmons peak show up as broad, smooth replicas of the peaks shifted to lower kinetic energy by an amount equal to the bulk plasmon energy $\hbar\omega p$. In the free-electron approximation, the bulk plasmon energy is related to the valence electron density as $(\hbar\omega_p)2=(\hbar^2e^2/m\pi)N\upsilon$, where m is the electron mass, e is the electron charge, and $\hbar$ is Plank's constant, the $N\upsilon$ valence electron density [6]. This peak experimentally can be observed in the REELS spectra.

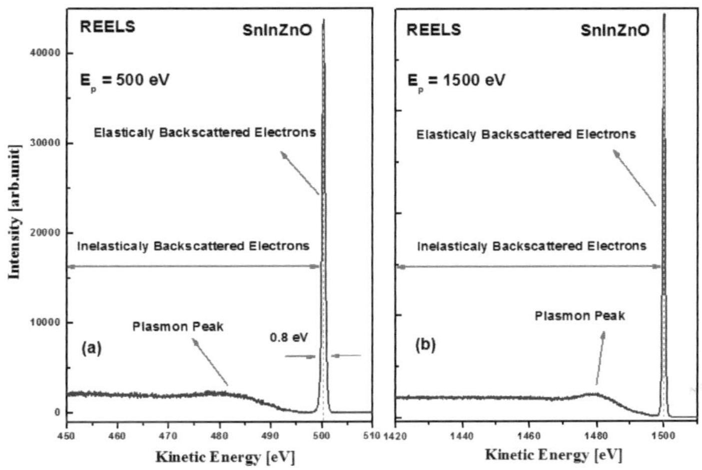

Figure 2. REELS spectra as a function of kinetic energy for different primary energies (Ep = 500 dan 1500 eV).

We made use of REELS to investigate the dependence of the band gap energy E_g of SnInZnO thin films with different primary energies. Figure 3 shows the REELS spectra with different primary energies. As can be seen in Fig. 3, the Plasmon loss peak form SnInZnO thin films located at loss energy of 23 eV and it was no changed correspond to the different primary energy. The REELS spectra were used to estimate the energy band gaps of the SnInZnO thin films by the method described in our previous work [7,8]. The band gap value of the probed region of the excited solid was estimated by the extrapolating the linear part of maximum negative slope to the background level. The onset of the loss signal refers to the band gap value. The result shows that the band gap of SnInZnO thin films were 3.17 and 3.47 eV for the primary energy of 500 and 1500 eV, respectively within the uncertainty of $\pm$ 0.1 eV. Previous study reported that the band gap value of ZnO, SnO_2, and In_2O_3 were 3.37, 3.60, dan 3.70 eV, respectively. Hence, the band gap of SnInZnO thin film is closed to the band gap of ZnO. Therefore, in the band structure of of SnInZnO thin film, there will be an upper band level states (known as conduction band) that derived from Zn 2p filled states and an lower band level states that derived from O 2p states (known as valence band) that are partially occupied. This result suggests the band gap value of SnInZnO thin film could be improved by changing the composition ratio of the ZnO. On the other hand, for SnInZnO, the inelastic mean free paths (λ) of electron energy of 500 eV and 1500 eV are 0.5 nm dan 2.2 nm, respectively [8]. Thus, the energy spectrum for 500 eV gives the surface sensitive information. On the other hand, the energy spectrum for 1500 eV gives the bulk sensitive information. By comparing the band gap values with different primary energy, it can be seen that the surface band gap of SnInZnO thin films is about smaller 0.3 eV than that of the bulk band gap. Overall, this result indicates that the band gaps are affected by the primary energy of electron.

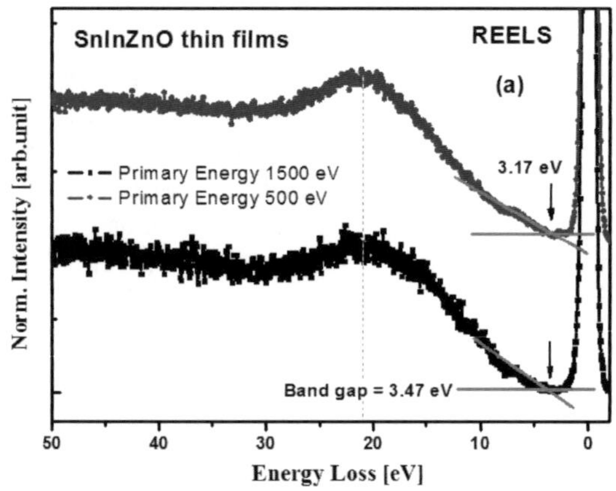

Figure 3. Reflection electron energy loss spectra of SnInZnO thin films in different primary energy of 500 and 1500 eV.

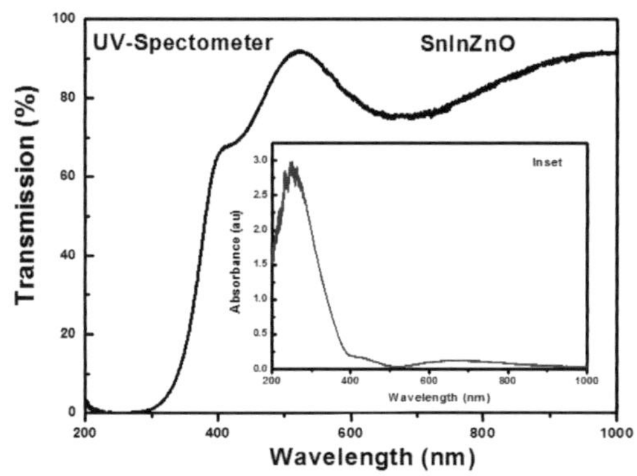

Figure 4. Transmittance spectra of SnInZnO thin films (inset: the absorbance of the film).

For comparison, the band gap was also measured by utilizing a UV-spectrometer. Figure 4 shows the transmittance and absorbance spectra of SnInZnO thin film. The measured transmittance spectra of the films shows that the film is transparent with the transmittance exceeding 80% all over the spectral range 300 to 1000 nm as shown in Fig. 4. The absorbance spectra were used to estimated the absorption coefficient, α(ω), by the relation [9]:

$$\alpha(\omega) = \left(\frac{1}{d}\right) ln\left(\frac{I_0}{I_t}\right) = \frac{2.303\,A}{d} \quad (1)$$

where d and A are the thickness of the films and absorbance, respectively, also I_0 and I_t are the intensities of incident and transmitted radiation, respectively. In the high absorption region, $\alpha(\omega) \geq 10^4$ cm^{-1}, Mott and Davis proposed the following relation for amorphous materials [9]:

$$\alpha\, \hbar\omega = A\, (\hbar\omega - E_{opt})^{m} \quad (2)$$

where E_{opt} is the optical band gap, ħω is incident photon energy, and m is the index which can have different values (1/2, 3/2, 2 and 3) corresponding to direct allowed, direct forbidden, indirect allowed, and indirect forbidden transitions, respectively. A is a constant called the band tailing parameter and is related to speed of light (c), the refractive index (n), the extrapolated D.C conductivity at T = 0 K, and the band tailing energy (E_e). It was noticed from Fig. 4 (a) that for m = 2, the extrapolation of the linear part gives the value of the direct band gap. The value shows that the direct band gap of SnInZnO thin films is about 3.08 eV. In addition, the extrapolation of the linear part of *(αhυ)* $^{1/2}$ versus *hυ* plot is shown in Fig. 4 (b), the indirect band gap is deducted as 3.48 eV. The values of direct band gap and indirect band gap agree with the reported value of other literatures [10,11]. A comparative study of the band gap by REELS spectra and UV-Spectrometer with the same previous finding is given in Table 1. It can be seen from the data in Table 1 that the band gap obtained from the preliminary analysis of the REELS spectra with the primary energy of 1500 eV is comparable to that of the optical band gap from indirect transitions. Similarly, the band gap from the REELS spectra with the primary energy of 500 eV closed to that of the optical band gap from direct transitions. These findings of this study suggest that the direct band gap from UV-Spectrometer can give the surface band gap information; meanwhile the indirect band gap gives the bulk band gap information.

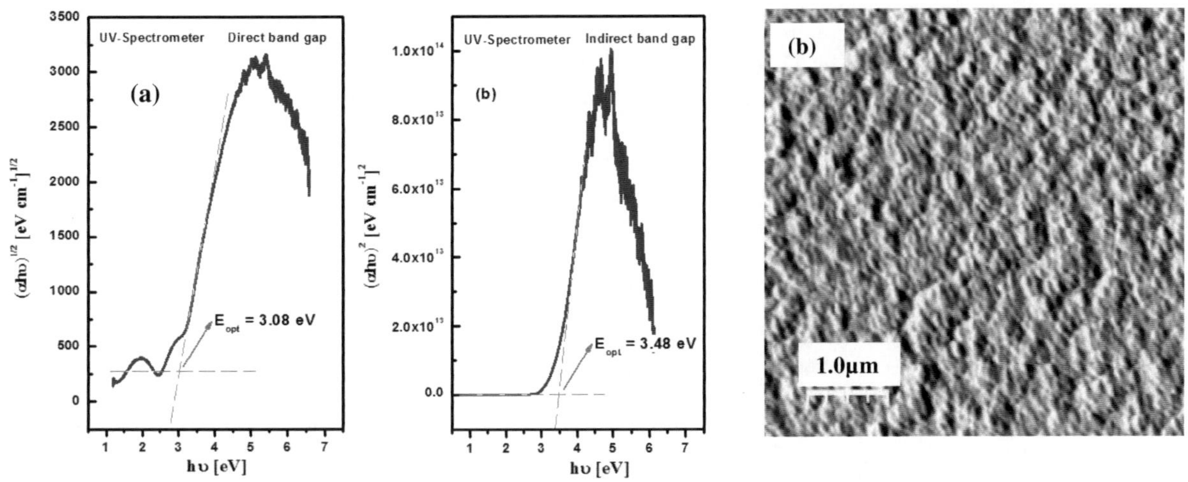

Figure 5. (a) Transmittance spectra of SnInZnO thin films (inset: the absorbance of the film), (b) Atomic force microscopy of SnInZnO thin films.

Table 1. A review of some published data on the optical energy gap and its comparison with the present findings.

Energy Gap (eV)	Method	Preparation technique	Ref.
3.50	Indirect band gap	Electron beam evaporation	[10]
3.00	Direct band gap	Co-sputtered DC Sputtering	[11]
3.44	Indirect band gap	Co-sputtered DC Sputtering	[11]
3.08	Direct band gap	RF magnetron sputtering	Present work
3.48	Indirect band gap	RF magnetron sputtering	Present work
3.17	REELS with E_p = 500 eV	RF magnetron sputtering	Present work
3.47	REELS with E_p = 1500 eV	RF magnetron sputtering	Present work

Figure 5 (b) shows the surface morphology and the root mean square (RMS) roughness of SnInZnO thin films measured using AFM. The roughness parameters are determined on each image obtained in the tapping mode (height image) and are defined as follows: Rq is the standard deviation of the Z values within a given area and is calculated using [12]

$$R_q = \left[\sum (Z_i - Z_{ave})^2 / N\right]^{1/2} \quad (3)$$

where Z_{ave} is the average of the Z values within the given area, Zi the current Z value and N the number of points within the given area. Ra is the mean roughness. This is the mean value of the surface fluctuations relative to the center plane and is calculated using [12]

$$R_a = 1/L_x L_y \int_0^{Ly}\int_0^{Lx} |f(x,y)| dxdy \quad (4)$$

where f (x,y) is the surface fluctuation relative to the center plane and Lx and Ly are the dimensions of the surface. The statistical parameters calculated in the roughness analysis mode, are listed in the Table 2 to give a clear view of the obtained data. From the Table 2, the RMS roughnesses of SnInZnO film for all samples are 5.407 nm.

Table 2. Statistical parameter of SnInZnO thin film by atomic force microscopy.

	Data obtained from AFM
RMS Roughness	5.407
Minimum Value	-10.1035
Maximum Value	10.1027
Peak to peak	20.2063
Roughness Average	4.6010
Average height	0.6184
Surface skewness	-0.0133
Surface Kurtosis	1.9559
Plane offset	0

4. Conclusion

In this study, we investigated the band gap of SnInZnO thin film by using different method REELS spectra and UV-Spectrometer. The REELS spectra showed that the band gap values are strongly affected by primary energy. In the visible light region, the transmittance coefficients SnInZnO thin film is approximately 80%. Furthermore, we found that the direct band gap from UV-Spectrometer analysis can give surface band information; meanwhile the indirect band gap gives the bulk band gap information. In addition, we examined the surface roughness by atomic force microscopy and the result shows that the roughness of the the RMS roughness of SnInZnO films is 5.407 nm. We believe that our results can be a good guide for those growing SnInZnO thin films with the purpose of device applications, which require TFTs good transparency and a proper the band gap values.

Acknowledgment

The authors also would like to thank to the Ministry of Research, Technology, and Higher Education, Indonesian Government in terms of Simlibtamas Hibah Penelitian. This research was also supported by the Basic Science Research Program through the National Research Foundation of Korea (NRF) funded by the Ministry of Education, Science and Technology (2012R1A1A2009590).

5. References

[1]K. Nomura, H. Ohta, A. Takagi, T. Kamiya, M. Hirono, and H. Hosono, Nature (London) **432**, 488 (2004).
[2] K. Nomura, H. Ohta, K. Ueda, T. Kamiya, M. Hirono, and H. Hosono, Science **300**, 1269 (2003).
[3] Damishi, H. C. Ma, J. J. Kim, and H. Y. Lee Thin Solid Films **520**, 3741 (2012).
[4] W. S. M. Werner, Surface and Interface Analysis, **31**, 141-76 (2001).
[5] R. F. Egerton, Rep. Prog. Phy., **72**, 016502 (2009).
[6] R. F. Egerton, *Electron Energy-Loss Spectroscopy in the Electron Microscopy*, 2nd Edition (Plenum, New York, 1996).
[7] Y. R. Denny, H. C. Shin, S. Seo, S. K. Oh, H. J. Kang, D. Tahir, S. Heo, J. G. Chung, J. C. Lee, and S. Tougaard, J. Elect. Spect. Rel. Phenom. **185**, 1822 (2012).
[8] Y. R. Denny, S. Lee, K. Lee, S. Seo, S. K. Oh, H. J. Kang, S. Heo, J. G. Chung, J. C. Lee, and S. Tougaard, J. Vac. Sci. Technol. A **31(3)**, 031508-1 (2013).

[9] D. L. Wood and J. Tauc, Phys. Rev. B **5**, 3144 (1972).
[10] H. A. Mohamed, J. Phys. D: Appl. Phys. **40**, 4234–4240 (2007).
[11] K. J. Saji, and M. K. Jayaraj, Thin Solid Films, **516** 6002–6007 2 (2008).
[12] G. Kavei, Y. Zare and A. Mohammadi gheidari, Scanning, **30**, 232 -230 (2008).

IOP Conf. Series: Materials Science and Engineering **343** (2018) 012007 doi:10.1088/1757-899X/343/1/012007

Influence of Sintering Temperature on Mechanical and Physical properties of Mill Scale based Bipolar Plates for PEMFC

Deni S. Khaerudini[1,2], Rina Berliana[1], Gatra B. Prakoso[1], Dita R. Insiyanda[1], Sagir Alva[2]

[1] Fuel Cell Group, Research Center for Physics, Indonesian Institute of Sciences, Bld. 440 Kawasan Puspiptek Serpong, Tangerang Selatan 15314, Banten, Indonesia

[2] Department of Mechanical Engineering, Universitas Mercu Buana, South Meruya No. 1, Jakarta 11650, Indonesia

deni.shidqi.khaerudini@lipi.go.id

Abstract. This work concerns the utilization of mill scale, a by-product of iron and steel formed during the hot rolling of steel, as a potential material for use as bipolar plates in proton exchange membrane fuel cells (PEMFCs). On the other hand, mill scale is considered a very rich in iron source having characteristic required such as for current collector in bipolar plate and would significantly contribute to lower the overall cost of PEMFC based fuel cell systems. In this study, the iron reach source of mill scale powder, after sieving of 150 mesh, was mechanically alloyed with the aluminium source containing 30 wt.% using a shaker mill for 3 h. The mixed powders were then pressed at 300 MPa and sintered at various temperatures of 400, 450 and 500 ^{o}C for 1 h under inert gas atmosphere. The structural changes of powder particles during mechanical alloying and after sintering were studied by x-ray diffractometry, scanning electron microscopy (SEM) with energy dispersive X-ray spectroscopy (EDX), microhardness measurement, and density – porosity analysis. The details of the performance variation of three different sintering conditions can be preliminary explained by the metallographic and crystallographic structure and phase analysis as well as sufficient mechanical strength of the sintered materials was presented in this report.

1. Introduction

Proton Exchange Membrane Fuel Cell (PEMFC) is one of the promising alternative energy technologies to be developed, because this technology has some advantages in term of high process efficiency, low operating temperature and producing electricity from electrochemical reactions in an environmentally friendly manner [1]. PEMFC consists of various components, among which bipolar plates allocate about 60 to 80% of PEMFC weight and 30 to 45% of their cost [2]. However, its use is not optimal yet, due to high weight and high production costs, which largely affected by bipolar plate.

Bipolar plate (BPP) is one the most significant part in PEMFC. The function of bipolar plate is the backbone of a PEMFC, facilitates water and thermal management, and provides conduits for reactant gases. Generally, there are two types of bipolar plate in PEMFCs; the carbon-based and the metal-based bipolar plate. The carbon-based bipolar plates have higher corrosion resistance and lower contact resistance, but they are considered as brittle and permeable to gases. Compared with the carbon-based bipolar plate, metal-based bipolar plates have higher mechanical strength, lower permeability, and much superior manufacturability. Due to their high strength, metallic BPP can be made of very thin sheets with a thickness of 0.1 mm or less [3-6]. However, the main handicap of the metal-based bipolar plates is the lack of ability to combat corrosion in the harsh acidic and humid environment of PEMFCs.

The unique combinations of properties provided by aluminum and its alloys make aluminum one of the most versatile, economical, and attractive metallic materials for a broad range of uses - from soft, highly ductile wrapping foil to the most demanding engineering applications. Aluminum alloys are the second after steels in use as structural metals. The applications of aluminum alloys are economical in many applications. These alloys, being formed by Fe_3Al or FeAl ordered phases, are characterized by very good high-temperature corrosion resistance in oxidizing and sulphidizing environments together with low cost of the constituents and lower density (5.7 g cm^{-3} for FeAl). Possible production routes of iron aluminides concern casting technology, hot working, or powder metallurgy processes. Appropriately alloyed and treated, aluminum can resist corrosion by water, salt, and other environmental factors, and by a wide range of other chemical and physical agents [7-9]. Therefore, the alloy of aluminium and mill scale is ones of the metallic bipolar plates that can be considered, due to the possibility to mass production and decreasing PEMFC volume and weight, which lowers the overall manufacturing cost. As an alternative one, waste-based materials can also be utilized as raw materials, for instance, mill scale. Mill scale is a waste material from the steelmaking industry and considered has rich iron content (about 70-75% Fe). Thus, the high content of Fe in mill scale is potentially used as raw material for manufacturing metallic-based bipolar plate. In this study, in order to reduce the cost as well as utilize the industrial mill scale waste, we report a study of the utilization of mill scale based bipolar plate with some variation addition of waste of aluminum for evaluation and justification on their microstructural, physical and mechanical properties [10].

2. Experimental

In the present study, the industrial mill scale (as received from PT Krakatau Steel, Cilegon, Indonesia) the chemical composition (wt.%) of Fe total: 74.24%, Fe metal: 0.20%, Fe_2O_3: 52.02%, SiO: 0.25%. CaO: 0.97%, MgO: balance (analyzed by XPS), was chosen as raw material. After sieving of 150 mesh and to prepare the substrate sample, the waste of aluminum material was added–with a composition of 30%wt Al addition with temperature variations of 450, 475 and 500 ^{o}C. Thus, the blends were mechanically alloyed using a shaker mill in a discontinues milling manner with milling speed of 600 rpm, delay time of 20 minutes and total milling time of 3 h. To prepare the green body, the milled powders were pressed with a loading of 300 MPa at room temperature into the mould with diameter of 20 mm and targeted thickness of of 3 mm. The green body was then sintered at those three variations sinter temperature for 1 h under inert gas atmosphere. After grinding and polishing, the samples obtained were microstructure analyzed by scanning electron microscopy (SEM Hitachi SU 3500) and with energy dispersive X-ray spectroscopy (EDX Horiba). Phase analysis was carried out by X-ray diffraction (XRD SmartLab, Rigaku) with a copper anode x-ray tube with 2θ from 10 to 90° using a 0.01^{o} step size and 0.20 second count time. To measure the hardness surface of sample, Vickers hardness was analyzed by Microhardness Tester (HV Leco LM100AT). The density and porosity analysis were carried out by Archimedes method.

3. Results and discussion

3.1. Phase characterization

The X-ray diffraction analysis was carried to identify the phases present in the powder of raw material and after the sintering process of alloyed sample, as showed in Figure 1. In this study, the bipolar plate samples were prepared by using the mill scale and 30 wt. % of aluminium. Mill scale is a waste material from the steelmaking industry and considered has rich iron content (about 70-75% Fe) and other oxide content. Therefore, the oxide phases that contained in mill scale can be observed due to the original of the raw mill scale content and during the heating process. Figure 1a shows the after milled

FeAl alloy powder for 3 h as well as their original raw materials of mill scale (Fe) and Al powder. From XRD peaks, it shows the existence of new phase Fe_3Al after 3 h milling process at 2θ value of about 65.899°. This Fe_3Al phase can be obtained due to some collision process between the ball milling (high-energy ball mills) and the powders during under the desired milling period and ball to powder ratio. Under such desired conditions, some eroded state of powder and size reduction will be occurred. Subsequently, the heat transfer of ball-powder process is also occurred due to the friction and collision, so that the new phase of Fe_3Al is formed.

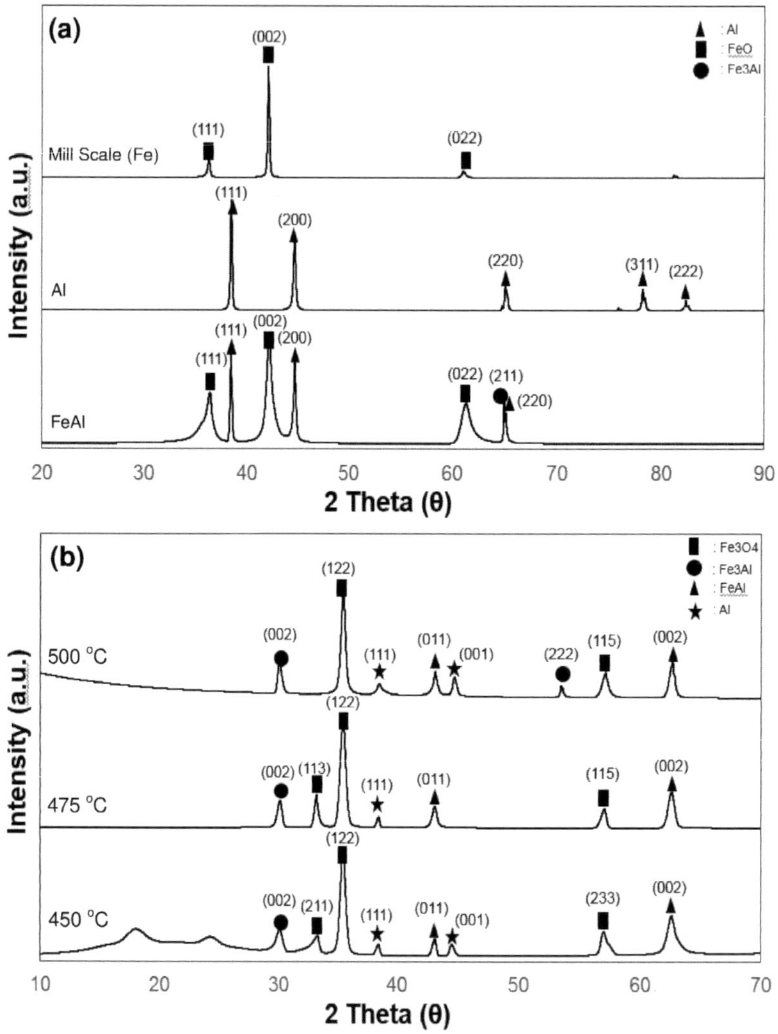

Figure 1. X-ray diffractogram of (a) mill scale (Fe), Al and FeAl powder; and (b) FeAl sample sintered at 450, 475 dan 500 °C showing Fe_3O_4, Fe_3Al, FeAl and Al phases

Figure 1b shows the XRD peaks of the bipolar plate samples after sintered at different temperatures of 450, 475 and 500 °C. After the sintering process, the different phases can be observed, including Fe_3Al, Fe_3O_4 and FeAl phases. Generally, the sintering process is a heating process under the recrystallization temperature or melting point. In this study, the variation of sintering temperatures was carried out, however, the Fe_3Al phase can be detected and formed on each variation of sintering condition. When the sinter temperature increases, these phases are continuously transformed into FeAl and Fe_3Al phases. Fe_3Al phase was formed due to the influence of alloying process of mill scale, which had sufficient Fe content, with the Al powder. In this case, the similar behaviour was observed i.e., the aluminium is melts; consequently the intermetallic Fe_3Al phase is emerged. The Fe_3Al phase has good mechanical and physical properties. Therefore, this material can be considered as a good raw material for bipolar plate purpose. From the XRD, after the sintering process, the magnetite Fe_3O_4 phase is also observed. This usually happened at high temperature. In this case, this occurs when the sintering temperature is increase, so that the Fe element (from the mill scale) tends to easily react with oxide that very reactive at high temperature condition. Other reason, the original content of mill scale itself, which contain some metal oxide as a by-product of the steel making industry. Therefore, the oxide phases that originally contained in the mill scale will be emerged during the sintering/heating process.

3.2. Vickers hardness

In this study, the hardness testing was carried out by Vickers micro-hardness method. This test was performed on the sintered samples in order to evaluate the Fe-Al bonding with respect of variations of sintering temperature of 450, 475 and 500 °C. The results showed that the obtained highest hardness value is 713.27 VHN for the sintered 475°C sample. From the graph, it also can be observed that the hardness value for two other different sintering temperatures at 450 and 500 °C are 421.79 and 384.26 HVN, respectively. At 500 °C, the hardness value shows decreases. This can be explained due to an over sintering condition which contributed by Al element. Another consideration is the pre-melting condition. This achieved when too high sinter condition on the necking process of the samples, then causing increases in porosity. Therefore, the pre-melting occurs so that the porosity increases at sinter condition of 500 °C. From the results, it is clearly indicated that the ideal sinter temperature should be less than 500 °C. From this study, the best one is at 475°C with hardness value of 713.27 VHN.

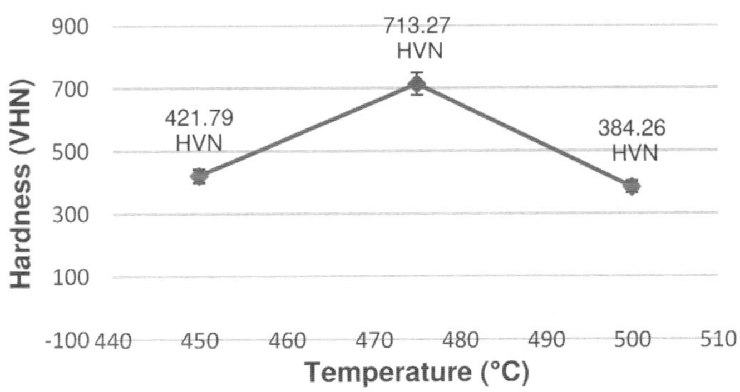

Figure 2. Vickers Hardness results with different of sintering temperature at 450, 475 and 500 °C.

3.3. Metallographic analysis

In this study, we further evaluated the microstructure of bipolar plate sample with the FE-SEM and EDX micrograph analysis, as shown in Figure 3. From the FE-SEM observation, from the surface view, indeed some void still can be observed. Further, indeed there are some inhomogeneous of the grain size, this can be explained due to the difference of melting point between the Fe (mill scale) and Al element. However, in term of hardness, as previously shown in Figure 2, the sintered 475°C is the highest one. This can be explained due to some better of Fe-Al compound bonding. Further, the point analysis is also indicated the existence of Fe_3O_4 and Fe_3Al phases, as also previously confirmed by XRD test result. However, one can be revealed from this observation is that the Fe_3Al phase is existed/overlapped on the surface of Fe_3O_4 phase.

Figure 3. FE-SEM metallography observation at x2000 magnification of FeAl sintered 475°C

Further analysis of EDX results are shown in Figure 4. One can be observed that the Fe and Al elements are distributed homogenously from the sample surface. When the Fe_3Al phases are being produced by powder metallurgy process under desired reactive sintering condition, it can be expected that inter-diffusion of iron and aluminium will be occurred and the ordered phases will be formed after being heated to sufficient temperature. Simultaneously, the densifying of the material by sintering of powder particles would happen. $FeAl_3$ phase is formed by the Fe and Al solid state reaction below the melting temperature of aluminium. After that, the remaining aluminium element melt and are enriched by iron. Consequently, other Fe-Al phases precipitate from the melt [11].

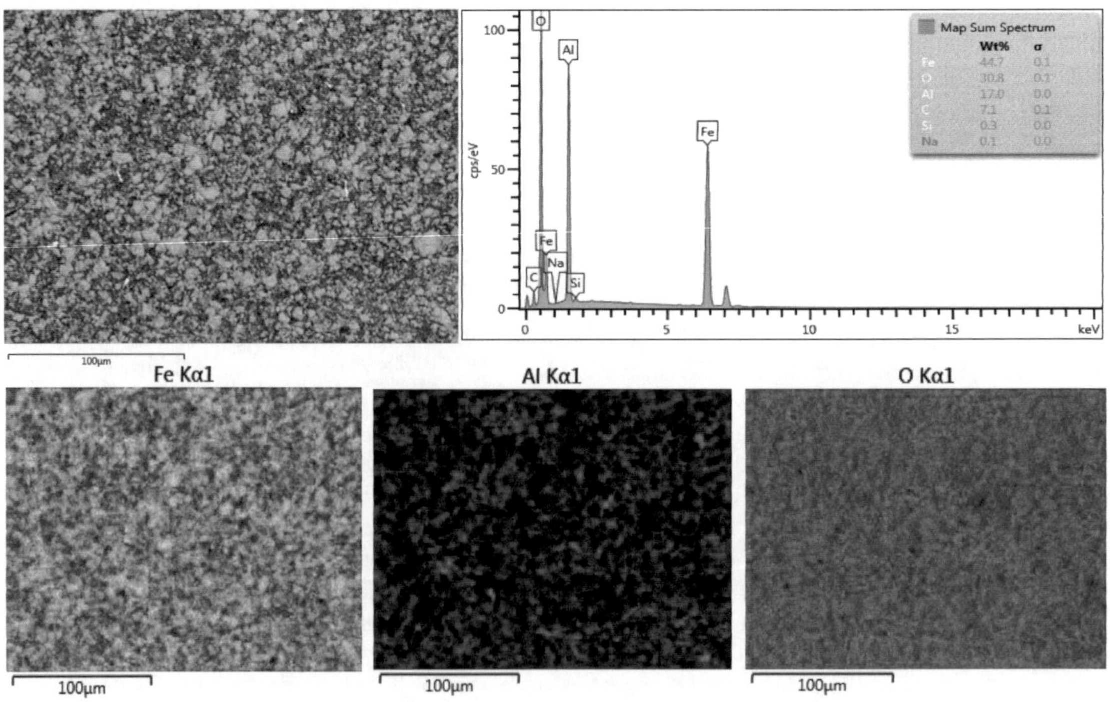

Figure 4. FE-SEM and EDX of FeAl sintered 475°C

The wt.% results data from EDX analysis is also evaluated, as shown in the insert figure of EDX peaks. From the results, in here we focused on the Fe and Al elements. The observed Fe element is 44.7 wt.% and Al element is 17.0 wt.%. After calculation, if we compared only between those two element, resulting on 72.4 wt.% Fe and 27.6 wt.% Al. Those results are relatively close with the designed alloying mixing sample of 70 wt.% mill scale (Fe) and 30 wt.% Al. This also indicated that the alloying process in this study is considered sufficient condition.

3.4. Density and Porosity Analysis

Figure 5 shows the density and porosity of Fe-Al samples. Figure 5a shows the density values with respect variation of sintering temperature. The obtained densities are 1.3495, 1.3793 and 1.3507 gr/cm^2 at 450, 475, and 500 °C, respectively. The density value corresponds to the DOE standard for the bipolar plate material should be less than 5 gr/cm^2. From Figure 5a, the density is decrease when the sinter temperature higher than 475°C. This can be explained due to effect of sintering temperature that causing some pre-melting phase when it's over from their ideal sinter condition which contributed mainly from Al element in the study. The increasing of sinter temperature also causing some re-grain growth in the necking area that established during sintering. This initialized by re-diffusion of the elements due to the heat transfer effect which resulting in rearrangement of the grain interface in the necking area. This causes increasing on their porosity as a consequence decrease their density and mechanical properties of the bipolar plates sample. In Figure 5b, it can be seen clearly that the porosity is increase at sinter temperature of 500 °C, i.e. 1.0479 %. This value is inappropriate due to the DOE standard on the desired porosity which should be less than 1 %. Meanwhile, at 450 °C, the obtained porosity is 0.8374 % and at 475 °C is 0.7250 %; those two conditions are still considered the suitable condition due to the DOE standard value.

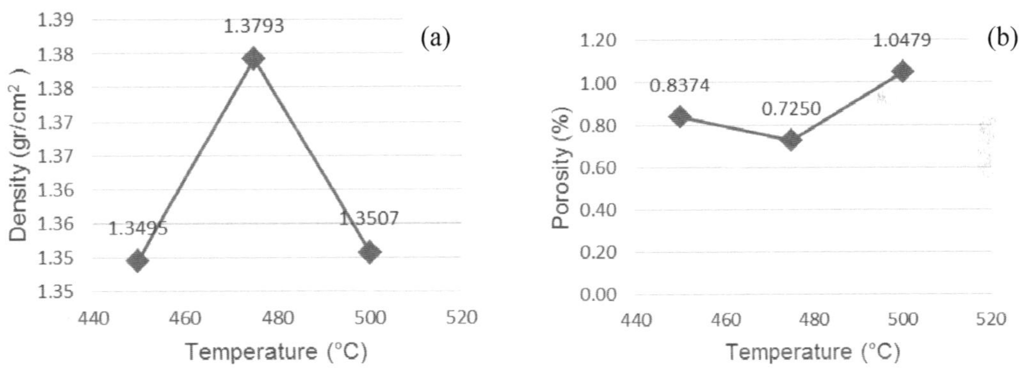

Figure 5. (a) Density and (b) Porosity of the bipolar plate sample after sintered at 450, 475 and 500°C

4. Conclusions

In this report, a series of sintered mill scale-based bipolar plate for PEMFC were developed by powder metallurgy method and the influence of various sintering condition on mechanical and physical properties were investigated. We fabricated it with aluminium addition of 30 wt.% and sintered at 450, 475 and 500 °C for 1h under inert gas atmosphere. The crystallographic analysis by XRD revealed that the obtained phases are magnetite (Fe_3O_4), Fe_3Al, FeAl and Al phases, while for the sintered sample the existence of aluminium (Al) phase is also can be detected. The hardness evaluation showed that the sample sintered 475 °C is much higher (713.27 VHN) than other two condition of sintered samples, as also confirmed by their density result (1.3793 g/cm^3). This is tentatively distributed to the not optimum and over sinter condition of mill scale-aluminium element. The SEM micrograph and EDX analysis showed the inhomogeneous of grain size morphology of the sintered sample due to the difference of optimum sintering condition of Fe and Al element. While the EDX analysis revealed the homogeneous distribution of Fe and Al element on the sample surface. The EDX analysis also confirmed the percentage of alloy composition that close to 70:30 wt.% of Fe:Al ratio. Those results suggest that the mill scale waste has some potential applications as bipolar plate that might be a suitable for commercialization of bipolar plate and PEMFCs technology after further research investigation and evaluation.

Acknowledgments
The authors would like to thank PT Krakatau Steel for raw material of mill scale support and the *Research Center for Physics* – LIPI for technical and facilities support.

References

[1] Wee J H 2007 *Renew. Sustain. Energy Rev.* **11** 1720
[2] Feng K, Li Z, Sun H, Yu L, Cai X, Wu Y, Chu P K 2013 *J. Power Sources* **222** 351
[3] Rajasekar S, Chetty R, Neelakantan L 2015 *Int. J. Hydrogen Energ* **40** 12413
[4] Zhou T –Y, Chen Y –S 2015 *J. Fuel Cell Sci. Technol.* **12** 051001
[5] Jung S –P, Lee C –I, Chen C –C, Chang W –S, Yang C –C 2015 *J. Power Sources* **283** 429
[6] Tawfik H, Hung Y, Mahajan D 2007 *J. Power Sources* **163** 755
[7] Kratochvíl, P 2008, *Intermetallics* **16** 587
[8] Hauild P, Siegl J, Málek P, Ima V 2009, *Intermetallics* **17** 680
[9] Judkins R R, Rao U S 2000, *Intermetallics* **8** 1347
[10] Lu C D, Mao P Z, Wen D H, Jia H 2013 *Key Eng. Mater.* **546** 164
[11] Novák P, Michalcová A, Marek I, Mudrová M, Saksl K, Bednarčík J, Zikmund P, Vojtěch D 2013 *Intermetallics* **32** 127

IOP Conf. Series: Materials Science and Engineering **343** (2018) 012008 doi:10.1088/1757-899X/343/1/012008

Local Structure and Magnetic Properties of $Fe_{50}Cr_{40}Si_{10}$ Nanocrystalline Alloys

K. Tarigan[1], D. Sebayang[1], S. C. Yu[2], and D. S. Yang[3]

[1]Department of Mechanical Engineering, Mercu Buana University, Jakarta 11650, Indonesia.
[2]Physics Department, Natural Collage, Chungbuk National University, Cheongju 361-763, Korea
[3]Physics Division, School of Science Education, Chungbuk National University, Cheongju 361-763, Korea

kontan.tarigan@mercubuana.ac.id

Abstract: The structural and the magnetic properties of nanocrystalline $Fe_{50}Cr_{40}Si_{10}$ alloys were prepared via mechanical alloying by using Fe, Cr, and Si elements with 1- to 24 hrs milling times. Structural analysis based on X-ray diffraction and Extended X-ray absorption fine structure spectroscopy. Concerning the magnetic behavior, the data obtained from a vibrating sample magnetometer at 300 K shows that both the magnetic saturation and the coercivity are dependent strongly on the milling time and the crystallite size. By adjusting the milling time, both appropriate structural transformation and magnetization values are obtained.

1. Introduction

In recently years, nanocrystalline materials have been the subject of scientific interest because of their attractive properties often superior to those of conventional polycrystalline materials and also amorphous alloys of the same composition. Their properties are quite different from those of the corresponding crystalline materials. Nanocrystalline materials are novel materials that are not only scientifically interesting but also holding great potential for a number of technological applications [1-3].

Nanocrystalline and amorphous magnetic materials have been studied for many applications in industrial products, including transformers, motors, and a wide variety of magnetic components in sensors, power electronics, electrical energy control/management systems, telecommunication equipment and pulse power devices [4-5]. The wide range of applications arises from the versatile nature of these materials which can provide fast magnetization reversal with minimal magnetic losses. One aspects of the versatility is that materials can be prepared without stoichiometric restrictions characteristic of crystalline materials and by post-fabrication heat treatment. In certain cases, alloys can be designed for specific applications. There are, however, constraints that limit the fabrication of these materials. In developing a magnetic material, two extreme cases are of interest from an application point of view: one is a material with as high permeability as possible and the other is a material with the saturation induction Bs as high as possible [6] is an important area of research and development [7]. The nanocrystalline supersaturated solid solutions and amorphous phase in the powders are obtained during mechanical alloying [8-9]. Nanocrystalline materials obtained by high-energy ball milling are of great interest since it is known that those materials may exhibit different electrical, magnetic, optical, and other physical properties in the nano-regime due to finite size effects [10].

Mechanical alloying (MA) is a non-equilibrium solid-state alloying technology for powders and can be used to synthesize novel alloys impossibly obtained by conventional technology. It is well known that mechanical alloying of elemental metal mixtures can generate equilibrium and non-equilibrium structures, including supersaturated solid solution, nanocrystalline, metastable compounds and amorphous solids [1,11]. MA has been shown to be capable of synthesizing a variety of equilibrium and non-equilibrium alloy phases starting from blended elemental or pre-alloyed powders. The non-equilibrium phases synthesized include supersaturated solid solutions, metastable crystalline and quasicrystalline phases, nanostructures, and amorphous alloys [1]. During the last years the mechanical alloying technique has been found to be very effective in producing powders with interesting properties. By this means it is possible to synthesize alloys or composite materials with highly dispersed components far away from thermal equilibrium state like amorphous or nanocrystalline materials [1-2].

Besides the incipient crystallisation of amorphous solids, mechanical alloying is nowadays one of the widely used preparation nanocrystalline structures. Mechanical alloying techniques involve the synthesis of materials by high-energy ball milling, in which elemental blends (or pre-alloyed powders, oxides, nitrides, etc) are milled to achieve alloys or composite materials. High-energy ball-milling offers indeed supplementary degrees of freedom in the choice of possible for synthesizing new materials and appears further as an attractive method of synthesis in view of its potential for large scale production. These techniques allow producing non-equilibrium structures/microstructures including amorphous alloys, extended solid solutions, metastable crystalline phases, nanocrystalline materials and quasi crystals. Last two decades, the various mechanical routes used in producing soft magnetic powders (ferrites and alloys) were reported [10-12]. Mechanical alloying is able to produce nanostructure materials with unique chemical, structural, electrical and magnetic properties, due to type of disorder created by the high density of defects and the small crystal size [13].

Nanocrystalline magnetic materials are intensively investigated because of their remarkable properties such as coercivity and saturation magnetization, which significantly differ from those of microcrystalline materials and are sensitive to the structure and microstructure. It is known that the interplay between the crystallite size and the magnetic properties i.e. saturation magnetization, coercivity of FeCr and FeCrSi alloys with nanometer crystallite size is not yet understood fully. This research constitutes a contribution to the comprehension of the relation between the physical properties (local structural and magnetic) of nanostructure some alloys elaborated by high energy ball milling method.

Although important progress has been made in the study of nanocrystalline magnetic materials prepared by mechanical alloying methods, there are numbers of open questions, whose solution could help to advance both their theory and practical applications. Among the challenges facing experts are those concerned with the correlation between structural and magnetic properties of materials at nanocrystallite size scales.

The main objective of this study is the investigation of the relationship between the structure and the magnetic properties of nanocrystalline alloys which were made by mechanical alloying. To understand the mechanism of solid state reaction by mechanical alloying on the atomic scale, the particle size of these alloys will be examine by the structural evolution by using X-ray diffraction (XRD) and Extended X-ray Absorption Fine Structure (EXAFS), and the magnetic properties are measured by using a vibrating sample magnetometer (VSM).

2. Experimental method

$Fe_{50}Cr_{40}Si_{10}$ nanocrystalline alloys were prepared by mechanical alloying using SPEX 8000 mixer with stainless steel balls and vial. The starting material were a mixture of pure Fe(53 um, 99 %), Cr(75 um, 99 %) and Si(105 um, 99 %) powders which were used commercial powders as the precursors. The weight ratio of balls-to-powder mixture was 5:1. $Fe_{50}Cr_{40}Si_{10}$ alloys were mixed and ground for different times 1-, 3-, 6-, 12-, and 24 hrs. The process was performed in Ar ambient to prevent oxidation during the alloying process. Magnetic measurements were carried out on VSM in magnetic field of 10 kOe. Structure data were obtained by using the X-ray diffractometer (XRD) Cu-K_α radiation. The data were analyzed using Match. Based on these data, crystallite sizes of the samples were estimated in terms of the Williamson-Hall plot. EXAFS data were collected with energy of 2.5 GeV, and a maximum current of 200 mA. EXAFS spectra were obtained at Fe K-edge (7112 eV) in the transmission mode at room temperature. The sample chamber was filled with pure nitrogen gas. Then the EXAFS data were analyzed using IFEFFIT software, an interactive program for XAFS analysis.

3. Results and discussions

Figure 1 shows the X-ray diffraction of the $Fe_{50}Cr_{40}Si_{10}$ powder. The XRD pattern of un-milled powder indicates the characteristic reflection of individual element constituent of the Fe, Cr, and Si elements. From 1- up to 24 h, Si patterns were disappeared, it resolved in Fe and Cr structure, only the bcc phase peaks are visible (A2 solid solution α-Cr (Fe, Si)).

Based on the line in Figure 1, that Fe and Si atoms dissolved into the Cr lattices, so that the intensity of the bcc phase diffraction peaks decreases with increasing of milling time.

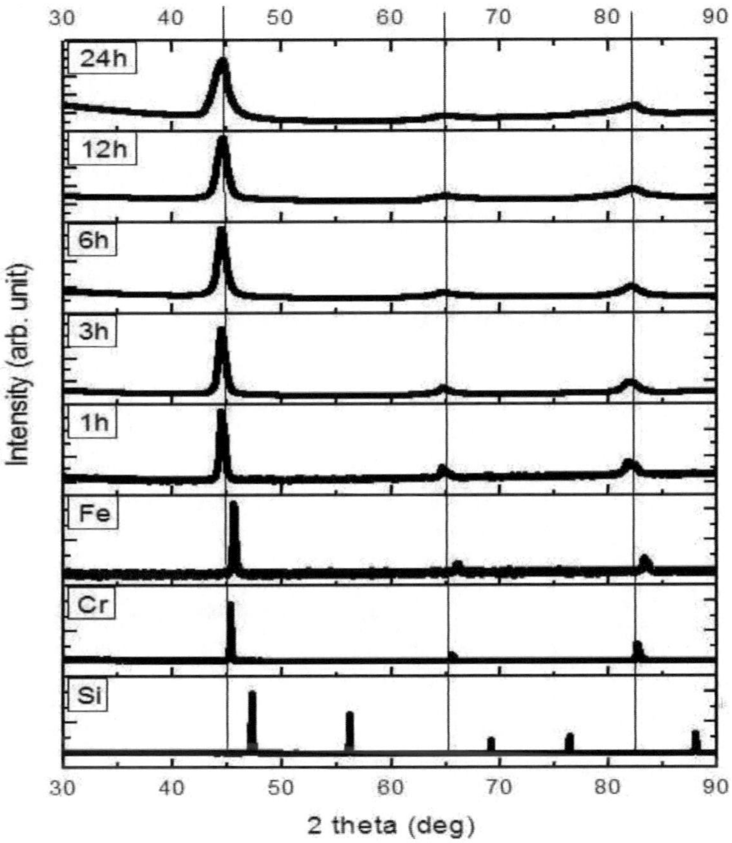

Figure. 1: The XRD of $Fe_{50}Cr_{40}Si_{10}$ of 1- to 24 h, respectively.

In Figure 2, the deformation is due to the replacement Cr atoms by Fe and Si atoms, which signals the formation of an alloy. Based the data, the crystallite size and the lattice strain can be evaluated from the intercept and slope the Williamson-Hall plot, $B\cos\theta = \frac{K\lambda}{D} + 2\epsilon\sin\theta$, where B is the PWHM in radians, D is the average crystallite size, ϵ is the strain, K is the shape factor, λ is the X-ray wavelength and θ is the Bragg angle. In general, both of crystallite size and lattice parameter are decreased.

In inset of Figure 2, the lattice parameter decreases for increasing milling time, except for 1 h milled which is the sample has not stabled yet. The peak after 3 h milled are broader, which is due to the deformation of the structure and variation in the crystallite size.

Figure 3 of the EXAFS give direct information about the variation of local structure. Its use to examine local structure around the Fe ions in the $Fe_{50}Cr_{40}Si_{10}$ alloys. Figure 3 shows the normalized near edge spectra for the processed sample were similar to each other but above the edge the spectra gradually changed. This suggests that electronic configuration for the Fe central atom was unchanged during the MA processing.

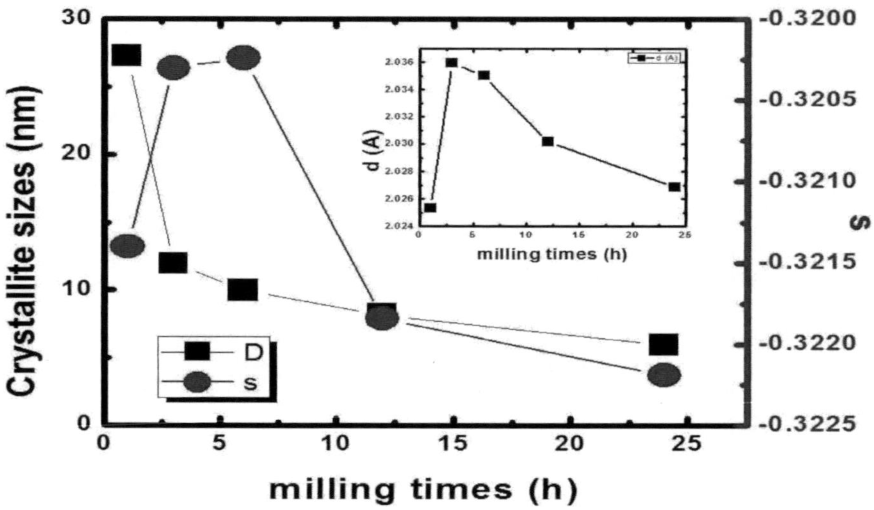

Figure 2. The crystallite size of $Fe_{50}Cr_{40}Si_{10}$ of milled powder for 1- to 24 h. Inset is the lattice parameter respect to milling times.

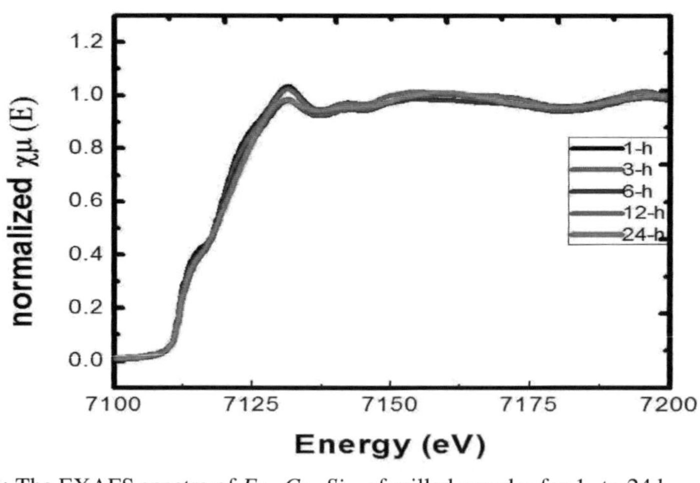

Figure 3: The EXAFS spectra of $Fe_{50}Cr_{40}Si_{10}$ of milled powder for 1- to 24 h

Figure 4 shows the EXAFS spectrum of $Fe_{50}Cr_{40}Si_{10}$ alloys for 1- to 24 h. The reduction of amplitudes is related to the disorder of local structure, and the variation of the phase is related to the change of chemical order. In Figure shows the significant change in the amplitude and the phase. Its mean there is a huge changed in local structure. The systemic variations of the amplitude and phase in the EXAFS spectra confirmed that alloying at atomic scale occurred during MA process.

The decrease of the amplitude before 12 h indicates that the fracture was dominant. After 12 h milling, the phase was shifted significantly. This indicates that the Fe and Si atoms were diffused into the Cr structure and new materials were produced during this process.

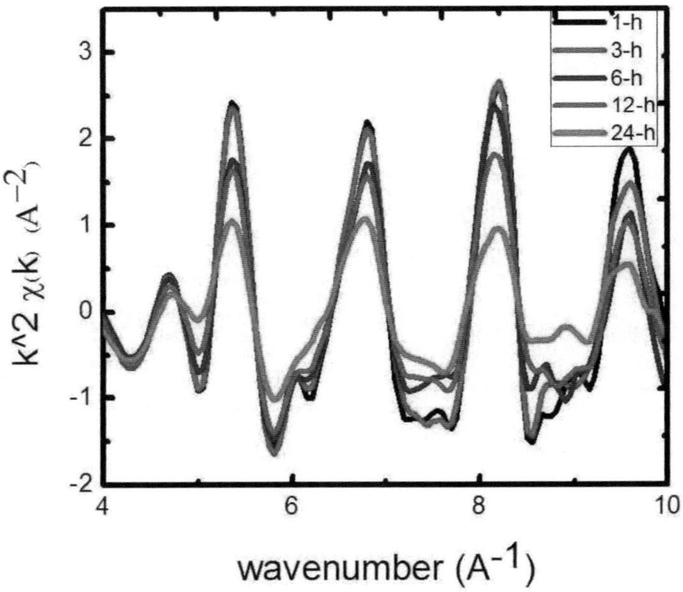

Figure 4. The k-weighted EXAFS spectra of milled powder for 1- to 24 h

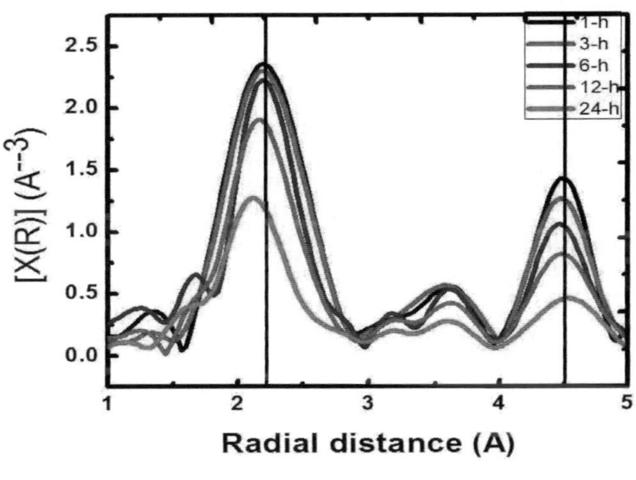

Figure 5: The Fourier transformation of EXAFS spectra of powder for 1- to 24 h

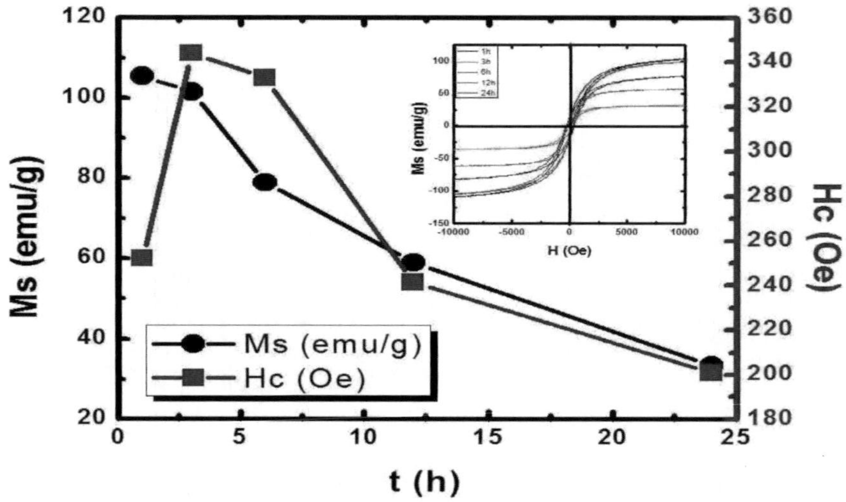

Figure 6. Variation of magnetization and Coercivity of $Fe_{50}Cr_{40}Si_{10}$ alloys with various milling times. The inset shows the variation of the hysteresis loop with magnetic saturation respected to the milling times.

Concerning the magnetic behaviors, both magnetization saturation *(Ms)* and coercivity *(Hc)* depended strongly on milling times, t_m. The *Ms* is decreased because of particle sizes is decreased and the transition from Fe, Cr, and Si grains to $Fe_{50}Cr_{40}Si_{10}$ grain. The *Hc* in beginning time is increased quickly because of fractures in process unstable yet. After 3 h milling times *Hc* is decreased continuously. The coercivity (*Hc*) is decreased continuously until 24 h milled as effect of parameter *d* is decreased. The variation of *Ms* and *Hc* can be seen in hysteresis loop in inset of the Figure 6.

Conclusion

The relatively new phase of $Fe_{50}Cr_{40}Si_{10}$ alloy is explicitly shown in the EXAFS spectra by the variation of amplitude and phase after 12 h milling times. The significant change of the structural phase revealed that new atom neighbors, the atom central Fe substituted by Cr and Si atoms were increased during the MA process. The magnetic saturation (*Ms*) is decreased in long milling times as effect of crystallite size is decreased. The coercivity (*Hc*) is decreased continuously until 24 h milled as effect of parameter *d* is decreased.

Acknowledgment

The research was supported by the Community Service Centre - Mercu Buana University (Grant No: 02-5/01/B-SPK/VIII/2016).

References

1. C. Suryanarayana, *Mechanical alloying and milling*, Progress in Materials Science 46 (2001) 1-184.
2. A. Djekoun et. al., *Structure and magnetic properties of Fe-rich nanostructured Fe_{100-X} Ni_X powders obtained by mechanical alloying,* Physics Procedia **2** (2009) 693–700.
3. C. Suryanarayana, C. C. Koch, *Nanocrystalline materials–Current research and future directions,* Hyperfine Interactions **130** (2000) 5–44.
4. R. Hasegawa, *Present status of amorphous soft magnetic alloys*, Journal of Magnetism and Magnetic Materials **215–216** (2000) 240-245.
5. Michael E. McHenry, Matthew A. Willard, David E. Laughlin, *Amorphous and nanocrystalline materials for applications as soft magnets,* Progress in Materials Science **44** (1999) 291-433.
6. R. Hasegawa, *Advances in amorphous and nanocrystalline magnetic materials,* Journal of Magnetism and Magnetic Materials **304** (2006) 187–191.

7. Ajay Gupta, *Nanocrystalline Soft-magnetic Alloys Produced by Controlled Crystallization of Amorphous Alloys,* PINSA **67, A**, No. 1 (2001) 31-46.
8. Ran Guang et. al., *Formation of nanocrystalline and amorphous phase of Al–Pb–Si–Sn–Cu powder during mechanical alloying*, Materials Science and Engineering *A* **416** (2006) 45–50.
9. Yifang Ouyang, Hongmei Chen, Xiaping Zhong, Yong Du, $(FeAl_3)_{1-x}Zr_x$ *amorphous alloys prepared by mechanical alloying,* Physica B **391** (2007) 380–384.
10. P. Kameli, H. Salamati, and A. Aezami, *Effect of particle size on the structural and magnetic properties of* $La_{0.8}Sr_{0.2}MnO_3$, Journal of Applied Physics **100**, 053914 (2006).
11. J.J. Ipus, et. al., *Structure and magnetic properties of Fe-Nb-B amorphous/nanocrystalline alloys produced by compaction of mechanically alloyed powders,* Journal of Applied Physics **107**, 073901(2010).
12. I. Chicinaş, *Soft magnetic nanocrystalline powders produced by mechanical alloying routes,* Journal of Optoelectronics and Advanced Materials Vol. **8**, No. 2, (April 2006) 439 – 448.
13. Gema González et. al., *Mechanical Alloying of FeCoCr*, Revista Latinoamericana de Metalurgia y Materiales 2011; **31** (1): 64-70.

IOP Conf. Series: Materials Science and Engineering **343** (2018) 012009 doi:10.1088/1757-899X/343/1/012009

Cure characteristics, crosslink density and degree of filler dispersion of kaolin-filled natural rubber compounds in the presence of alkanolamide

I Surya[1,2], N Hayeemasae[3] and M Ginting[1]

[1]Department of Chemical Engineering, Faculty of Engineering, Universitas Sumatera Utara, Medan, Sumatera Utara, Indonesia
[2]Sustainable Energy and Biomaterial Center of Excellence, Faculty of Engineering, Universitas Sumatera Utara, Medan, 20155, Indonesia
[3]Department of Rubber Technology and Polymer Science, Faculty of Science and Technology, Prince of Songkla University, Pattani Campus, 94000, Pattani, Thailand

indradanas@yahoo.com

Abstract. The effects of alkanolamide (ALK) addition on cure characteristics, crosslink density and degree of filler dispersion of kaolin-filled natural rubber (NR) compounds were investigated. The kaolin filler was incorporated into NR compounds with a fixed loading, 30.0 phr. The ALK was prepared from Refined Bleached Deodorized Palm Stearin (RBDPS), a waste product of cooking oil production, and diethanolamine. The ALK is an oily material and added into the filled NR compounds as a rubber additive at different loadings, 0.0, 3.0, 5.0 and 7.0. The kaolin-filled NR compounds with and without ALK were vulcanized using a semi-efficient vulcanization system. It was found that ALK decreased the scorch and cure times and improved filler dispersion of the kaolin-filled NR compounds. The higher the ALK loading, the shorter were the scorch and cure times. It was also found that ALK increased the crosslink density of kaolin-filled NR compound up to 5.0 phr of loading. Due to its oily properties, The ALK acted as an internal plasticizer which decreased the minimum torque and improved the degree of kaolin dispersion in NR phases. The higher the ALK loading; the lower the minimum torque and better the filler dispersion.

1. Introduction

The rubber compounds are prepared by mixing a base rubber and rubber additives. The base rubber or rubber gives rubbery characteristics to the compound and rubber additives plasticizers reduce viscosity, enhance deformability and alter properties; curing agent crosslinks the rubber chains; and accelerators enhance the action of curing agent to speed up the resultant crosslinking formation.

Rubber additives can be divided into curative and noncurative additives. Curative additives affect the cure characteristics of a rubber compound. Non-curative additives do not affect the rate and nature of the curing process. Fillers are non-curative additives. The main purpose of utilizing them is to enhance the mechanical properties of rubber vulcanisates such as tensile strength, resistances to abrasion, tearing and flexing or cheapen the cost of products. Fillers can be divided into black and nonblack fillers. The black filler is carbon black (CB), and non-black fillers are calcium carbonate, kaolin clay, precipitated silica, talc, barite, wollastonite, mica, precipitated silicates, fumed kaolin and diatomite. Of these, the three most widely used, by volume and by functionality, are calcium carbonate, precipitated silica, and kaolin. For many applications, kaolin and the other non-black fillers are not satisfactory alternatives to CB, because of their lower reinforcement levels. The basic problem is their surface chemistry which is more polar and hydrated than CB, and which cause them to be difficult to wet, disperse and interact with hydrocarbon rubbers. Some methods have been introduced to improve the reactivity of nonblack fillers with the rubber phases. The most popular one is the utilising of a silane-coupling agent. The coupling agent modifies the surface of polar fillers. The modified polar fillers provide chemically active surfaces that can participate in vulcanization, providing coupling bonds between silane and both the polar fillers and the rubber phases [1-3]. Those products show significant improvement in performance compared to their base materials.

Another alternative method to overcome the weakness of polar fillers, such as kaolin, is the using of alkanolamide (ALK). The ALK is prepared by reacting Refined Bleached Deodorized Palm Stearin (RBDPS) with diethanolamine with a certain condition. This study reports the effects ALK loading on cure characteristics, crosslink density and degree of filler dispersion of kaolin-filled NR compounds.

2. Materials and Methods

2.1. Materials

NR grade SMR-L was obtained from Guthrie (M) Sdn. Bhd., Seremban, Malaysia. Other compounding ingredients such as kaolin, sulphur, zinc oxide, stearic acid, N-isopropyl-N'-phenyl-p-phenylenediamine (IPPD) and Mercapto Benzothiazolyl disulfide (MBTS) were supplied by Bayer (M) Ltd., Sdn. Bhd., Petaling Jaya, Selangor, Malaysia. All materials were utilized as supplied. The ALK was prepared in the laboratory using RBDPS and diethanolamine [4]. The chemical formula of ALK is $CH_3(CH_2)_{14}CON(CH_2CH_2OH)_2$.

2.2. Compounding

A sulphur-accelerated curing system was used for compounding. The recipe for the preparation of the kaolin-filled NR compounds is given in Table 1. The compounding procedure was done in accordance with the American Society for Testing and Material (ASTM) – Designation D 3184 – 80. Compounding was done on a two-roll mill. Table 1 also shows the designation and composition of the NR-based recipes used in this study.

Table 1. The composition and designation of the kaolin-filled NR compounds

Ingredients	Content (phr)*	Designation				
		A/0.0 (Control)	B/1.0	C/3.0	D/5.0	E/7.0
SMR L	100.0	100.0	100.0	100.0	100.0	100.0
Zinc oxide	5.0	5.0	5.0	5.0	5.0	5.0
Stearic acid	2.0	2.0	2.0	2.0	2.0	2.0
IPPD	2.0	2.0	2.0	2.0	2.0	2.0
MBTS	1.5	1.5	1.5	1.5	1.5	1.5
Sulphur	1.5	1.5	1.5	1.5	1.5	1.5
Kaolin	30.0	30.0	30.0	30.0	30.0	30.0
ALK	0.0 – 7.0	0.0	1.0	3.0	5.0	7.0

*parts per hundred parts of rubber

2.3. Cure characteristics

The cure characteristics of the kaolin-filled NR compounds with and without ALK were obtained using a Monsanto Moving Die Rheometer (MDR 2000) which was employed to determine the scorch time (ts_2), cure time (t_{90}), minimum torque (M_L), maximum torque (M_H) and torque difference (M_H—M_L) according to ISO 3417. Samples of the respective compounds were tested at 150°C.

2.4. Crosslink density

Swelling tests on the kaolin-filled NR vulcanisates were performed in toluene in accordance with ASTM D471-12a. The cured test pieces (30 mm × 5 mm × 2 mm) were weighed using an electric balance and swollen in toluene until equilibrium, which took 72 hours at room temperature. The samples were taken out from the liquid, the toluene was removed from the sample surfaces and the weight was determined. The samples were then dried in the oven at 70 °C until constant weights were obtained. The swelling results were used to calculate the molecular weight between two crosslinks (M_c) by applying the Flory-Rehner equation [5-6].

$$M_c = \frac{-\rho_r V_s V_r^{1/3}}{\ln(1-V_r)+V_r+\chi V_r^2} \tag{1}$$

$$V_r = \frac{1}{1+Q_m} \tag{2}$$

where ρ is the rubber density (ρ_r of SMR-L = 0.92 g/cm^3), V_s is the molar volume of the toluene (V_s = 106.4 cm^3/mol), V_r is the volume fraction of the rubber in the swollen specimen, Q_m is the weight increase of the vulcanisate in toluene and χ is the interaction parameter of the rubber network–solvent (χ of SMR-L = 0.393). The crosslink density is given by;

$$V_c = \frac{1}{2M_c} \tag{3}$$

3. Results and Discussion

3.1. The effects of ALK on cure characteristics of kaolin-filled NR compounds

The effects of ALK addition on cure characteristics of the kaolin-filled NR compounds are shown in Table 2. The cure characteristics are the scorch time, cure time, minimum torque, maximum torque and torque difference. As shown in Table 2, the scorch times of kaolin-filled NR compounds with ALK were lower than that of without ALK (control compound). The higher the ALK loading, the lower were the scorch times. As shown in Table 1, rubber compounds used MBTS as the accelerator. It is functionally classified as a primary accelerator which causes a scorch delay to rubber compounds [7-8]. Whenever ALK was added to the filled-rubber compounds, the additive decreased the mass ratios of both rubber and non-rubber constituents. Therefore, ALK decreased the mass ratio of MBTS and consequently, decreased the effect of scorch safety.

Table 2. The cure characteristics of kaolin-filled NR compounds with and without ALK

Compounds	Cure characteristics				
	ts_2, min.	**t_{90}, min.**	**M_H, dN.m**	**M_L, dN.m**	**M_H−M_L, dN.m**
A/0.0	5.94	9.89	5.34	0.39	4.95
B/1.0	4.45	7.61	6.49	0.13	6.36
C/3.0	3.71	6.31	6.59	0.11	6.48
D/5.0	2.98	5.40	7.57	0.09	7.48
E/7.0	1.47	3.38	7.21	0.03	7.18

IOP Conf. Series: Materials Science and Engineering **343** (2018) 012009 doi:10.1088/1757-899X/343/1/012009

As shown in Table 2, the addition of 1.0 phr of ALK decreased the cure time of control compound. A lower cure time means a higher cure rate and hence, a cure enhancement phenomenon was observed. In this case, the ALK acted as an additional accelerator in the curing process [8-9] and it was considered as a secondary accelerator for the kaolin filled NR compounds. The cure enhancement was due to the amine presence in ALK, together with zinc oxide and fatty acid, they activated the MBTS-accelerator more pronouncedly [10] and hence, improved the cure rate. It was also seen that increases the ALK loading caused in a more pronounced the cure enhancement and it was simply due to a higher amount of amine in the kaolin-filled NR compounds.

As shown in Table 2, the addition of 1.0 phr of ALK decreased the minimum torque of control compounds. Further increases the ALK loading decreased the minimum torque. The minimum torque relates to the viscosity of a rubber compound qualitatively [3, 11]. A lower minimum torque means a lower viscosity and vice versa. The ALK decreased the viscosity of the filled NR compounds because of its waxy properties [4, 8] which plasticised, softened and decreased the viscosity of the filled NR compounds, respectively. The higher ALK loading the lower was the viscosity.

As shown in Table 2, the addition of 1.0 phr of ALK increased the maximum torque. The maximum torque relates to the value of stock modulus which increased in this case. It was due to the nature of rubber - filler interaction including intercalation and exfoliation [12]. The increases in maximum torque were more pronounced whenever the ALK loadings were further increased up to 5.0 phr and decreased beyond the loading. Therefore, the intercalation, exfoliation, and rubber - filler interaction were further increased also up to 5.0 phr of ALK. In this case, the ALK acted as a compatibilizer in the kaolin-filled NR compounds.

As shown in Table 2, the addition of 1.0 phr of ALK increased the torque difference (M_H—M_L) value. The value was further increased by the addition of ALK up to 5.0 phr and decreased beyond the loading. Torque difference relates to the degree of crosslink density of a rubber compound [4, 13]. The greater the value, the higher is the crosslink density. The increases of torque difference or crosslink density up to the optimum loadings of ALK (5.0 phr) was due to the acting of ALK not only as an additional accelerator but also as a compatibilizer. As an accelerator, the amine constituent activated chemically both the rubber and elemental sulphur during curing reaction. Together with the others curatives, the amine formed intermediate complexes which attached the available elemental sulphur to rubber chains more efficient [8], causing in a higher degree of cure. As a compatibilizer, the oily properties of ALK decreased the viscosity of the filled NR compounds and hence, improved the kaolin dispersion or rubber-filler interaction. The rubber – filler interaction is defined as additional physical crosslinks [9], and together with sulphide crosslink contribute to total crosslink density of a filled rubber vulcanisates [8-9].

The decrease in torque difference (beyond the optimum loading) might be due to the excessive amount of ALK which formed some oily layers. The layers covered and coated the kaolin and other curatives and hence, decreased the total crosslink.

3.2. The effect of ALK on crosslink density of kaolin-filled NR compounds

The effect of ALK on the total crosslink density of the kaolin-filled NR vulcanisates is shown in Figure 1. The total crosslink density was determined by the Flory—Rehner approach (Equation 1). The additions of up 5.0 phr of ALK into the kaolin- filled NR compound increased the total crosslink density and further increase the ALK loadings decreased the total crosslink density. This observation is in line with the result of the torque difference as presented in Table 2. It surely confirms that torque difference indicates the degree of crosslink density of the filled NR compounds.

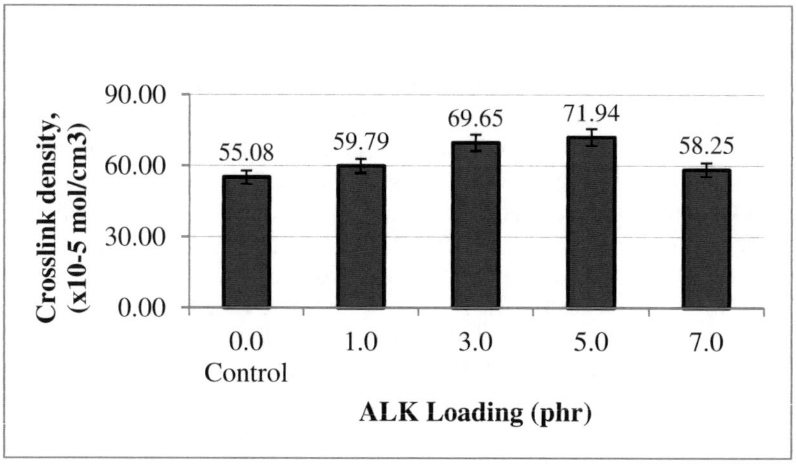

Figure 1. The effect of ALK loading on crosslink density of the kaolin-filled NR vulcanisates

3.3. The effect of ALK on degree of filler dispersion of kaolin-filled NR compounds

Based on the cure characteristics, the degree of kaolin dispersion in NR phases in the presence of ALK was determined qualitatively by using Equation 4 [9, 14,15].

$$L = \eta_r — m_r \quad (4)$$

where: $\eta_r = [M_{Lf}/ M_{Lg}]$, and $m_r = [M_{Hf}/ M_{Hg}]$; where M_{Lf} and M_{Hf} are the minimum and maximum torques of the filled compounds and M_{Lg} and M_{Hg} are the minimum and the maximum torques of the unfilled rubber compound. A lower value of L, at a particular kaolin loading, means a better degree of filler dispersion. The minimum and maximum torques of unfilled NR compounds were 0.05 and 4.85, respectively.

Table 3 shows the values of L (degree of kaolin dispersion) of kaolin-filled NR compounds. The L values of kaolin-filled compounds with ALK were lower than that of control compound. The higher was the ALK loading, the lower was the L value. It was due to the plasticizing effect of ALK which decreased viscosity the filled NR compounds and hence, improved the kaolin dispersion.

Table 3. The values of L for kaolin-filled NR compounds

Dispersion parameter	Kaolin-filled NR compounds with ALK				
	0.0 phr	**1.0 phr**	**3.0 phr**	**5.0 phr**	**7.0 phr**
Minimum torque	0.39	0.13	0.11	0.09	0.03
Maximum torque	5.34	6.49	6.59	7.57	7.21
η_r	5.57	1.86	1.57	1.29	0.43
m_r	1.09	1.33	1.35	1.55	1.48
$L = \eta_r — m_r$	4.48	0.53	0.22	- 0.26	- 1.05

4. Conclusion

From this study, the following conclusions were drawn:

1. Alkanolamide acted as an additional accelerator in the curing process of kaolin-filled natural rubber compounds. It decreased scorch and cure times. The higher the alkanolamide loading, the lower were the scorch and cure times.

2. Alkanolamide acted as a compatibilizer in the kaolin-filled NR compounds. It increased the maximum torque and torque difference of the kaolin-filled natural rubber compounds up to an optimum loading. A 5.0 phr of alkanolamide was the optimum loading.
3. Alkanolamide also acted as internal plasticiser which decreased the minimum torque and improved filler dispersion of kaolin in natural rubber phases. Increases the alkanolamide loading further increases in degree dispersion of kaolin.

5. Acknowledgments

The authors would like to thank Universiti Sains Malaysia for providing the research facilities for carrying out the experiment and for making this research work possible.

6. References

[1] Ranney M *et al* 1971 *Rubber Chem Technol* **44**(4) 1080-92.
[2] Wagner M 1974 *Rubber Chem Technol* **47**(4) 697-716.
[3] Pinter PE and Mc Gill CR 1978 *Rubber World Intern* **177** 30-37.
[4] Surya I *et al* 2013 *Polym Test* 2013 **32**(8) 1313-21.
[5] Flory PJ and Rehner JJr 1943 *J Ch Ph* **11** 521-26.
[6] Flory PJ 1950 *J Ch Ph* **18**(1) 108-11.
[7] Long H 1985 *Basic Compounding and Processing of Rubber, Rubber Division* American Chemical Society Inc USA.
[8] Surya I and Ismail H 2016 *Polym Test* **50** 276-82.
[9] Surya I *et al* 2014 *Polym Test* **40** 24-32.
[10] Surya I *et al* 2015 *Polym Test* **42** 208-14.
[11] Ismail H *et al* 2002 *Polym Test* **21**(5) 565-69.
[12] Teh P *et al* 2004 *Eur Polym J* **40**(11) 2513-21.
[13] Surya I and Ismail H 2016 *Plast Rubber and Compos* **45**(7) 287-93.
[14] Lee B 1979 *Rubber Chem Technol* **52**(5) 1019-29.
[15] I Surya *et al* 2017 *IOP Conf Ser Mater Sci Eng* **223** 012006.

Synthesis of Composite Adsorbent from Calcium Carbonate and *Cocos Nucifera* Carbon Powder Crosslinked with Biopolymer Matrix

M A Kamaruddin[1,2], M M A Bakri[2], F A Norashiddin[1], M H Zawawi[3] and M R R A Zainol[4]

[1]Environmental Technology Division, School of Industrial Technology, Universiti Sains Malaysia, 11800 Penang, Malaysia

[2]Center of Excellence Geopolymer & Green Technology (CEGeoGTech), Universiti Malaysia Perlis, Perlis, Malaysia

[3]College of Engineering, Universiti Tenaga Nasional, Jalan Serdang, Seksyen 11, Selangor, 43650 Bandar Baru Bangi, Malaysia

[4]School of Civil Engineering, Universiti Sains Malaysia, Engineering Campus, 14300 Nibong Tebal, Penang, Malaysia

anuarkamaruddin@usm.my

Abstract. The use of coconut shell based adsorbent for removing various pollutants from wastewater offers attractive advantages such as exceptional adsorption capacity with larger surface area, low-cost and biocompatibility for a wide range of pollutants. The same goes for calcium carbonate based adsorbent, which provides better removal for metals ions through precipitation method. In fact, recycling of waste material is considered environmentally preferable and is supported by public opinion and government policy. However, because of unit operations mechanisms and effectiveness are different to one another in downstream process, it is quite difficult to provide good adsorbent that exhibits dual attributes of hydrophobic and hydrophilic characters. This paper provides brief process for sintering of composite adsorbent for separation and purification of industrial wastewater application. Characterization was performed by physical and chemical approach. Results prove that the inclusion of biopolymer as composite binder improved mechanical properties of the composite adsorbent.

1. Introduction

Generally, activated carbon has been the adsorbent of choice rather than the others because of their physical and/or chemical properties that can easily be tailored in order to optimize their performance [1]. Activated carbon has a very porous structure with a large internal surface area ranging from 500 to 2000 m^2/g that makes it a good adsorbent in the adsorption process [2]. Broad application of activated carbon is due to its chemical and mechanical stability, high degree of surface reactivity and high adsorption capacity [3]. Common types of activated carbon are granular (GAC) and powdered (PAC) form. Granular activated carbon is produced from hard materials that may be used to remove pollutants from wastewater with no requirement to separate the activated carbon from bulk fluid.

Large particle size of GAC presents a smaller surface area thus suitable for diffusion (gas and odor) purpose. Meanwhile, PACs may be obtained from crushed ground carbon particles that pass designated sieve aperture. Typical size of PAC granules are less than 1.0 mm with average diameters of 0.15 to 0.25 mm [4].

Despite overwhelming anticipation in textile wastewater treatment, the biggest barrier of commercial activated carbon is the prohibitive cost and difficulty associated with regeneration [5, 6]. Moreover, large scale application for activated carbon will only raise the operational cost of a treatment plant due to the expensive cost of activated carbon over the years (M. A. Kamaruddin, Yusoff, & Ahmad, 2011). Therefore, the drawbacks of commercial activated carbon have resulted in a growing research interest searching for low-cost alternative precursors for activated carbon as cheaper and renewable are the trends among scientists in the recent years [7].

In the field of wastewater treatment, limestone is commonly employed to remove certain types of heavy metals such as manganese; as discussed by Aziz and Smith [8]. This study attempted to use limestone particles to remove manganese from water containing 53.9% $CaCO_3$ and 5.2% $MgCO_3$, gravel, crushed bricks or without solid media at different pH values. They found that at final pH value of 8.5, the removal of Mn using limestone, crushed bricks, gravel, aeration and settlement without solid media were 95%, 82%, 60% and less than 15% respectively, in exploring the capabilities of limestone in purifying wastewater.

Although abundant literature data for the application of coconut shell based activated carbon and limestone particle in wastewater treatment, there are still gap of knowledge to be further explored such as enhancement of adsorption capacity through modification of *Coco Nurfira* carbon (CSAC) and calcium carbonate powder (LP) with the presence of sodium alginate as biopolymer binder. In fact, there is very limited information on the assessment of sodium alginate composite based adsorbent for in terms of their physical and chemical characteristics. Thus, this study attempts to provide brief supporting data for the potential of composite preparation by the combination between hydrophobic and hydrophilic precursors.

2. Materials and Methods

2.1 Composite preparation

Briefly, the composite was prepared based on [9, 10]. The composite adsorbent preparation involved several physical and mechanical procedures and small modification was adapted in order to suit the different characteristics of *Coco Nufira* carbon (CSAC) and calcium carbonate powder (LP). First, known amount of CSAC and LP was introduced into 500 mL beaker. A known volume of distilled water was then added and stirred with mechanical stirrer (Ika, Germany) for 15 minutes with 50 rpm speed.

Next, the beaker was heated to 80 °C on a hot plate and alginate powder was slowly added into the beaker. This process was maintained for 15 minutes until a homogeneous condition between CSAC, LP and alginate was reached. The mixture was then transferred via peristaltic pump into the beads injector. Figure 1 illustrates the schematic diagram of the experimental set up. The mixture was then extruded from the injector by controlling the opening of control valve. Finally, the release composite beads were immersed in $CaCl_2$ solution for 12 hours to achieve hardened beads. Figure 1 illustrates the schematic diagram of the experimental set up.

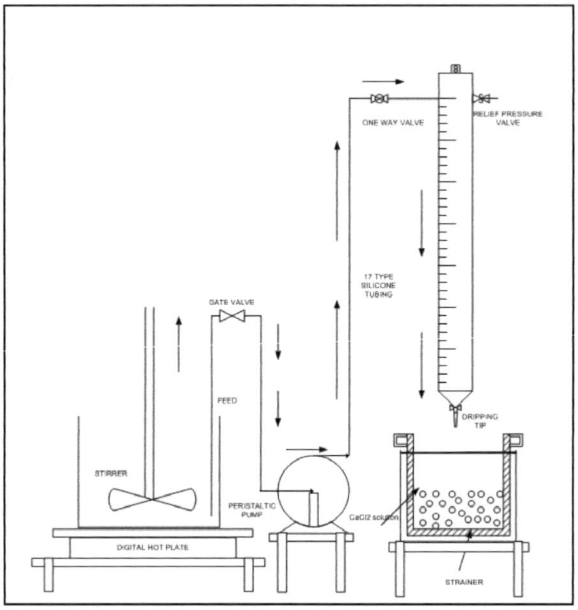

Figure 1. Schematic flow of composite preparation

2.2 Physical and Chemical Characteristics Analysis

X-ray powder diffraction (XRD)
X-ray powder diffraction (XRD) was used to support the SEM image attained. XRD is the most common technique used to determine the mineral composition of composite adsorbent that exists in the crystalline state. The equipment used was D2 Phase (Bruker). It is based on Brag Equation as follows:

$$\eta\lambda=2d\sin\theta \quad (1)$$

Thermogravimetry Analysis
Thermal analysis or Thermogravimetry Analysis (TGA) was carried out using TGA Analyzer (DTG 60 H Shimadzu, Japan). Principally, 5 mg of sample was loaded into a crimp cell whereby the pan holder was used to minimize the potential reaction with the sample. The furnace chamber was then raised and degassed for several minutes. During inert atmosphere exertion in the chamber, a small amount of sample was heated up to 110 °C for complete dehydration and followed by decomposition at 850 °C for 7 minutes to determine volatile matters. Oxidizing state was achieved once the sample was cooled down to 600 °C and this process was maintained until there was no change in the weight of sample.

Fourier Transform Infrared (FTIR) Spectroscope
An FTIR transmission spectrum of the sample was carried out by using an FTIR spectroscope (Model Perkin Elmer 2000, USA). The purpose of FTIR measurement is to study the surface chemistry of the raw materials and the composite adsorbent by exploring the functional groups present in the sample surfaces. Prior to the analysis, the sample was ground into a fine form that can pass a 75 µm sieve aperture and dried at 110 °C for 12 hours. Next, the solid sample was mixed with 400 mg of potassium bromide (KBr) using impact mortar and pestle. A translucent round disk of 12.7 mm internal diameter and 1 mm thickness was retrieved by using a bench press with 8 tons of pressure. The translucent disk was then placed into the infrared spectrometer. Finally, with the aid of software (Spectrum version

5.0.2) in the FTIR spectroscope system, the wave number in the range of 400-4000 cm^{-1} was eventually recorded.

Results and Discussions

X-ray Diffraction (XRD)

Figure 2 illustrates important peaks that emerged from the XRD analysis were carbon nitride 84.1%, calcium peroxide (6.4%), iron nitride (2.8%), potassium oxide (5.4%) and magnesium zinc (1.3%). These peaks were observed in the range of 24° to 48°. However, except the peak intensity, the presence of AG ($Na_2C_6H_7O_6$) was not observed due to the weakened condition after forming within the composite adsorbent [11]. In the amorphous phase, normally X-rays will be scattered in many directions leading to a large bump distributed in a wide range (2 Theta) instead of high intensity narrower peaks. In this case, plenty of remotely narrow peaks emerged which suggested the degree of amorphous and large particle size that could be observed from Figure 2 (b).

The tricarbon tetranitride peak was observed at 30° with 41.4% intensity. Meanwhile, the formation of hydrogen (28.7%) element suggested sufficient paths for oxygen bonding during the adsorption process. This reaction could initiate hydrogen bonding with carboxylic, hydroxyl as well as amines groups for metal sorption, respectively. Overall, the composite adsorbent suggested the amorphous condition rather than crystalline because of the interference of irregular base line with noise and pulsed shape being the highly noisy line is the highly amorphous phase as manifested from Figure 2.

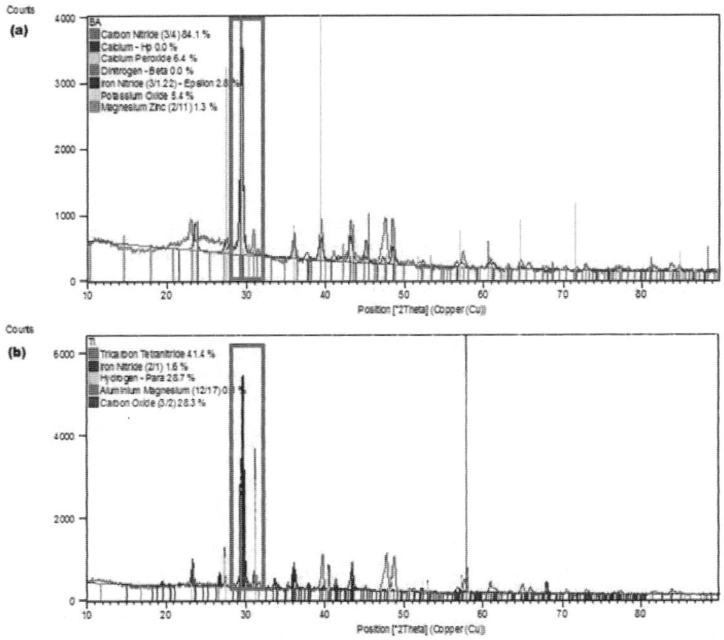

Figure 2. XRD patterns of (a) Composite adsorbent before adsorption and (b): composite adsorbent after adsorption

Thermogravimetry Analysis

Figure 3 shows the TGA plot of the composite adsorbent before and after adsorption process. As shown in the figure, gradual weight loss occurred for the dried composite adsorbent during heating between 0-200 °C, which was attributed by moisture loss (5.01%). The next heating step to 400 °C caused a weight loss of 27.71% of initial dry mass, indicating volatile matter combustion [12]. Further increased heating rate to 600 °C caused significant loss of the initial dry sample to 79.30% which largely represented by the fixed carbon. Finally, the ash content showed 7.96% from total weight loss. However, the transition of weight loss for the spent composite adsorbent after adsorption indicates slight changes in terms of weight %. From the figure, only 1.45% of moisture loss can be observed during thermal analysis. This condition was due to the oversaturated condition of the composite adsorbent. In fact, this phenomenon proved the water bearing capacity of the adsorbent with the presence of sodium alginate as the core binder [13]. Further increased heating from 200 to 600 °C led to reduction weight loss of 14.36 and 62.88% than original weight of the sample which represent the volatile matter and fixed carbon contents. The weight loss of the composite adsorbent as shown in the figure suggested the matrix decomposition of carbon that is rich with fixed carbon and ash as partial constituents in the adsorbent [14].

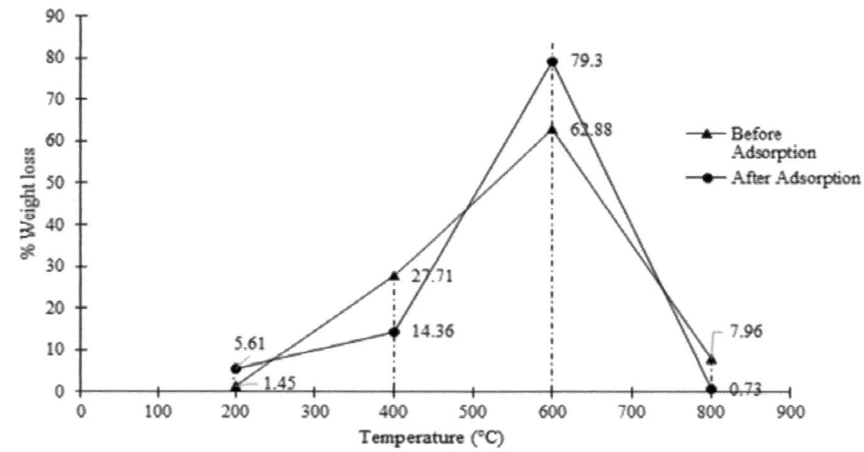

Figure 3. TGA of composite before and after trial adsorption

Fourier Transformation Infrared (FTIR) Spectroscope

The adsorption characteristics of a composite adsorbent are also influenced by the surface functional groups, which can be determined from the Fourier Transform Infrared (FTIR) spectroscopy analysis. Figure 4 illustrates the FTIR spectrums of the composite adsorbent before and after undergoing adsorption process. Based on FTIR spectrums (Figures 4.19 (a) and (b)), the surface chemistry of the composite adsorbent before and after adsorption of the precursors showed some differences due to the modification of many functional groups of the precursors during the gelation process. In fact, the most dominant peak observed were C≡N stretching nitriles of 2358 cm^{-1} that conformed to significant presence of alginate as the binding agent for the precursors (Figure 4.9 (a)).

The FTIR spectrums also revealed that the functional groups in the precursors such as CSAC and alginate (AG) have been shifted to C=O stretching alpha (1726 cm^{-1}), which belongs to unsaturated esters. This functional group also indicated that the shifted spectrums from aliphatic amines to C=O stretching provides the exchange path during the adsorption course. In fact, those medium peaks that have been detected from the precursors were mostly belonging to general carbonyl groups, in which stretching were from 1665 to 1769 cm^{-1}. Next, another significant peak observed for the composite adsorbent before the adsorption was at band 1025 cm^{-1}. The spectrum belongs to aliphatic amines group that stretch from C-N, which implied important absorption bands for grafting purpose of alginate in the gelation process.

Figure 4 (b) shows the FTIR spectrums of the composite adsorbent after the adsorption process. From the figure, there was significant H stretching vibrations of hydroxyl functional groups including hydrogen bonding, which was detected at a bandwidth of 3409 cm^{-1}. The location of the detected hydrogen-bonded O-H group is usually in the range of 3200–3750 cm^{-1} for alcohols and phenols and related to adsorbed water [15]. However, the disappeared medium C=O stretching bond was found shifted to 1408 cm-1 and belong to C–C stretch aromatics rings. The shift to lower wave numbers in comparison with AG samples explained the interaction of the regular long chain with the sodium ions [16]. Table 1 summarizes detected bandwidth of composite adsorbent in this work with respect to previous works.

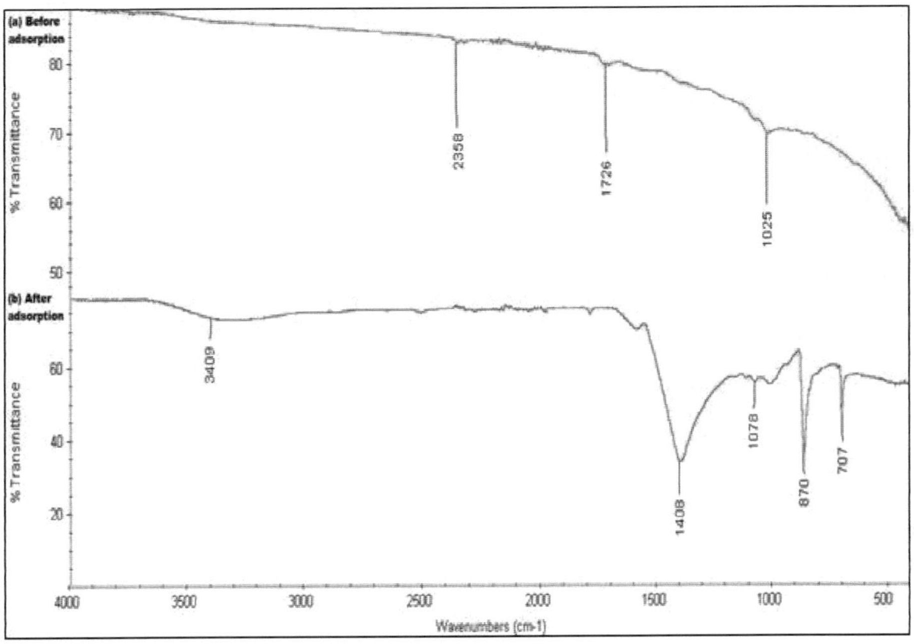

Figure 4. FTIR spectrums of composite adsorbent (a) before and (b) after adsorption

Conclusions

Based on the preliminary analysis, the results have demonstrated the successful synthesis of composite from hydrophobic and hydrophilic conjugates through biopolymerization technique. From the XRD analysis, the composite adsorbent was categorized as having amorphous structure, thereby plenty of narrower peaks were detected in a wide range of 2 theta. TGA analysis has shown that water bearing capacity as a result from preparation condition was responsible for the high moisture content of the composite. FTIR results indicated the presence of functional groups such as hydroxyl, carbonyl and alkyl on the surfaces of the composite adsorbent prepared. Further studies for the effectiveness of the composite shall be expanded in near future.

Acknowledgments

This research is supported by Solid Waste Management Cluster (SWAM), Grant No. 1001/CKT/870023 and 304/PTEKIND/6315062 RU Short Term Grant from Universiti Sains Malaysia. Authors gratefully acknowledge contribution in this paper whom directly or indirectly towards the outcome of the present research.

References

[1] Mezohegyi, G., van der Zee, F. P., Font, J., Fortuny, A., & Fabregat, A. (2012). Towards advanced aqueous dye removal processes: a short review on the versatile role of activated carbon. *Journal of Environmental Management, 102*, 148-164.

[2] Ribas, M. C., Adebayo, M. A., Prola, L. D., Lima, E. C., Cataluña, R., Feris, L. A., . . . Calvete, T. (2014). Comparison of a homemade cocoa shell activated carbon with commercial activated carbon for the removal of reactive violet 5 dye from aqueous solutions. *Chemical Engineering Journal, 248*, 315-326.

[3] Trikha, R., & Sharma, B. K. (2014). Studies on factors affecting fluoride removal from water using passive system. *Journal of Environmental Chemical Engineering, 2*(1), 172-176.

[4] Salleh, M. A. M., Mahmoud, D. K., Karim, W. A. W. A., & Idris, A. (2011). Cationic and anionic dye adsorption by agricultural solid wastes: A comprehensive review. *Desalination, 280*(1), 1-13.

[5] Hung, Y. T., Aziz, H. A., Amr, S. S. A., Yeh, R. Y. L., Liu, L. H., & Kamaruddin, M. A. (2013). Chemicals and allied products. *Water Environment Research, 85*(10), 1474-1500. doi: 10.2175/106143013X13698672322426

[6] Hung, Y. T., Aziz, H. A., Yusoff, M. S., Kamaruddin, M. A., Yeh, R. Y. L., Liu, L. H., . . . Fu, Y. P. (2015). Chemical waste and allied products. *Water Environment Research, 87*(10), 1312-1359. doi: 10.2175/106143015X14338845155822

[7] Kamaruddin, M. A., Yusoff, M. S., & Aziz, H. A. (2014). Preparation and Characterization of Alginate Beads By Drop Weight. *International Journal of Technology, 5*(2), 121-132.

[8] Aziz, H. A., & Smith, P. G. (1996). Removal of manganese from water using crushed dolomite filtration technique. *Water Research, 30*(2), 489-492.

[9] Kamaruddin, M. A., Yusoff, M. S., & Ahmad, M. A. (2011). Optimization of durian peel based activated carbon preparation conditions for ammoniacal nitrogen removal from semi-aerobic landfill leachate.

[10] Kamaruddin, M., Yusoff, M., Aziz, H., & Alrozi, R. (2017). Statistical Determination for Eco-Friendly Composite Adsorbent Preparation for the Removal of COD, Colour and Cu (II) from Textile Effluent. *Journal of Scientific & Industrial Research, 76*, 506-511.

[11] Lv, X., Jiang, G., Xue, X., Wu, D., Sheng, T., Sun, C., & Xu, X. (2013). Fe 0-Fe 3 O 4 nanocomposites embedded polyvinyl alcohol/sodium alginate beads for chromium (VI) removal. *Journal of hazardous materials, 262*, 748-758.

[12] Lemlikchi, W., Sharrock, P., Mecherri, M., Fiallo, M., & Nzihou, A. (2012). Treatment of Textile Waste Waters by Hydroxyapatite Co-Precipitation with Adsorbent Regeneration and Reuse. *Waste and biomass valorization, 3*(1), 75-79.

[13] Donati, I., & Paoletti, S. (2009). Material properties of alginates *Alginates: Biology and applications* (pp. 1-53): Springer.

[14] Shen, Y.-S., Wang, S.-L., Tzou, Y.-M., Yan, Y.-Y., & Kuan, W.-H. (2012). Removal of hexavalent Cr by coconut coir and derived chars–The effect of surface functionality. *Bioresource technology, 104*, 165-172.

[15] Ahmad, A., & Hameed, B. (2009). Reduction of COD and color of dyeing effluent from a cotton textile mill by adsorption onto bamboo-based activated carbon. *Journal of hazardous materials, 172*(2), 1538-1543.

[16] Daemi, H., & Barikani, M. (2012). Synthesis and characterization of calcium alginate nanoparticles, sodium homopolymannuronate salt and its calcium nanoparticles. *Scientia Iranica, 19*(6), 2023-2028.

Phase study of titanium dioxide nanoparticle prepared via sol-gel process

Alaba Oladeji Araoyinbo[1], Mohd Mustafa Al Bakri Abdullah[1], Mohd Arif Anuar Mohd Salleh[2], Nurul Nadia Abdul Aziz[1] and Azwan Iskandar Azmi[1]

[1]Faculty of Engineering Technology, Department of Mechanical Engineering Technology, Universiti Malaysia Perlis, Sungai Chuchuh, 02100 Padang Besar Utara, Perlis, Malaysia

[2]Center of Excellence Geopolymer and Green Materials (CEGeoGTech), School of Materials Engineering, Universiti Malaysia Perlis (UniMAP), Perlis, Malaysia

alaba@unimap.edu.my.

Abstract. In this study, titanium dioxide nanoparticles have been prepared via sol-gel process using titanium tetraisopropoxide as a precursor with hydrochloric acid as a catalyst, and ethanol with deionized water as solvents. The value of pH used is set to 3, 7 and 8. The sols obtained were dried at 100 °C for 1 hr and calcined at 350, 550, and 750 °C for 3 hrs to observe the phase transformation of titanium dioxide nanoparticle. The samples were characterized by x-ray diffraction and field emission scanning electron microscope. The morphology analysis is obtained from field emission scanning electron microscope. The phase transformation was investigated by x-ray diffraction. It was found that the pH of the solution affect the agglomeration of titanium dioxide particle. The x-ray diffraction pattern of titanium dioxide shows the anatase phase most abundant at temperature of 350 °C. At temperature of 550 °C the anatase and rutile phase were present. At temperature of 750 °C the rutile phase was the most abundant for pH 3, 7 and 8. It was confirmed that at higher temperature the rutile phase which is the stable phase are mostly present.

Keywords: anatase, nanoparticle, rutile, sol-gel, titanium dioxide.

1. Introduction

Titanium dioxide (TiO_2) is one of the most current nanomaterials that have been researched a great deal of times due to its unique properties. Titanium dioxide is also referred to as titanium (IV) oxide or titania. Titanium dioxide is categorized as nanoparticle because it is a particle that has one or more dimensions of the order of 100 nm or less [1]. Titanium dioxide which is also a known semiconductor shows unique characteristics, e.g, easy processing control, inexpensive, and non-hazardous. These benefits make titanium dioxide a material of choice in solar cells, chemical sensors for hydrogen gas evolution, as pigments, self-cleaning surfaces and environmental purification applications [2]. The photocatalytic activity of TiO_2 is one of its most unique features, and it is mostly determined by properties such as the crystalline phase, crystallite size, and specific surface area [3].

Titanium dioxide is observed to be mainly in the crystalline or amorphous forms in which the crystalline polymorphos, specifically, is anatase, rutile and brookite. The active crystallite phases of titanium dioxide are rutile and anatase [4, 5]. The anatase and rutile phase are tetragonal structure, but brookite phase is an orthorhombic structure [6]. The anatase phase is known to exhibit a higher photocatalytic behavior, while the rutile phase is the most stable phase [7]. The crystalline structure reveals physical properties such band gap which governs their applications and uses [8, 9]. The enlargement of TiO_2 surface area employs the use of different methods which includes microemulsion, hydrothermal crystallization and sol–gel by reducing the particle size down to nanoscale [10]. Sol-gel is one of the most successful methods for preparing nano-sized metallic oxide materials and has been mostly applied for the preparation of titanium dioxide nanoparticles. Titanium dioxide nanoparticles prepared via sol-gel method are highly crystalline and have smaller crystallite size compare to the other methods [11]. In this present study, the nano-sized TiO_2 was prepared from organometallic precursor-titanium tetraisopropoxide (TTIP). Titanium alkoxide is not hazardous or toxic to the surroundings, and is stable and provides an easy approach in the production of titanium dioxide nanoparticles [12]. In this research, the prepared sample will be characterized by using techniques like X-ray diffraction (XRD) and Field emission scanning electron microscope (FE-SEM). Processing conditions, such as chemical concentration, the pH, calcinations time and temperature have a great influence on the particle size and phase purity of the final powder.

2. Materials and methods

Titanium tetraisopropoxide (TTIP, 98+ % solution Acros Organics), Hydrochloric acid (HCl,1.0 M solution, R&M Chemicals), Ethanol (CH_3CH_2OH, 95 %, solution, HmbG Chemicals), Deionized water (H_2O).

2.1 Preparation of Titanium Dioxide Nanoparticle by Sol-Gel Process

Titanium isopropoxide (TTIP) was dissolved in absolute ethanol and stirred for 30 min using magnetic stirrer. In this step, to investigate the effect of pH upon the samples properties, a mixture of deionized water and hydrochloric acid (HCl) was added to the solution dropwise to adjust the pH range. The concentration of acid was 1 M and the Table 1 shows the amount of other materials used. The pH was measured using litmus paper to prepare the different sols at pH 3, 7 and 8. A homogeneous solution was obtained after stirring for 2 hrs at room temperature with a magnetic stirrer. A sol was formed after stirring, and then the sols were aged for 24 hrs to form the gel. In order to obtain the nanoparticle, the gel was dried at 100 °C for 1hr to evaporate water and organic material. The dried gel was calcined in the LT furnace at 350, 550 and 750 °C at heating rate of 10 °C/min and for 3 hrs soaking time to obtain the desired TiO_2 nanoparticles.

Table 1: The amount of materials.

Amount of TTIP(ml)	Amount of ethanol (ml)	Amount of H_2O (ml)	Amount of HCl (ml)
7	60	2	1

IOP Conf. Series: Materials Science and Engineering **343** (2018) 012011 doi:10.1088/1757-899X/343/1/012011

2.2 Characterization Method
The morphology of the samples was observed by using Field emission scanning electron microscope (FE-SEM, Nova NanoSEM 450). The phase analysis was obtained by utilizing X-ray diffraction (XRD, BRUKER D5 Phaser).

3. Result and Discussion
Figure 1 shows the morphology of titanium dioxide at pH 3. The diameter of the particles that was observed was 0.10, 0.22, 0.24, and 0.27 µm. The titanium dioxide particles exhibited irregular morphology due to the agglomeration of the primary particle with an average diameter of 0.22 µm. These particles were classified as fine particles that combined together with a coarse particle in the range between 100 and 2500 nm [13]. The diameter of the particle that was observed from FE-SEM morphology analysis was large because the fine particle easily bond together and formed the coarse particle. According to titanium dioxide manufacturers association (2012), the primary particles are strongly bound together by chemical bonds to form aggregates. These aggregates tend to agglomerate via Van-der Waals attractive forces to form particles in the micron (µm) size range. From the observation at 1KX magnification, the crack was formed after calcination [14]. Due to this problem, the cracks observed after calcination leads to an increase of the crack opening displacement [15].

Figure 2 shows the morphology of titanium dioxide at pH 7. The titanium dioxide powder shows agglomeration with fine powder and irregular shape. The diameter of the particle that was observed by FE-SEM morphology analysis was 0.10, 0.27, 0.31, 0.36, 0.40 and 0.49 µm. The titanium dioxide particles exhibited irregularly morphology due to the agglomeration of an average diameter 0.32 µm. These particles were classified as fine particle because it is between the fine particle ranges [13]. The large crack occurs on the surface of titanium dioxide particle at pH 7 after calcination at 550 ^{o}C . This crack was clearly seen at 10 KX magnifications.

Figure 3 shows the morphology of titanium dioxide at pH 8. The particle showed the soft agglomerated fine powder with irregular shapes and fused together. The diameter of these particles observed was 0.09, 0.10, 0.14, 0.18, 0.22 and 0.40 µm. These particle sizes were smaller than the particle at pH 3 and 7. These particles were classified as fine particle because the average diameter of titanium dioxide was 0.18 µm is within the range of fine particle size. The fine crack was clearly seen at 1KX magnification. This crack occurs after the calcination process.

Based on the analysis from the Figure 1, 2 and 3, all the samples of titanium dioxide powder from FE-SEM analysis reveals that the crystallites are in micron meter size. These particle sizes are larger than those previously reported. All sample of titanium dioxide particle was classified as a fine particle. The sample of titanium dioxide for pH 3 at temperature 550 ^{o}C shows higher agglomeration than the sample of titanium dioxide for pH 7 and 8. The particles agglomerates resulting in an increase of particle size as the calcination temperature increases. Consequently, the growth of titanium dioxide nanoparticles is accelerated at higher calcination temperature. The effect of pH also contributes to the changes in particle size. It was observed that the morphology of titanium dioxide greatly depends on pH values. The higher the pH value, the size of the particles becomes smaller and agglomerate.

IOP Conf. Series: Materials Science and Engineering **343** (2018) 012011 doi:10.1088/1757-899X/343/1/012011

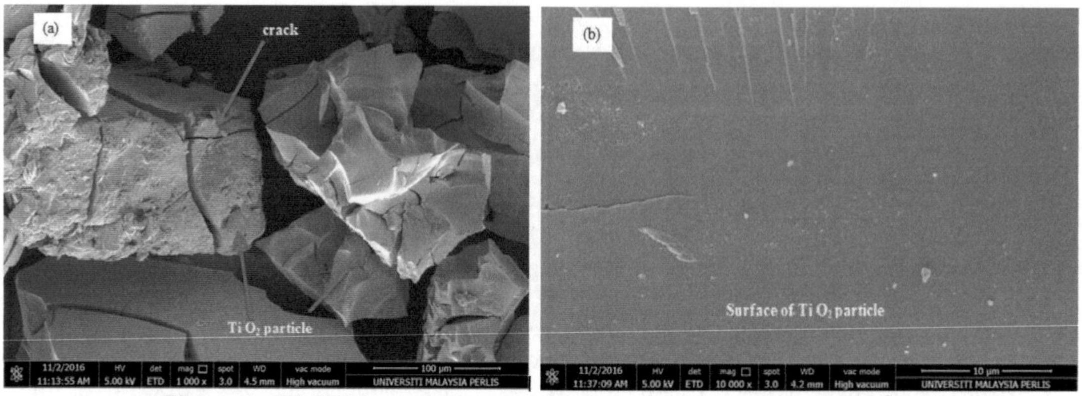

Figure 1: FE-SEM morphology of titanium dioxide powder at 550 °C for pH 3
(a) 1KX magnification, (b) 10 KX magnification

Figure 2: FE-SEM morphology of titanium dioxide powder at 550 °C for pH 7
(a) 1KX magnification, (b) 10 KX magnification

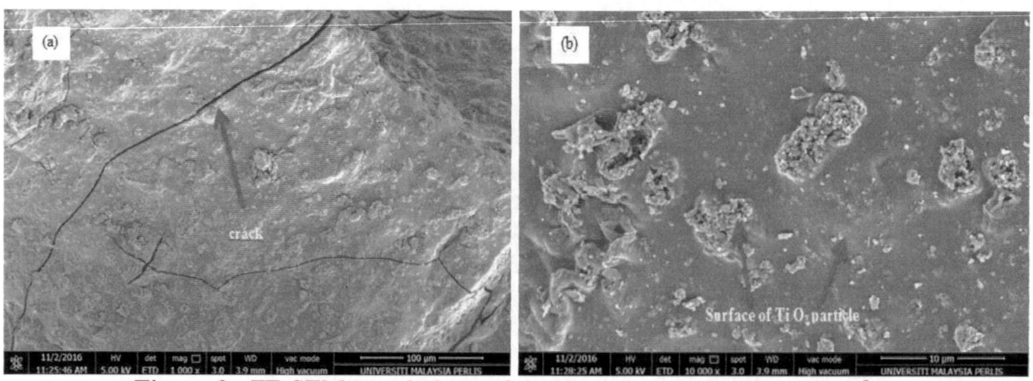

Figure 3 : FE-SEM morphology of titanium dioxide powder at 550 °C for pH 8
(a) 1KX magnification, (b) 10 KX magnification

IOP Conf. Series: Materials Science and Engineering **343** (2018) 012011 doi:10.1088/1757-899X/343/1/012011

3.1 Phase Analysis

The phase analysis of titanium dioxide nanoparticle was obtained by using X-ray diffraction (XRD). The XRD patterns were recorded in the range 2θ = 10 – 90° using step size 0.002 ° and scan rate at 5 °/min. Based on the Figure 4, the X-ray diffraction patterns of titanium dioxide particles dried for sample pH 3 shows an amorphous phase but there are some peaks present due to some unintentional contaminations resulting from the preparation of the sols. While, for the sample pH 7 and 8 were largely amorphous as no distinct peak was observed from the XRD pattern.

Figure 5 shows the XRD pattern of titanium dioxide particles calcined at 350 °C for pH 3, 7 and 8. At the pH value of 3, 7 and 8, anatase peaks of TiO_2 powder after calcination was obviously obtained. The amorphous phase of titanium dioxide particle dried for pH 3 crystallizes to the anatase phase after calcination at temperature of 350 °C. X-ray diffraction patterns of titanium dioxide particles calcined for pH 3 shows the sharp peak which is anatase peak located at 2θ = 25. 50 °, 38.22 °, 48.30 °, 54.54 °,55.02 °, 63.18 °, 70.14 °, 75.66 ° and 83.10 ° are respond to the (101), (004), (200), (105), (211), (204), (220), (215) and (312) lattice plane of TiO_2 at 350 °C. The titanium (Ti) element presence at 2θ = 23. 05° corresponds to the (101) for all pH. The amorphous phase of titanium dioxide particle dried for pH 7 and 8 crystallizes to the anatase phase after calcination at 350 °C The X-ray diffraction patterns of titanium dioxide particles calcined for pH 8 shows the additional anatase peak located 2θ = 58.32 ° respond to the (222) lattice plane of TiO_2 at 350 °C.

Figure 6 shows the XRD pattern of titanium dioxide particles calcined at 550 °C for pH 3, 7 and 8. Definitely phase transformation occurred from anatase phase to thermodynamically more stable rutile phase with increasing calcination temperature [16]. As the temperature is increased to 550 °C, the anatase and rutile peak of TiO_2 powder was obtained with decreased intensity. The rutile phase has its peak at 2θ = 27.39 °, 36.15 ° and 41.53 ° which corresponds to (110), (101) and (111) lattice plane of TiO_2. The anatase phase retains its peak location with further decrease in intensity.

Figure 7 shows the XRD pattern of titanium dioxide particles calcined at 750 °C for pH 3, 7 and 8. At calcination temperature of 750 °C the rutile phase was noticed as the dominant phase. X-ray diffraction patterns of titanium dioxide particles calcined for pH 3, 7 and 8 shows the rutile peak located at the same location which are at 2θ = 27.70 °, 36.30 °, 39.37 °, 41.53 °, 44.29 °, 54.59 °, 57.04 °, 62.88 °, 62.27 °, 68.18 °, 70.10 °, 76.71 °, 82.70 °, and 84.55 ° are responds to the (110), (101), (004), (111), (210), (105), (220), (204), (204), (400), (220), (215), (312) and (312) lattice plane of TiO_2 at 750 °C.

IOP Conf. Series: Materials Science and Engineering **343** (2018) 012011 doi:10.1088/1757-899X/343/1/012011

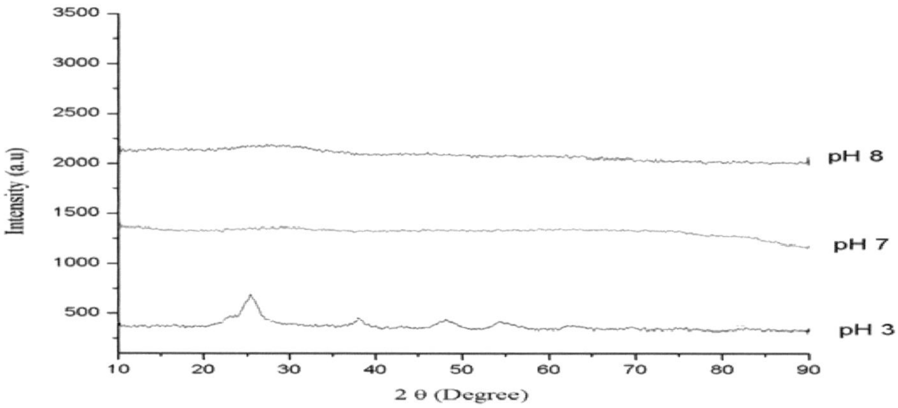

Figure 4: XRD pattern of titanium dioxide particles dried at 100 °C for pH 3, 7 and 8.

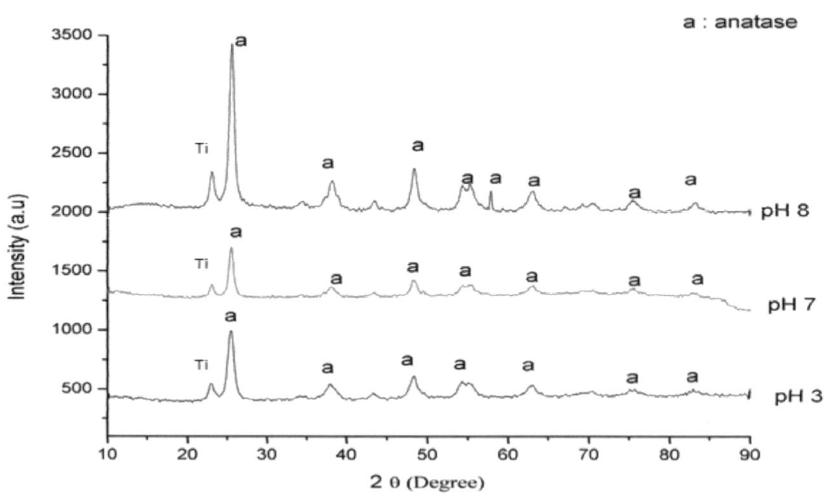

Figure 5 : XRD pattern of titanium dioxide particles calcined at 350 °C for pH 3, 7 and 8.

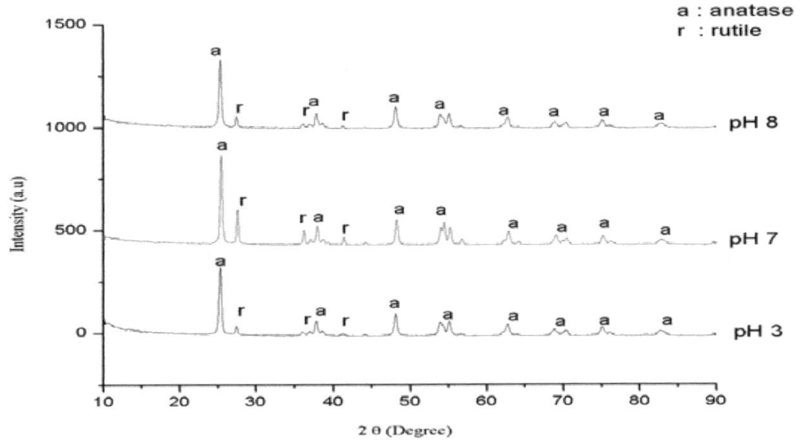

Figure 6: XRD pattern of titanium dioxide particles calcined at 550 °C for pH 3, 7 and 8.

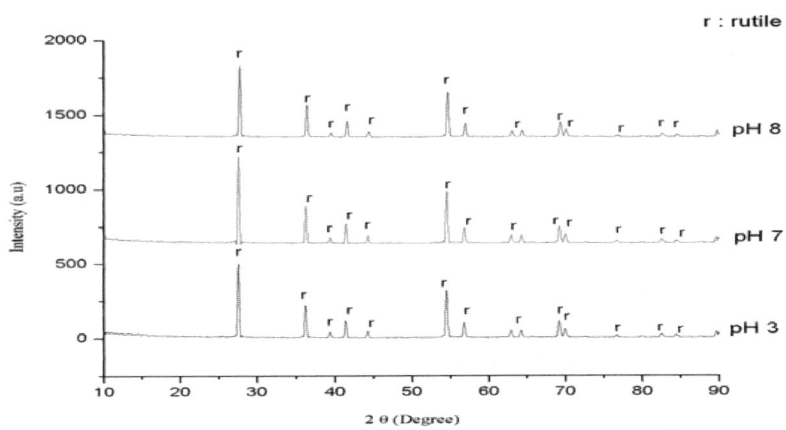

Figure 7 : XRD pattern of titanium dioxide particles calcined at 750 °C for pH 3, 7 and 8.

4. Conclusions

Titanium dioxide nanoparticles were successfully synthesized via sol-gel process using titanium tetraisopropoxide (TTIP) as the precursor, hydrochloric acid (HCL) as the catalysts, and ethanol with deionized water as a solvent. The synthesized titanium dioxide powders were characterized by field emission scanning electron microscopy (FE-SEM) and X-ray diffraction (XRD). The morphology was studied using FE-SEM. The phase transformation was investigated by X-ray Diffraction (XRD).The morphology analysis from FE-SEM confirms an agglomeration of particles from the sol-gel method by different pH value. At pH 3, 7 and 8 the phase transformation from XRD pattern shows the phase of titanium dioxide transform from anatase to rutile after calcination. The titanium dioxide phase for pH 3 before calcination shows had a peak but it was amorphous. Whereas, the titanium dioxide phase for pH 7 and 8 before calcination shows the amorphous phase. It was observed that pH and temperature plays an important role toward the phase transformation of titanium dioxide.

Acknowledgments

I would like to express my gratitude to Universiti Malaysia Perlis (UniMAP) and Center of Excellence Geopolymer and Green Materials (CEGeoGTech) for giving me the opportunity to be involved in this project.

References

[1] Zhao J, Wang X, Sun T and L Li 2005 In situ templated synthesis of anatase single-crystal nanotube arrays *Nanotechnol.* **16** 2450-2454.

[2] Hoffmann M R, Martin S T, Choi W and D. W. Bahnemann 1995 Environmental applications of semiconductor photocatalysis *Chem. Rev.* **95** 69-96.

[3] Li Z, Hou B, Xu Y, Wu D and Sun Y 2005 Hydrothermal synthesis, characterization, and photocatalytic performance of silica-modified titanium dioxide nanoparticles *J. Colloid Interface Sci.* **288** 149-154.

[4] Joseph K, Antony R and Vishwanathan B 2009 Effect of surface area pore volume and particle size of P25 titania on the phase transformation of anatase to rutile *Indian J. Chem Technol* **48** 1378-1382.

[5] Gang L, Xuewen W, Zhigang C, Hui-Ming C and Gao Q L 2009 The Role of crystal phase in determining photocatalytic activity on nitrogen doped TiO_2 *J. Colloid Interface Sci.* **329** 331-338.

[6] Mahshid S, Askari M and Ghamsari M S 2007 Synthesis of TiO_2 nanoparticles by hydrolysis and peptization of titanium isopropoxide solution *J. Mater Process Tech.* **189** 296-300.

[7] Mark H F, Othmer D F, Overberger C G and Seaborg G T 1983 Encyclopedia of Chemical Technology 3rd Edn. John Wiley and Sons New York 139-145.

[8] Koelsch M, Cassaignon S, Minh C T T, Guillemoles J F and Jolivet J P 2004 Electrochemical comparative study of titania (anatase, brookite and rutile) nanoparticles synthesized in aqueous medium *Thin Solid Films* **451** 86–92.

[9] Tang H, Prasad, K, Sanjines R, Schmid P E and Levy F 1994 Electrical and optical properties of TiO_2 anatase thin films. *J. Appl. Phys.* **75** 2042–2047.

[10] Gupta S M and Tripathi M 2012 A review on the synthesis of TiO_2 nanoparticles by solution route *Cent. Eur. J. Chem.* **10**(2) 279–294.

[11] Vijayalaxmi R and Rajendran V 2012 Synthesis and characterization of nano-TiO_2 via different methods. *Archives Appl. Sci. Res.* **4**(2) 1183–1190.

[12] Pathan H.M, Sun-Ki M, Desai J.D, Kwang-Deog J and Oh-Shim J 2007 Preparation and characterization of titanium dioxide thin films by SILAR method *Mater. Chem. Phys.* **97** 5-9.
[13] OMICS International 2016 Nanoparticles http://www.omicsonline.org/nanoparticles.php
[14] Titanium Dioxide Manufacturers Association 2012 About Titanium Dioxide http://www.cinkarna.si/si/files/default/tio2/tdma_about_tio2__20120914.pdf
[15] Mahe M, Heintz J M, Rodel J and Reyndes P 2008 Cracking of titania nanocrystalline coatings *J. Eur. Ceram. Soc.* **28** 2003-2010.
[16] Farhadian A K and Bagheri-Mohagheghi M 2013 Transition from anatase to rutile phase in titanium dioxide (TiO_2) nanoparticles synthesized by complexing sol–gel process: effect of kind of complexing agent and calcinating temperature *J Sol-Gel Sci Technol* **65** 329–335.

Corrosion Protection Of Mild Steel In Sea Water Using Chemical Inhibitor

Alaba O. Araoyinbo [1], Mohd Arif Anuar Mohd Salleh[2], Muhammad Zulerwan Jusof [1]

[1]Faculty of Engineering Technology, Universiti Malaysia Perlis (UniMAP), Perlis, Malaysia

[2]Center of Excellence Geopolymer and Green Materials (CEGeoGTech), School of Materials Engineering, Universiti Malaysia Perlis (UniMAP), Perlis, Malaysia

alaba@unimap.edu.my.

Abstract. The effect of sodium nitrite as a corrosion inhibitor of mild steel in sea water (i.e ASTM standard prepared sea water and sea water obtained from a local river) was investigated, using the weight loss technique. Different amount of sodium nitrite were prepared (i.e 2 % to 10 %) in the inhibition of the mild steel corrosion in sea water exposed to irradiation condition from sunlight exposure. The cut samples of mild steel were exposed to these corrosive media and the corresponding weight loss subsequently obtained was recorded at intervals of 1 to 4 weeks. It was observed that corrosion rate increases with the time of exposure to the corrosive medium exposed to sunlight and that sodium nitrite that was used at the chemical inhibitor was able to retard the corrosion rate of mild steel if the appropriate concentration is applied. The results obtained from the weight loss analysis shows that the optimum percentage of sodium nitrate in sea water that gives the optimum corrosion inhibition of mild steel is 4 %.

1. Introduction

Corrosion is known to be a problem normally associated with material deterioration in the field of oil and gas, construction, and other engineering fields. Mild steel find used in most structural shapes such as beams, bars and pipes [1]. Due to the level of the damage caused by corrosion in various engineering field, there is the need to retard the corrosion rate, if not to prevent it completely [2]. The different methods that have been reported for the prevention control of corrosive activity comprise the separation of the metal interphase from corrosive media or by altering the environment. This method of corrosion prevention includes but not limited to anodic protection, cathodic protection, coating and inhibitors [3]. Inhibitors generally are substances when added in the right concentrations to the environment decrease the corrosion rate considerably. To be fully effective the inhibitors required should be present above a certain minimum concentration.

Corrosion inhibitors can be classified as anodic, cathodic or mixed according to their influence on the electrochemical reaction involving the metals, and their interaction with the environment. Corrosion inhibitors have been in use for several years and the most familiar examples of their applications are in paints and coatings on metals where nitrate, chromate, phosphate, benzoates, borates and oxides are incorporated [4].

It is known that inhibitors could function in several ways to control corrosion activity by adsorption of the film onto the surface of a corroding material, thereby inducing the formation of a thick corrosion product, or by changing the characteristics of the environment resulting in reduced aggressiveness. This work is to study the potency of different concentrations of sodium nitrite in corrosion inhibition of mild steel in sea water obtained from the local river exposed to sunlight [5].

2. Materials and methods

The main material used for this research work is mild steel (i.e. AISI/SAE 1020). The mild steel was purchased locally from a metal fabrication workshop at padang besar , uniciti alam. The XRF analysis of the purchased mild steel used is shown in Table 1:

Table 1. X-ray fluorescence spectrometer of as-received mild steel

Element	C	Si	Mn	Cu	Fe
Weight percent (%)	0.203	0.152	0.627	0.204	Balance

2.1 Sample preparation

Sea water

1. The sea water was prepared according to the ASTM G1 standard 35g/ 1 liter.
2. An original sea water was also obtained from a local river in Pantai Irama, Bachok, Kelantan, Malaysia.

Mild steel

The test specimens of each mild steel to be tested, were measured to obtain fixed and regular dimensions, and cleaned to remove grease and oxidation films that could be present on its surface. The mild steel was cut to a sample dimension of 3.5 cm diameter and 7.5 cm length. The sample was weighed before and after the experiment. Each of the test samples is marked and its initial weight is recorded.

2.2 Solution Preparation

Five different solutions were prepared, each having different concentrations of sodium nitrite to be used as the inhibitor in both sea water. The percentages of the sodium nitrite in the prepared corrosive media are as follows; 2 %, 4 %, 6 %, 8 %, and 10 % respectively.

Eighteen test samples were carefully arranged into each of the corrosive media prepared and left for 4 weeks shown in Figure 1 before being transferred and exposed to sunlight. For every week the corrosion behaviour was observed. After the corrosion testing each sample was removed from the corrosive media, dried and weighed to determine the weight loss.

Figure 1: The immersion of mild steel in sea water for 4 weeks before field exposure

3. Results and Discussion

The weight loss for the mild steel without sodium nitrite inhibitor is shown in Table 2 and Figure 2. It was observed that the rate of corrosion activity increases with time as the weight loss increases which can lead to serious material failure when in use in the engineering field.
The data from one week immersion test shows that the increase in the weight loss started at 0.78 for the original seawater (sea water from the local river) and 1.04 for ASTM prepared sea water and it continued for the subsequent weeks. The data observed for week four analyses for the immersion test shows the weight loss of mild steel in original sea water is higher than the ASTM prepared standard. This is probably due to the presence of other contaminants that could speed up the corrosion activity in the local river. For week four the weight loss in original sea water is 3.35 g while the weight loss in ASTM prepared standard is 2.64 g.

Table 2: The results for comparison without inhibitor

Time (weeks)	Weight loss	
	Sea water (original)	Sea water (based on ASTM)
1	0.78	1.04
2	1.82	1.80
3	2.54	2.32
4	3.35	2.64

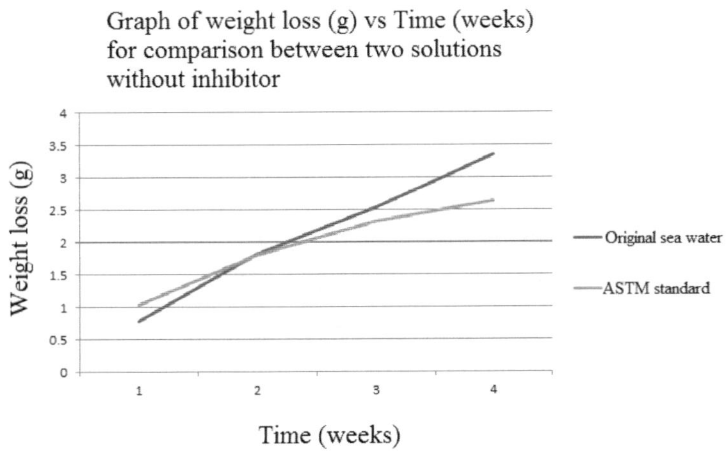

Figure 2: Mild steel weight loss without inhibitor

Figure 3 shows the weight loss that was recorded against the percentage of sodium nitrite in the corrosive media for two different solutions exposed to sunlight.
Based on the graph, for comparison between both sample solutions it was observed that the original sea water sourced from a local river with 2 % sodium nitrite concentration has the highest weight loss and for the ASTM Standard 10 % concentration of inhibitor has the highest weight loss from the immersion test. This can be summarized that the 2 % concentration of inhibitor cannot support and protect the mild steel from rust. Then, the 8 % and 10 % concentration of inhibitor for both samples also could not protect the mild steel from corroding and the best result of decreasing weight loss was 4 % and 6 % concentration of the sodium nitrite inhibitor. The percentage was enough to protect and slow the rate of weight loss in the mild steel.

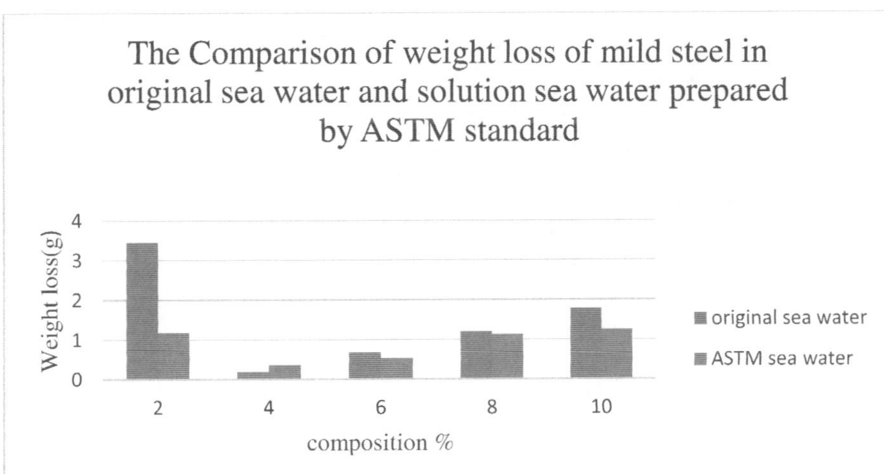

Figure 3: The graph show the comparison weight loss between two sea water conditions

IOP Conf. Series: Materials Science and Engineering **343** (2018) 012012 doi:10.1088/1757-899X/343/1/012012

The surface morphology as observed by optical microscope in Figures 4 (a), (b), (c), (d) and (e) for 2 %, 4 %, 6 %, 8 % and 10 % concentration of sodium nitrite inhibitor in sea water. Rust was observed from the morphology to indicate corrosion activity while the concentration with the most effective inhibitor shows minimal presence of rust on the surface.

(a) 2 % concentration

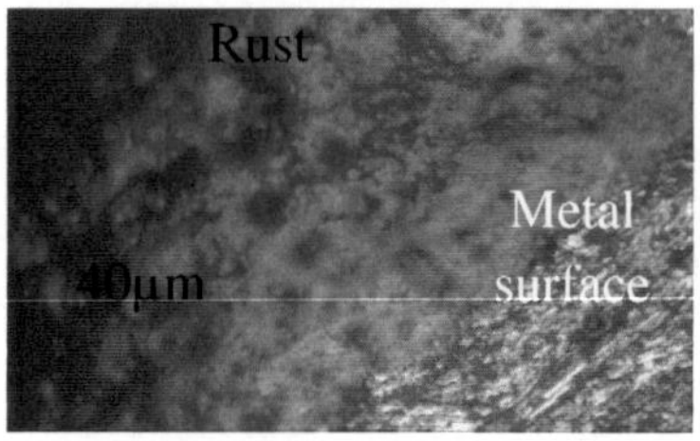

(b) 4 % concentration

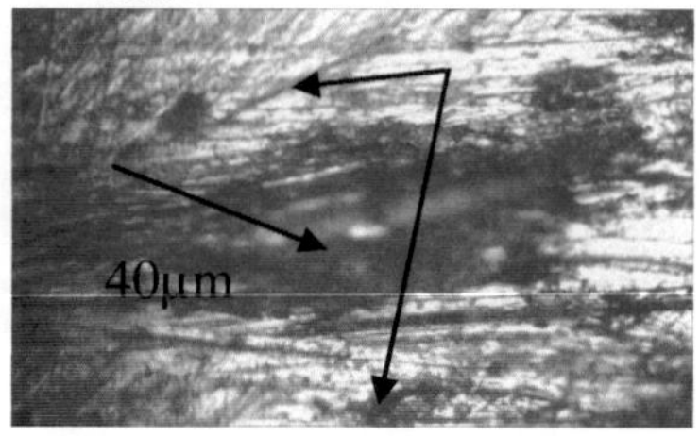

(c) 6 % concentration

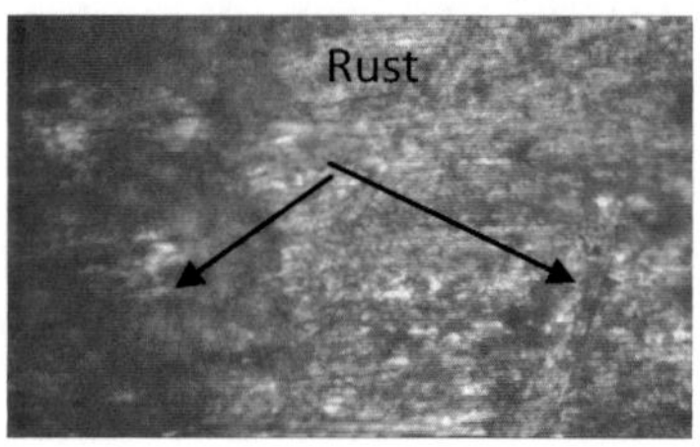

(d) 8 % concentration

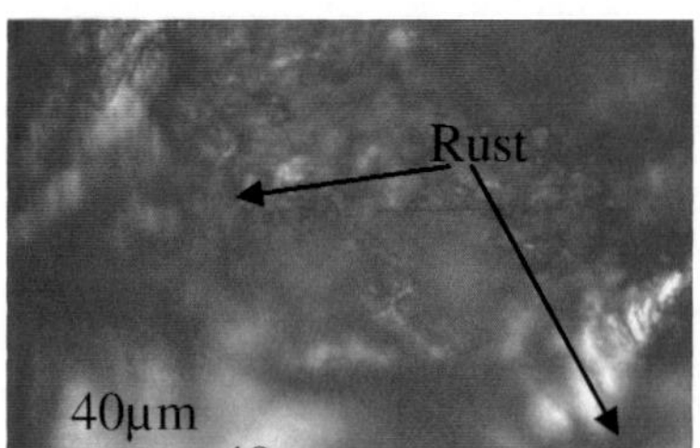

(e) 10 % concentration

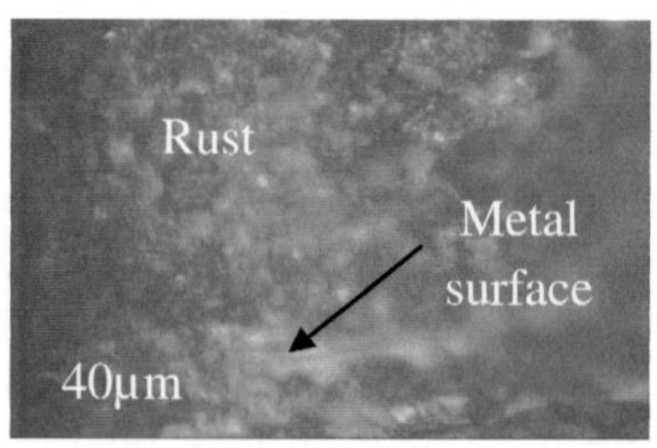

4. Conclusions

Mild steel corrodes when exposed to any aggressive or marine environment which can lead to material failure and shortened life span of engineering materials in service. Sodium nitrite inhibitor provides an alternative method of inhibiting the corrosion behavior when the right dosage or concentration is applied. The corrosion rate was observed to increase with the time of exposure to the corrosive medium being inhibited or not inhibited. Sodium nitrite is shown in this experiment as a suitable inhibitor which can retard the corrosion rate of mild steel if the appropriate concentration is used in sea water. The optimum percentage composition of sodium nitrite required in sea water for optimum corrosion inhibiting performance of mild steel is 4 %.

Acknowledgments

I would like to express my gratitude to Universiti Malaysia Perlis (UniMAP), my supervisor Dr. Alaba Oladeji Araoyinbo, and my project coordinator Assoc. Prof. Dr. Mohd Mustafa Al Bakri Abdullah for giving me the opportunity to be involved in this project.

References

[1]Rastoyi R B, Singh M M, Singh K and Yadav M 2005 Organolic dithiohydrazodicarbonamides as corrosion inhibitors for mild steel-dimethyl sulphoxide containing HCl, Portugaliae *Electrochim. Acta* **22** 315 - 332.

[2] Dubey A K and Singh G 2007 Corrosion inhibition of mild steel in sulphuric acid solution by using polyethylene glycol methyl ether (PEGME) *Electrochim. Acta* **25** 221–235.

[3] Akpan I A and Offiong N A O 2013 Inhibition of mild steel corrosion in hydrochloric acid solution by ciprofloxacin drug. *Int. J. Corros.* 1–5.

[4] Montemor M F 2014 Functional and smart coatings for corrosion protection: A review of recent advances *Surf. Coat. Tech.* **258** 17–37

[5] Venereau E, Diveu C, Grimaud L, Ravon E, Froger J, Preisser L and Gascan H 2010 Definition and characterization of an inhibitor for interleukin-31. *J. Biol. Chem.* **285**(20) 14955–14963.

[6] Adamu M, Umoru L E and Ige O O 2014 Effect of toluene and dioctylphthalate on the rebar corrosion of medium carbon steel in seawater and cassava fluid *J. Miner. Mater. Charact. Eng.* **2** 1–7.

[7] Chinwko E C, Odio B O, Chukwuneke J L and Sinebe J E 2014 Investigation of the effect of corrosion on mild steel in five different environments *Int. J. Sci. Tech. Res.* **3**(7) 306–310.

IOP Conf. Series: Materials Science and Engineering **343** (2018) 012013 doi:10.1088/1757-899X/343/1/012013

Effect of Alkali Concentration on Fly Ash Geopolymers

Siti Fatimah Azzahran Abdullah[1], Liew Yun-Ming[1], Mohd Mustafa Al Bakri[1, 2], Heah Cheng-Yong[1], Khairunnisa Zulkifly[1] and Kamarudin Hussin[1,2]

[1]Centre of Excellence Geopolymer & Green technology (CeGeoGTech), Universiti Malaysia Perlis (UniMAP), 02600, Arau, Perlis, Malaysia

[2]Faculty of Engineering Technology, Universiti Malaysia Perlis (UniMAP), 02100, Padang Besar, Perlis, Malaysia

sitifatimahazzahran@yahoo.com

Abstract. This paper presents the effect of NaOH concentration on fly ash geopolymers with compressive up to 56 MPa at 12M. The physical and mechanical on fly ash geopolymer are investigated. Test results show that the compressive strength result complied with bulk density result whereby the higher the bulk density, the higher the strength. Thus, the lower water absorption and porosity due to the increasing of NaOH concentration.

1. Introduction

Geopolymer was developed by Davidovits [1] in 1978 describing a family of mineral binders which have amorphous to semi-crystalline polymeric silicon-oxygen-aluminum framework structure. Davidovits [2] also proposed a terminology of polysialates as shown in Figure 1 by taking into consideration the Si-to-Al ratio defined as the following: "geopolymer is a polymer which is not original polycondensation, but by so-called "geopolymerization" reaction as result of the alkali activation of aluminosilicate materials.

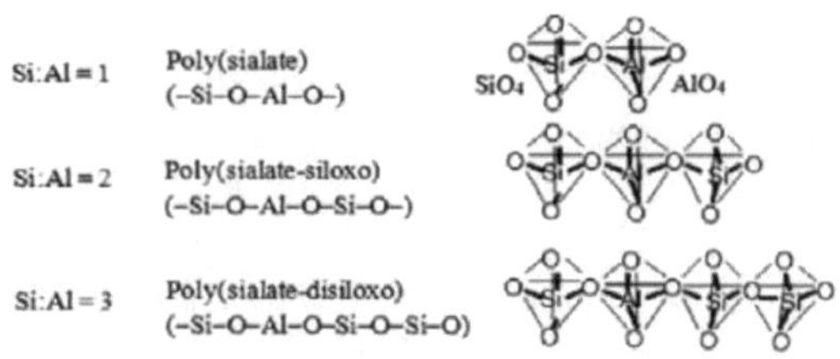

Figure 1: Davidovits terminology on geopolymer [2].

A three-dimensional aluminosilicate network created by the empirical formula Na, Kn [3] n·wH_2O where M is K, Na or Ca cation, n is the degree of poly-condensation and z is the 1, 2, 3 or more than 3 is formed. The network is configured of SiO_4 and AlO_4 tetrahedrons united together by oxygen bridges. Chains or rings united by Si–O–Al bridges are formed. Positive ions (Na^+, K^+, Ca^{2+}) must compensate the negative charge of Al that is present in the IV-fold coordination [2].

IOP Conf. Series: Materials Science and Engineering **343** (2018) 012013 doi:10.1088/1757-899X/343/1/012013

Geopolymer can be categorized as an inorganic polymer that created by the reaction between sodium hydroxide (NaOH) or potassium hydroxide (KOH) and sodium silicate (Na_2SiO_3) or potassium silicate (K_2SiO_3) with solid aluminosilicate materials [4]. A strong alkaline solution induces the dissolution of those solid raw materials producing aluminate and silicate species which linked by means of a polycondensation reaction to form the three-dimensional-cross-linked polysialate structure ([-(Si-O)z-Al-O-]n) [5]. It is necessary to dissolve and release the certain amounts of silica and alumina as to increase the surface hydrolysis the particles of the raw materials [6]. Sodium hydroxide (NaOH) and sodium silicate (Na_2SiO_3) are commonly used in the studies.

According to previous studies [7], it was found that with the higher concentration of NaOH solution that is above 10M, a lower rate of polymerization took place and the strength decreased. In addition, higher alkaline activator content gave a higher compressive strength. Molarity of NaOH solution plays a vital role in the strength of geopolymer. The NaOH concentration will affect the formation of aluminosilicate geopolymer network and change the geopolymer structure of the material since some losses in mechanical properties may occur due to the free OH^- in the alkali-activated matrix and consequently changes the properties of the end product [6, 8]. Previous studies [3, 9, 10] reported that the type of activator affected the properties of its alkali-activated and recommended sodium hydroxide since it is commonly used. Thus, in this study, the effect of NaOH concentration on the physical and mechanical properties of fly ash geopolymers was investigated.

2. Experimental Method

2.1 Raw Materials

Fly Ash (FA) was used as aluminosilicate source. It was collected from a local coal power plant Manjung, Perak. The chemical composition of fly ash as determined by X-Ray Fluorescence (XRF) analysis is given in Table 1. It was a class F fly ash due to the calcium oxide (CaO) content was less than 10% according to the ASTM C-618.

Table 1. Chemical composition of fly ash (FA).

Compound	SiO_2	Al_2O_3	Fe_2O_3	TiO_2	CaO	SrO	SO_3	K_2O	MnO	PbO	NiO
Mass (wt%)	52.3	25.9	10.52	2.92	4.90	0.654	0.05	1.81	0.032	0.046	0.027

The NaOH used was caustic soda flakes with the brand name of Formosoda-P purchased from the Formosa Plastic Corporation, Taiwan. The technical grade liquid Na_2SiO_3 was purchased from South Pacific Chemicals Industries Sdn.Bhd. (SPCI), Malaysia. The liquid Na_2SiO_3 is colourless and dissolves readily in water. It has SiO_2/NaO modulus of 3.20.Stock solution of NaOH (4M-14M) was prepared and allowed at room temperature. The NaOH and sodium silicate were mixed at ratio 2.5 to prepare the alkali activator. This activator solution was prepared 24 hours before use.

2.2 Preparation of Fly Ash Geopolymers

The fly ash was mixed with alkali activator at ratio of 2.0 to produce the fly ash geopolymers paste. The fly ash geopolymers were then cured at room temperature for 24 hours, followed by 24 hours at 60°C. After curing, the geopolymers samples were kept in room temperature until the day of testing at 7 and 28 days.

2.3 Testing and Analysis Methods

Before conducting the compressive strength, the bulk density was measured by measuring the dimension and weight. For water absorption and porosity testing, three samples from each series of NaOH concentration were oven dried at a temperature of 105°C for 24 hours and its weight was determined as weight dry (W_d). The samples were then immersed in water for 24 hours as the saturated weight in water (W_w) and its saturated surface dry weight as the saturated weight in air (W_a). The following equations (1) and (2) were used to calculate the water absorption and porosity values, respectively, in percentage.

$$\% \text{ Porosity} = \frac{(W_a - W_d)}{(W_a - W_w)} \times 100\% \quad (1)$$

$$\% \text{ Water Absorption} = \frac{(W_a - W_d)}{W_d} \times 100\% \quad (2)$$

The compressive strength of fly ash geopolymers was tested using Mechanical Tester as according to ASTM C 109/C 109M-08. Three samples were tested to obtain an average value of strength. All the tests were conducted after 7 and 28 days. A JSM-6460LA model Scanning Electron Microscopy (JEOL) was used to analyze in the morphology of fly ash and fly ash geopolymers. For microstructural analysis of fly ash geopolymers, fly ash geopolymers were the fracture surfaces after the compressive test.

3. Results and Discussion

3.1 Bulk Density, Porosity and Water Absorption

The bulk density, water absorption and total porosity values of the fly ash geopolymers at different NaOH concentration after 7 and 28 days are presented in Figure 2. The density of fly ash geopolymers varied from the range 1.63 until 1.89 g/cm^3. In general, the bulk density did not change significantly with varying NaOH concentration. Even so, it could be clearly seen that the highest bulk density was recorded for NaOH concentration at 12M, the bulk density which is 1.89 g/cm^3 after 7 days and 1.87 g/cm^3 after 28 days. The bulk density slightly dropped with increasing NaOH concentration up to 14M as achieved the optimum concentration at 12M as shown in Figure *3*. The concentration of 14M shows a clearer difference between 7 and 28 days due to a reduction in the resulted weight. Increasing the NaOH concentration may result in paste of a lower density compared to other molarity due to the mixing problems. Density is known to affect the mechanical strength of geopolymers [11].

From Figure 2 (b) and Figure 2 (c), the results showed that water absorption and porosity values decrease when using the higher NaOH concentration. In particular, with NaOH concentration was increased up to 12M, the lowest porosity percentage of 24.64% after 7 days and 22.74% after 28 days were recorded. The porosity is related to the structure and water absorption of the geopolymers [10]. At 4M, fly ash geopolymers showed the highest water absorption of 32.22% after 7 days and at 12M, the lowest water absorption of 12.51% after 7 days and 11.89% after 28 days were collected. The results complied with the bulk density measurement in Figure 2 (a). It is observed that the bulk density of 4M and 8M almost the same density values although the porosity shows the quite difference. This is due to the low strength and porous morphology of the fly ash geopolymer which reduce the porosity at 8M. The higher bulk density samples had lower porosity and thus lower water absorption. The result showed that higher NaOH concentration in the mix gives better reactivity with fly ash resulting in denser microstructure [12].

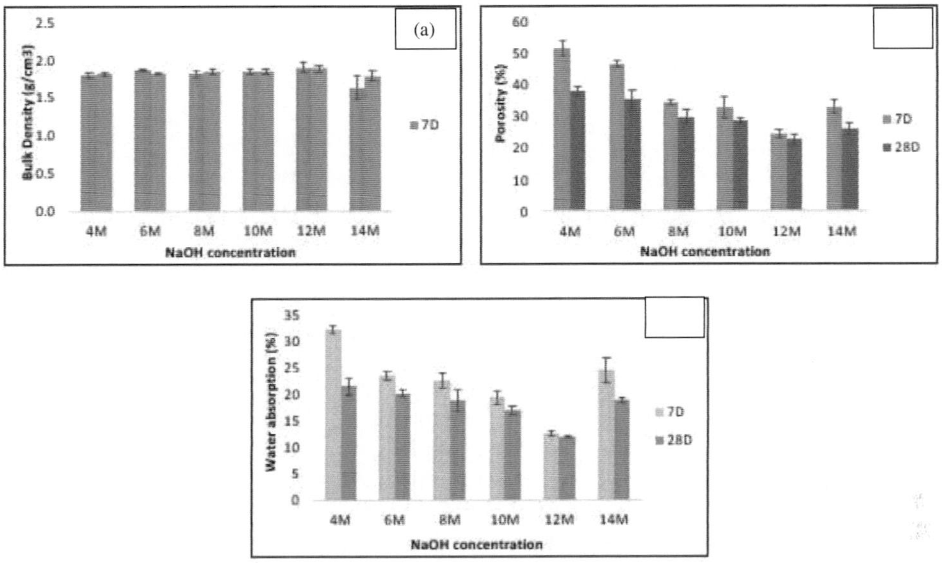

Figure 2: Bulk density of fly ash geopolymers at varying NaOH concentration, (b) porosity (%) of fly ash geopolymers after 7 and 28 days, (c) water absorption (%) of fly ash geopolymers.

3.2 Compressive Strength

Figure 3 presents the compressive strength of fly ash geopolymers at different NaOH concentration in the range of 4M to 14M after 7 and 28 days. Initially, compressive strength gain of fly ash geopolymers for all NaOH concentration slightly fluctuated after 7 days. At 10M, the compressive strength was recorded the highest (29 MPa) after 7 days. This might due to the geopolymers were still at the early stage of strength development [13]. After 28 days, the compressive strength development showed an increasing trend with increasing NaOH concentration up to 12M (56 MPa) due to the high concentrations of NaOH can increase the dissolving rate of silicon and aluminium atoms in raw materials [14]. Beyond 12M, the strength reduced (52 MPa for 14M). The strength reduction at 14M was most probably due to a low of geopolymerization formation [15].

Palomo et al. [16] also proposed that alkaline activator with 12M NaOH solution indicated better result compared to 18M NaOH. At low NaOH concentration of 12M, it meant that there are sufficient Na^+ and OH^- ion for complete geopolymerization reaction and hence the highest compressive strength. The increased strength is related to the structure and density of the geopolymers. The compressive strength result complied with bulk density, porosity and water absorption values. Higher bulk density, lower porosity and percentage of water absorption contributed to higher compressive strength [12]. Thus, the compressive strength with respect to the increase of NaOH concentration was the factor that controls the performance of the fly ash geopolymers. The increased of strength at high NaOH concentration was supported by Pimraksa et al. [17] and Bakkali et al.[18]. Bakri et al. [15] observed the molarity of NaOH is one of the factors that affect the early age compressive strength of fly ash geopolymers.

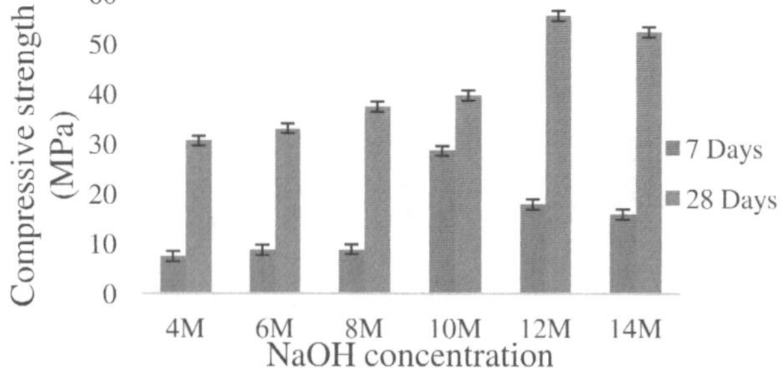

Figure 3: Compressive strength (MPa) of fly ash geopolymers after 7 and 28 days.

3.3 Microstructure Analysis

Figure 4 reveals the SEM micrographs of fly ash whereas Figure 5 shows the SEM micrograph of fly ash geopolymers with various NaOH concentration after 7 and 28 days. Fly ash particles are spherical in shape with smooth surface. After alkali activation, the fly ash particles dissolved and participated in the geopolymerization reaction. Some fly ash particles can still be clearly seen in the geopolymer matrix. However, the smooth surface could no longer be seen with some agglomerated particles on its surface due to alkali attack. According to Bakri et al. [19], fly ash reacted with the NaOH concentration produced the roughness of the surface.

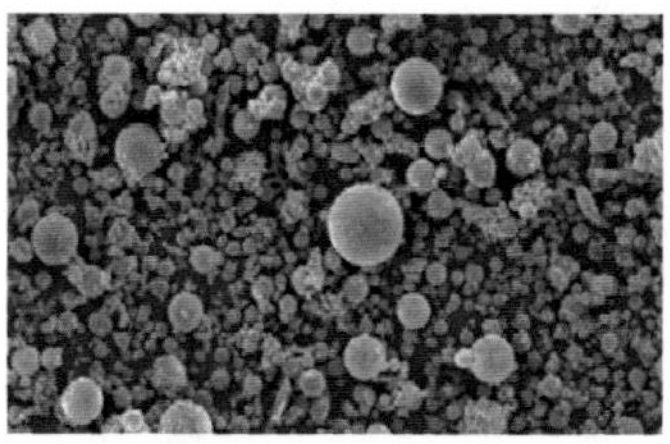

Figure 4: SEM micrograph of fly ash.

Moreover, images of fly ash geopolymers showed the combination of unreacted fly ash particles and also heterogeneous elements. It showed in Figure 5 that the materials are heterogeneous and contain a large proportion of unreacted fly ash after 7 days. However, images of fly ash geopolymers showed the smooth geopolymer matrix and less the unreacted fly ash particles after 28 days. The smooth geopolymer matrix presented in different shapes of the pure fly ash and the fly ash geopolymers. Besides, the various NaOH concentration resulted in presence of pore and less the unreacted fly ash since the spherical shape rarely existed. With increasing NaOH concentration, the unreacted fly ash particles were less abundant and the matrices looked rather dense. This was because of the better ability to dissolve fly ash particles which resulted in better geopolymerization [7].

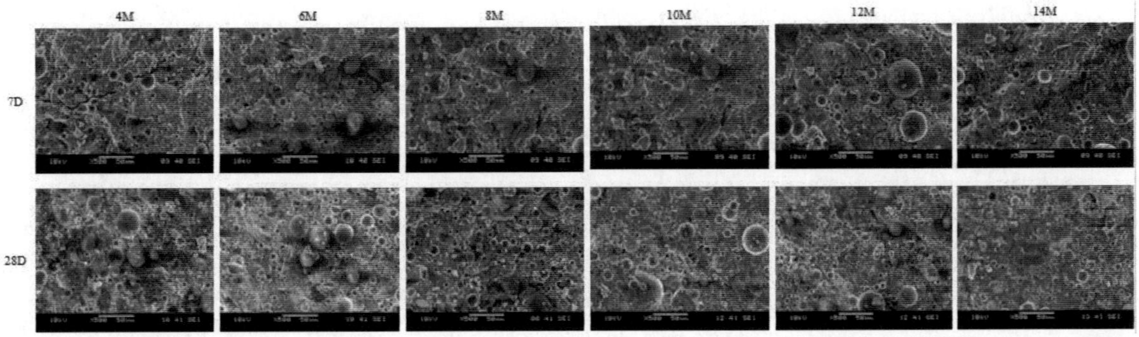

Figure 5: SEM micrograph of fly ash geopolymers after 7 and 28 days with varying NaOH concentration.

4. Conclusion

The molarity concentration of sodium hydroxide (NaOH) in fly ash geopolymers gives a major effect on the geopolymerization process and on the strength of the resulting geopolymer. When the NaOH concentration increased up to 12M, the highest strength of 56 MPa was achieved thus lowering the water absorption and porosity. In SEM images, the higher concentration of NaOH solution showed less the unreacted fly ash particles and smooth geopolymer matrix with fewer pores.

Acknowledgments

The authors acknowledge the funding provided by Fundamental Research Grant Scheme (FRGS/1/2015/TK05/UNIMAP/02/2) sponsored by Ministry of Education Malaysia.

References

[1] Cheng, T.W. and J.P. Chiu, *Fire-resistant geopolymer produced by granulated blast furnace slag.* Minerals Engineering, 2003. **16**(3): p. 205-210.

[2] Davidovits, *Geopolymeric Reaction in the Economic Future of Cementsand Concretes.* 1999: p. 111-121.

[3] Heah, C.Y., H. Kamarudin, A.M. Mustafa Al Bakri, M. Bnhussain, M. Luqman, I. Khairul Nizar, C.M. Ruzaidi, and Y.M. Liew, *Kaolin-based geopolymers with various NaOH concentrations.* International Journal of Minerals, Metallurgy and Materials, 2013. **20(3)**: p. 313-322.

[4] Barrie, E., V. Cappuyns, E. Vassilieva, R. Adriaens, S. Hollanders, D. Garcés, C. Paredes, Y. Pontikes, J. Elsen, and L. Machiels, *Potential of inorganic polymers (geopolymers) made of halloysite and volcanic glass for the immobilisation of tailings from gold extraction in Ecuador.* Applied Clay Science, 2015. **109-110**: p. 95-106.

[5] Catauro, M., F. Papale, G. Lamanna, and F. Bollino, *Geopolymer/PEG hybrid materials synthesis and investigation of the polymer influence on microstructure and mechanical behavior.* Material Research 2015. **18(4)**: p. 698-705.

[6] Görhan, G. and G. Kürklü, *The influence of the NaOH solution on the properties of the fly ash-based geopolymer mortar cured at different temperatures.* Composites Part B: Engineering, 2014. **58**: p. 371-377.

[7] Hanjitsuwan, S., S. Hunpratub, P. Thongbai, S. Maensiri, V. Sata, and P. Chindaprasirt, *Effects of NaOH concentrations on physical and electrical properties of high calcium fly ash geopolymer paste.* Cement and Concrete Composites, 2014. **45**: p. 9-14.

[8] Hamidi, R.M., Z. Man, and K.A. Azizli, *Concentration of NaOH and the effect on the properties of fly ash based geopolymer.* Procedia Engineering, 2016. **148**: p. 189-193.

[9] Abdulkareem, O.A., A.M. Mustafa Al Bakri, H. Kamarudin, I. Khairul Nizar, and A.e.A. Saif, *Effects of elevated temperatures on the thermal behavior and mechanical performance of fly ash*

geopolymer paste, mortar and lightweight concrete. Construction and Building Materials, 2014. **50**: p. 377-387.

[10] Al Bakri, A.M.M., H. Kamarudin, O.A.K.A. Karem, C.M. Ruzaidi, A.R. Rafiza, and M.N. Norazian, *Optimization of alkaline activator/fly ash ratio on the compressive strength of manufacturing fly ash-based geopolymer.* Applied Mechanics and Materials, 2012. **110-116**: p. 734-739.

[11] Wang, H., H. Li, and F. Yan, *Synthesis and mechanical properties of metakaolinite-based geopolymer.* Colloids and Surfaces A: Physicochemical and Engineering Aspects, 2005. **268**(1-3): p. 1-6.

[12] Thokchom, S., P. Ghosh, and S. Ghosh, *Effect of Water Absorption, Porosity, and Sorptivity on Durability of Geopolymer Mortars.* ARPN Journal of Engineering and Applied Sciences, 2009. **4**.

[13] Posi, P., C. Teerachanwit, C. Tanutong, S. Limkamoltip, S. Lertnimoolchai, V. Sata, and P. Chindaprasirt, *Lightweight geopolymer concrete containing aggregate from recycle lightweight block.* Materials and Design, 2013. **52**: p. 580-586.

[14] Onutai, S., T. wasanapiarnpong, S. Jiemsirilers, S. Wada, and P. Thavorniti, *Effect of Sodium Hydroxide Solution on the Properties of Geopolymer based on Fly Ash and Aluminium Waste Blend.* Suranaree Journal Science Technology, 2012. **21**: p. 9-14.

[15] A. M Mustafa Al Bakri, H. Kamarudin, M. Bnhussain, I. Khairul Nizar, A. R Rafiza, and Y.Zarina, *Microstructure of different NaOH molarity of fly ashbased green polymeric cement.* Journal of Engineering and Technology Research, 2011. **3(2)**: p. 44-49.

[16] A.Palomo, M.T.B. Varela, M.L.Granizo, F.Puertas, T.Vazquez, and M.W.Grutzeck, *Chemical stability of cementitious materials based on metakaolin.* Cement and Concrete Research, 1999. **1999**: p. 997-1004.

[17] Pimraksa, K., P. Chindaprasirt, A. Rungchet, K. Sagoe-Crentsil, and T. Sato, *Lightweight geopolymer made of highly porous siliceous materials with various Na2O/Al2O3 and SiO2/Al2O3 ratios.* Materials Science and Engineering: A, 2011. **528**(21): p. 6616-6623.

[18] Bakkali, H., M. Ammari, and I. Frar, *NaOH alkali-activated class F fly ash: NaOH molarity, Curing conditions and mass ratio effect.* Journal Materials Environment Science, 2016. **7**: p. 397-401.

[19] Bakri, A.M.M.A., H.Kamarudin, M.BnHussain, I.K. Nizar, and W.I.W.Mastura, *Mechanism and Chemical Reaction of Fly Ash Geopolymer Cement- A Review* Asian Journal of Scientific Research, 2011. **5**: p. 247-253.

IOP Conf. Series: Materials Science and Engineering **343** (2018) 012014 doi:10.1088/1757-899X/343/1/012014

Friction and Braking Application of Unhazardous Palm Slag Brake Pad Composite

Norizzahthul Ainaa Abdul Khoni[1,2], Che Mohd Ruzaidi Ghazali [1,3] and Mohd Mustafa Al Bakri Abdullah [1,2]

[1] Center of Excellence Geopolymer and Green Technology, School of Materials Engineering, Universiti Malaysia Perlis, Pusat Pengajian Jejawi 2, Taman Muhibbah, 02600 Jejawi, Perlis, Malaysia.

[2] Faculty of Engineering Technology, Uniciti Alam Campus, Universiti Malaysia Perlis, 02100, Sungai Chuchuh, Padang Besar, Perlis, Malaysia

[3] Faculty of Fundamental Science,University Malaysia Terengganu 21030, Kuala Terengganu Malaysia.

Abstract. This paper reveals new alternative friction materials for brake pads. Palm slag was studied as new friction materials in brake pads but its much harder made it difficult to be applied. As a way to reduce the hardness, tire dust was including as purpose on stabilizing the hardness of brake pads. The palm slag was sieves to get desired size that is 150 μm, 300 μm and 600 μm. The percentage weight of materials used are 20% graphite, 20% aluminium oxide, 20% steel fiber, 20% polyester resin and another 40% are varied between tire dust and palm slag. All of materials were blend and compress by using hot pressed machine. The composites properties that were examined are density, porosity, hardness, compressive strength, microstructure analysis and wear rate. The composition of 30% palm slag, 10% tire dust and larger size of filler give better result of mechanical properties and less wear rate of brake pads composites. Then, palm slag can be used in producing of non asbestos brake pads.

1. Introduction

Utilization of friendly materials into production of brake pads was discovered since 117 years ago [1]. The friction materials in pads of brake has important impact to the performance in system of brake and also on health [2]. The reason of replacement ingredient in existing brake pad is the disease that afflicts humans as example cancers, mesothelioma and asbestosis. One of the materials which cause the disease is asbestos. Theoretically, asbestos was chosen once time ago because of its goodness in frictional properties and thermal resistance [1, 3]. Asbestos was blend with other ingredients in the matrix. The use of asbestos was banned caused by its carcinogenic nature give harmful to human health [4].

Besides that, to develop better qualities for new brake pad comparable with existing brake pad, there are government rules that restrict to use materials in brake pad for application on road [1]. Generally, filler in brake pad has significant in plays a role which provide an optimization of friction, stabilization and minimization of wear, under operating variable such as sliding speed, braking force, braking duration and temperature [5]. Nowadays, brake pad was made from aramid [6], palm ash [7], coconut [8, 9], banana peel [10], ceramic [5] ,copper [11] and others.

In this paper, the physical, mechanical and wear test of variation percentage to weight ratio and size of filler for brake pad composites were studied experimentally.

2. Materials and Methods

2.1 Experimental Procedures

In this study, palm slag is use together with tire dust in brake pads composites. Five different compound were produced which contain of phenolic resin as binder, Graphite as lubricant, alumina as abrasive, steel fiber as reinforcement, palm slag and tire dust as filler. Weight balance was used to weight the materials. Dry materials were mixed in pulverizer to obtain homogeneous mixture. The mixture was mixed with polyester and put into hot pressed mold. The composites were pressed at 150^0C with pressure 30 MPa.

Table 1. The ingredients used in palm slag composite.

Materials	C1	C2	C3	C4	C5
Phenolic resin	20%	20%	20%	20%	20%
Palm slag	-	10%	20%	30%	40%
Tire Dust	40%	30%	20%	10%	-
Graphite	10%	10%	10%	10%	10%
Steel fiber	20%	20%	20%	20%	20%
Alumina	10%	10%	10%	10%	10%

2.2 Characterization of palm slag composites

The measurement of density was carried out by using pycnometer equipment. The porosity test using sample dimension of 25mmx25mmx7mm and was left in desiccator for a day. The weight of sample was taken before it was test in oil at 90^oC temperature for 8 hours. Then, the sample was left in oil until the oil was cool down to room temperature.

Hardness test was conducted by using shore D hardness Test. The dimension of samples is 15mm diameter with 6mm thick. The measurement penetration of a specified indentor into the materials was under specified conditions of force and time.

Compression Test of the composites were conducted according to ASTM D695 standards The result of compressive strength was obtained from universal testing machine that retrieves data through software of Test Expert II. There are three variation sizes: 150μm, 300 μm and 600 μm of composites that was tested. The sample with dimension 25mm x 7mm x 7mm were subjected to compressive force, loaded continuously until failure.

Microstructure analysis was carried out using SEM instrument model JEOL JSM-6460LA. Composite samples that have been tested on friction test were observed.

Wear Test on brake pad composites were conducted by using polisher machine with applied 10N load. The samples have dimensions 15 mm in diameter and 6 mm in height. The sample was set up on pin on the surface of stainless steel wheel and a wheel speed of 100 rpm. The weight loss of the composites was measured after friction test with sliding distance of 1 km. The samples were weighed before and after testing to determine weight loss within an accuracy of 0.0001 mg. Wear rate for the brake pad composites were determined by dividing wear volume and sliding distance. Calculation of the wear rate was listed in equation (1) and (2) below:

$$\text{Wear Volume, } m^3 = \frac{\text{Weight Before-Weight After}}{\text{Density, } \rho} \quad (1)$$

$$\text{Wear Rate} = \frac{\text{Wear Volume, } m^3}{\text{Sliding distance, } m} \quad (2)$$

3. Result and Discussion

The brake pad composites which have more tire dust show lowest density than brake composites with more palm slag as shown in Table 2, Table 3 and Table 4. Figure 1(a) and Figure 1 (b) indicates that formulation C5 has the highest porosity and density properties which mean that there should be high stability to the higher temperature because of its attribution to hydrophobic nature of the pores presented in the matrix of the composite through lubricious effect in hydrophobic group [12]. Porosity in composites will absorb energy and heat. Basically, low percentage of porosity will result in higher wear rate. Brake pad composites must contain at least small amount of porosity as an advantage to reduce the effect on the wear rate by water and oil [1].

Table 2: Density and Porosity for 150μm size of filler

Physical Properties	C1	C2	C3	C4	C5
Density (g/cm^{-3})	1.7527	1.9166	2.0667	2.2159	2.2231
Porosity (%)	22.94	27.82	31.61	34.87	36.23

Table 3: Density, Porosity for 300μm size of filler

Physical Properties	C1	C2	C3	C4	C5
Density (g/cm^{-3})	1.7612	1.9311	2.1421	2.1922	2.2413
Porosity (%)	23.21	28.06	32.34	35.18	37.08

Table 4: Density, Porosity for 600μm size of filler

Physical Properties	C1	C2	C3	C4	C5
nsity (g/cm^{-3})	1.7817	1.9402	2.1931	2.2232	2.2513
·osity (%)	25.02	28.89	33.67	35.98	37.97

IOP Conf. Series: Materials Science and Engineering **343** (2018) 012014 doi:10.1088/1757-899X/343/1/012014

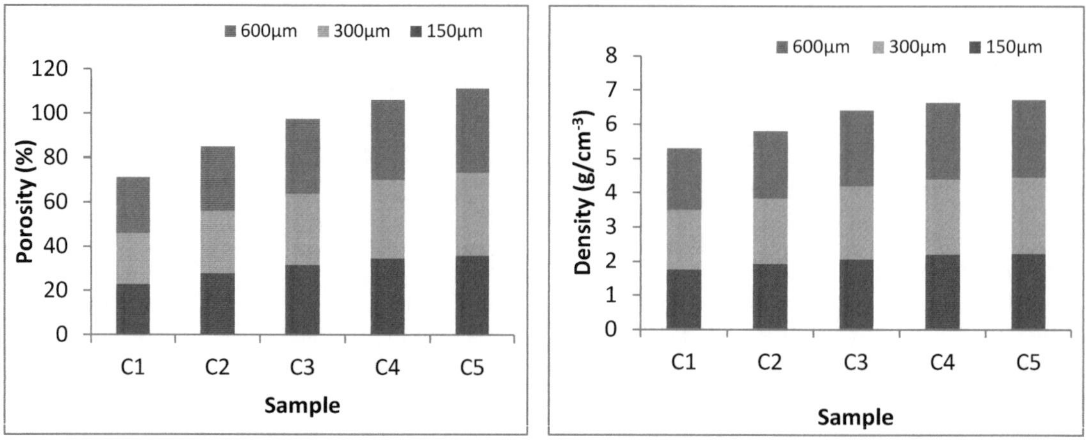

(a) (b)

Figure 1.Physical properties of different size of filler brake pad composites (a) Density analysis of the samples (b) Porosity analysis of the samples

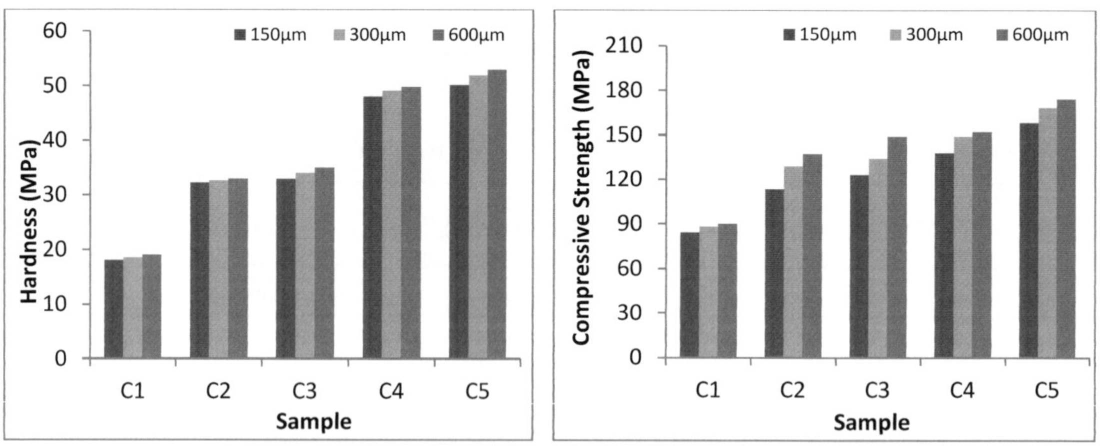

(a) (b)

Figure 2. Mechanical properties of different size of filler for brake pad composites (a) Hardness properties (b) Compressive strength

In this study, the hardness properties of composites shows increase in value when the percentage of palm slag was increase. It was observed that the composites have more palm slag materials will result in higher hardness as shown in Figure 2 (a). The correlation between size of filler and hardness shows that the larger size of filler will increase the value of hardness. The size of filler, 600µm gives higher value of hardness than 150µm and 300µm.

Figure 2(b) shows the result of compressive which correlate with size of filler. The formulation C5 exhibits higher strength to resist applied load and more capability to hold compressive force. Formulation C5 have higher compressive strength caused it contain more palm slag materials.

Surface of friction composites was analyzed through scanning electron microscope to view scratched on surface after friction test. Figure 3 shows SEM micrograph of worn surface. Some materials from the composite have come out of the composites. Figure 3(i) illustrate the brake pad composite with 150μm size of palm slag and tire dust which shows more depth of scratch than other size of filler. It also can be seen that all of the ingredient such as, graphite, aluminium oxide, steel fiber, tire dust and palm slag have been distributed well in the matrix binder. The existence of Al and O have given part of formation of aluminium oxide [12].The worn surface of brake pad composite in Figure 3(ii) have less materials come out and more scratch on surface. The brake pad composites Figure 3(iii) have greater size of filler than others, 600μm, has some tiny plateaus on the surface which responsible to low wear rate [1]. The wear debris that stick on surface of composites (Figure 3) showed the platelets in microstructure which can be form during the materials transfer from composites to disc or disc to composites.

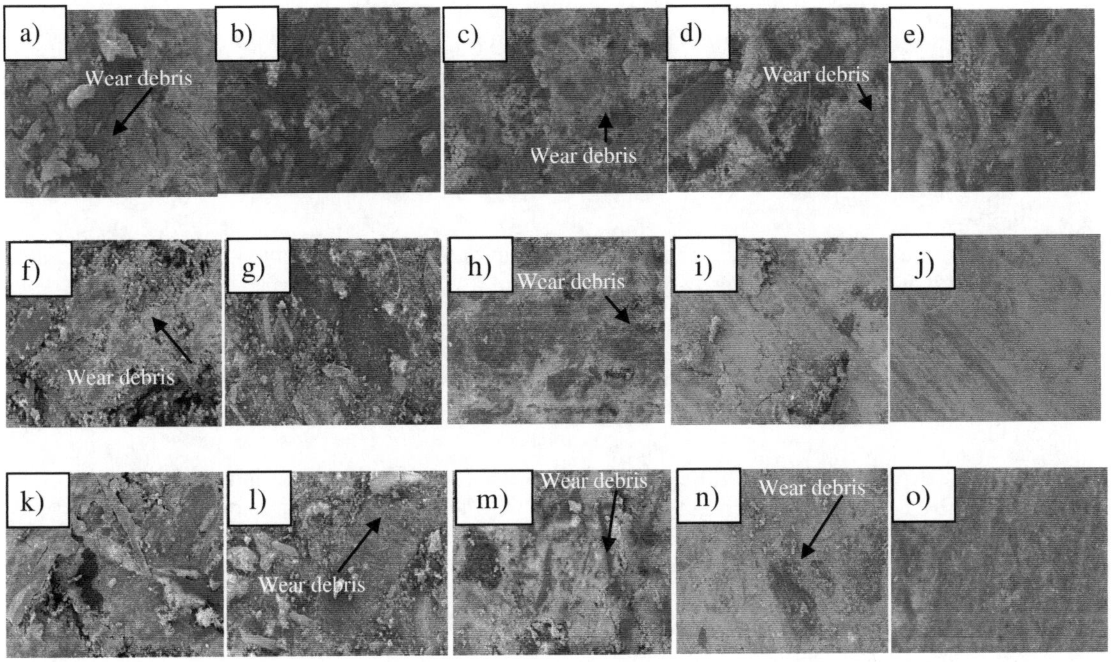

Figure 3. Surface Morphology of Brake Pads of (i) 150μm (a-e) ,(ii) 300μm (f-j) and (iii) 600μm (k-o).

Figure 4 shows wear rate of palm slag composites which varied with percentage and size of filler. Composites of formulation C5 have higher wear rate because it does not have tire dust. After addition tire dust into the composite, formulation C2, C3 and C4 have achieved low wear rate. It can be seen that formulation C4 contributed to higher wear resistance.

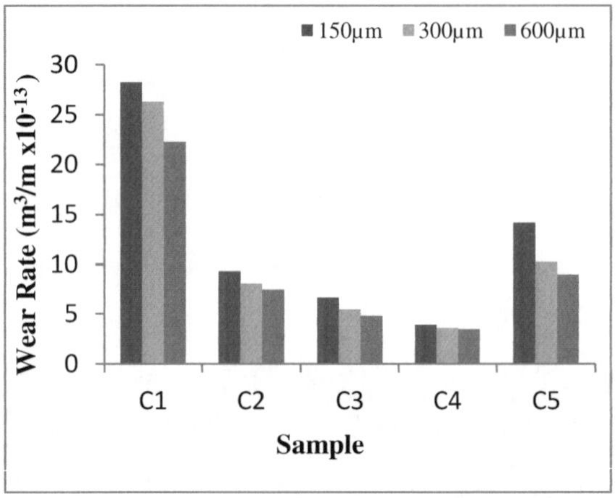

Figure 4. Wear Rate of brake pad composites with variation in size of filler.

4. Conclusions

Palm slag and tire dust can be used as friction materials in brake pad composites reinforced with palm slag and tire dust causing its properties below:

i. Brake pad composites reinforced with palm slag and tire dust show better properties. Hardness and compressive strength is in stability range.
ii. Morphology on surface composites shows hard bonding properties between matrix and filers. The worn surface indicates that the increasing in size of filler will result in low formation of wear debris.
iii. Formulation of composites with tire dust has less wear rate than composites without palm slag. The larger size of filler also decreases the wear rate of the brake pad composites.

5. References

[1] A L Craciun, et al. 2017 *Composites Materials for Friction and Braking Application. Materials Science and Engineering*

[2] Ahmad Murad, et al. 2016 *Thermal Performance of disc brakes using genuine and non-genuine brake pads in Kuwait. International Journal of Applied Engineering Research* **11** 5806-10

[3] R. Dineshkumar, E V V Ramanamurthy, and C. Krishnapavanteja. 2017 *Development of Friction Material by using precast prefired (pcp f) blocks* Materials Science and Engineering **197**

[4] Sadiq Sius LAWAL, Katsina Christopher BALA, and A.T. ALEGBEDE. 2017 *Development and Production of Brake Pad from sawdust composite.* Leonardo Journal of Sciences, **30** 47-56

[5] Patnaik, A., Mukesh Kumar, Bhabani K. Satapathy, and Bharat S. Tomar. 2010 *Performance Sensitivity of Hybrid Phenolic Composites in Friction Braking : Effect of Ceramic and Aramid Fibre Combination* Wear. **269** 891-9

[6] Singh, T. and A. Patnaik. 2015 *Performance assessment of lapinus aramid based brake pad hybrid phenolic composites in friction braking.* Archives of Civil and Mechanical Engineering. **15** 151-61

[7] Ruzaidi, et al. 2011 *Morphology and Wear Properties of Palm Ash and PCB Waste Brake Pad. International Conference on Asia Agriculture and Animal*

IOP Conf. Series: Materials Science and Engineering **343** (2018) 012014 doi:10.1088/1757-899X/343/1/012014

[8] A L Craciun, C Pinca-Bretotean, and D.U.a.A. Josan. 2017 *Tribological Properties of Nonasbestos Brake Pad Material by using Coconut Fiber. International Conference on Applied Sciences*

[9] Rudramurthy, Chandrashekara. K, Ravishankar R, and Abhinandan. S. 2014 *Evaluation of the Properties of Eco-friendly Brake Pad Using Coconut Shell Powder as an Filler Materials. International Journal of Research in Mechanical Engineering & Technology* **4** 98-106

[10] Idris, et al. 2015 *Eco-friendly asbestos free brake-pad: Using banana peels. Journal of King Saud University - Engineering Sciences* **27** 185-192

[11] Gilardi, R., Luigi Alzati, Mamadou Thiam, Jean-Francois Brunel, Yannick Desplanques, Philippe Dufrenoy, Sanjeev Sharma, and Jayashree Bijwe. 2012 *Copper Substitution and Noise Reduction in Brake Pads: Graphite Type Selection.* **5** 2258-69

[12] Sanjeet Kumar, Deepak Kumar, and J. Jain. 2017 *A Study on Mechanical and Tribological Behavior of Brake Pad Materials.* **737** 162-7

IOP Conf. Series: Materials Science and Engineering **343** (2018) 012015 doi:10.1088/1757-899X/343/1/012015

Effect of Sodium Hydroxide Molarity on Physical, Mechanical and Thermal Conductivity of Metakaolin Geopolymers

Nur Ain Jaya[1], Liew Yun-Ming[1], Mohd Mustafa Al Bakri Abdullah[1, 2], Heah Cheng-Yong[1], Kamarudin Hussin[1, 2]

[1]Center of Excellence Geopolymer & Green technology (CeGeoGTech), Universiti Malaysia Perlis (UniMAP), 02600, Arau, Perlis, Malaysia

[2]Faculty of Engineering Technology, Universiti Malaysia Perlis (UniMAP), 02100, Padang Besar, Perlis, Malaysia

ainjaya@gmail.com

Abstract. In the present work, the effect of different sodium hydroxide (NaOH) molarity (6M, 8M, 10M, 12M and 14M) on the physical, mechanical and thermal conductivity of metakaolin geopolymers (MkGPs) was investigated. Geopolymers were prepared by activating the metakaolin with a mixture of NaOH with sodium silicate (Na_2SiO_3). The products obtained were characterized after 28 days of curing. The density, porosity, compressive strength and thermal conductivity (TC) were determined. In general, the NaOH molarity has a significant effect on the compressive strength of the MkGPs. The highest compressive strength was 14.6 MPa achieved with 10M of NaOH solution. The thermal conductivity of MkGPs measured in this work was low in the range between 0.71-0.97 W/mK. NaOH molarity had a significant effect on compressive strength but a marginal effect on thermal conductivity of MkGPs. The thermal conductivity was mainly affected by the bulk density and thus the total porosity. The results showed that the geopolymer can be considered to be used as the thermal insulating material.

1. Introduction

In 1978, Professor Joseph Davidovits has developed a new class of material called geopolymer [1]. Geopolymers are ceramic-like materials produced by geopolymerisation reaction that uses aluminosilicate materials and an alkaline activating solution as the primary raw materials [1-4]. Depending on the raw materials selection, geopolymers exhibit excellent compressive strength [5], fire resistance [4, 6], toxic waste immobilisation [7, 8] and ability to form insulating foams [9]. The alkaline activator for geopolymer comprises of soluble alkali metal and they influence the kinetics of geopolymerisation [10, 11]. The degree of concentration of the alkaline activator greatly affects the geopolymerisation process as excessive alkalinity can adversely affect the strength of the geopolymer [12]. The concentration of NaOH is very important for the dissolution ability of the aluminosilicate source as low molarity causes insufficient of Na^+ ions to be reacted with Si^{3+} and Al^{4+} to complete the geopolymerisation reaction. However, too high concentration results in a faster setting due to reaction products rapidly forming around the unreacted particles preventing further reaction [13].

The composition of geopolymer greatly affects the mechanical and thermal properties. Thermal conductivity is important when considering thermal performance as geopolymer has the ability to be thermally insulating materials due to its amorphous structure [14, 15]. In other words, it is a physical property that determines how much heat will flow in a material. Lower TC means better thermal insulation [16-18]. Density and the degree of porosity are expected to control the thermal conductivity and compression strength of the geopolymer [19, 20]. For geopolymer to be categorized as insulating material, certain requirement set by functional classification of RILEM such as thermal conductivity must be met [21].

The current work focus on the effect of NaOH molarity on the physical (bulk density, true density and total porosity), mechanical (compressive strength) and thermal (thermal conductivity) properties of sodium-based geopolymers synthesized from MK. The alkali activator was prepared using different molarity of NaOH and Na_2SiO_3 solution.

2. Experimental method

2.1. Materials

Metakaolin was used as the Si-Al source material. The metakaolin was obtained by calcining the kaolin, purchased from Associated Kaolin Sdn. Bhd. Malaysia, at 850 °C for 6 h in the furnace. The chemical composition of metakaolin determined by X-ray fluorescence (XRF) spectrometer is shown in Table 1. The particle size distribution of metakaolin was obtained using a Malvern particle size analyser. The mean particle size $d_{(50)}$ was 9.465 μm. The sodium hydroxide (NaOH) powder was of caustic soda micropearls and 99% purity with the brand name of Formosoda-P, made in Taiwan. The sodium silicate (Na_2SiO_3) solution was supplied by South Pacific Chemicals Industries Sdn. Bhd., Malaysia, with a chemical composition of 30.1% SiO_2, 9.4% Na_2O and 60.5% H_2O.

Table 1. Chemical composition of metakaolin as determined by XRF analysis

Chemical	Wt (%)
	Metakaolin
SiO_2	55.7
Al_2O_3	38.6
Fe_2O_3	2.03
TiO_2	0.78
CuO	0.03
ZrO_2	0.04
K_2O	2.43
MnO_2	0.04
LOI	1.74

2.2. Preparation of metakaolin geopolymers (MkGPs)

NaOH solution was prepared in a volumetric flask using distilled water. The alkaline activator solution was formed by mixing Na_2SiO_3 and NaOH solutions at a ratio of 0.24 until a clear solution was obtained. The sodium hydroxide molarity used in this study was 6M, 8M, 10M, 12M and 14M. In order to produce geopolymer paste, metakaolin was mixed with the prepared activator solution. Then the paste was poured into steel moulds with a dimension of 50 ×50 × 50 mm. The moulded samples were sealed with a thin film to prevent moisture loss. The samples were cured at the room temperature for 24 hours and in the oven at 60 °C for another 24 hours.

2.3 Test and analysis method

The bulk density of the MkGPs was obtained as the ratio between the mass of the sample and its geometrical volume. The true density of the metakaolin geopolymer was determined by helium pycnometer technique (Multipycnometer, Quantachrome). The total porosity of the MkGPs was calculated according to the following equation:

$$Total\,porosity\,(\%) = \left(1 - \frac{bulk\ density}{true\ density}\right) \times 100$$

The compressive strength of metakaolin geopolymers was measured in accordance to ASTMC109 by using Instron machine series 5569 Mechanical Tester.

Room temperature thermal conductivity was measured on cylindrical specimens (diameter of 4 mm and height of 4 mm) by using a Hot-Disk thermal constant analyzer (Hot-Disk AB Uppsala, Sweden). At least 3 measurements were performed to ensure reproducibility.

The microstructural analysis was carried out using JSM-6460 LA model Scanning Electron Microscope (JEOL).

3. Result and discussion

3.1. Density and porosity

The bulk density and the total porosity of MkGPs are presented in Figure 1. The measured densities were in the range between 1.49 to 1.64 g/cm^3 for MkGPs with NaOH molarity ranged from 6M to 14M. In general, increasing NaOH concentration increases the bulk density of geopolymers. The lowest density is observed at 8M NaOH molarity with 1.49 g/cm^3. After 8M, the bulk density is increased with increasing NaOH molarity.

The porosities of the MkGPs were ranged between 18% and 24%. The highest porosity of 24% was recorded at 8M NaOH solution. This was probably contributed to the lowest bulk density recorded for geopolymers with 8M of NaOH solution. Further increase in the NaOH molarity from 10M to 14M decreased the porosity as the MkGPs become denser as shown by the density values. In general, density is inversely proportional to the porosity [22].

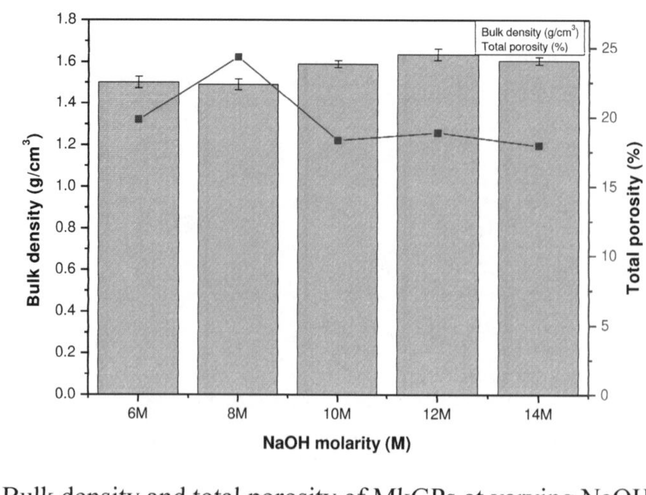

Figure 1: Bulk density and total porosity of MkGPs at varying NaOH molarity.

3.2. Compressive strength

The concentration of NaOH is of great importance affecting the geopolymer properties particularly the compressive strength development. At low molarity (6M), the compressive strength was very low with only 1.3 MPa. This was because of the lower dissolution ability of metakaolin at low NaOH molarity which causes insufficient Al^{3+} and Si^{4+} ions to allow for complete geopolymerisation process. The compressive strength increased towards an optimum of 14.6 MPa when the NaOH molarity increased from 8M to 10M (Figure 2). The optimum strength at this molarity is likely due to the sufficient Al^{3+} and Si^{4+} ions released from the metakaolin to participate in geopolymerisation with the optimum alkaline medium. The concentration of NaOH solution affects the dissolution ability of the metakaolinite particulates as the higher concentration provides the better dissolving ability of raw materials and accelerate the condensation of the monomer. Hence, the bonding strength of the geopolymer increases [23]. On due course, a dense MkGPs were obtained as shown by the increasing density (Figure 1) with the increasing NaOH molarity. However, at a certain NaOH threshold concentration, the compressive strength of geopolymer decreased [24]. Proper NaOH concentration is needed to facilitate and activate the system in order to develop a proper mechanical property as the compressive strength. As refer to Figure 2, the strength of MkGPs reduced when activated with 12M and 14M of NaOH solution.

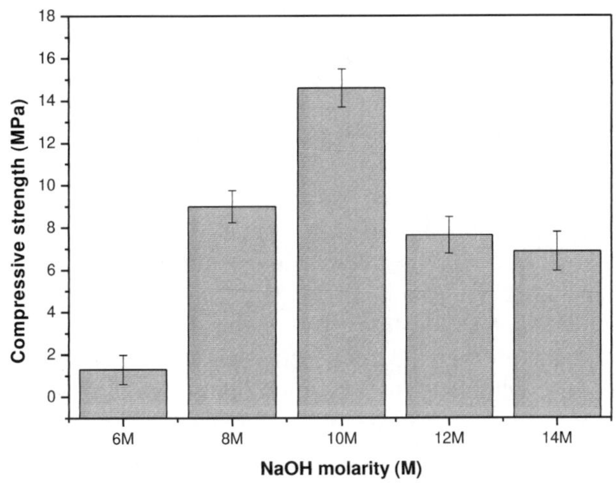

Figure 2: Compressive strength of MkGPs at varying NaOH molarity.

3.3. Thermal conductivity (TC)

The thermal conductivity value was ranged from 0.71 to 0.97 W/mK for MkGPs activated with 6M to 14M of NaOH solution. The lowest value of 0.71 W/mK was recorded at 8M and the highest value of 0.97 W/mK at 14M. The thermal conductivity of MkGPs did not vary significantly with NaOH molarity except for geopolymers activated with 8M NaOH solution. This might most probably due to the highest porosity of geopolymer with 8M NaOH solution (Figure 3). Porosity has the main influence on lowering the TC. The correlation between the TC and compressive strength was not significant because the highest strength was not necessarily attributed to the lowest TC.

Duxson *et al.* [18] obtained a lower value (0.8 W/mK) of TC when activated the geopolymer with potassium-based alkaline activator. The lower value of TC obtained in this work was assumed to be lower because the TC of sodium-based geopolymer is doubled of the potassium-based geopolymer. According to Gouloure *et al.* [25], geopolymers are promising to be used as thermal insulation materials because of the low TC (< 1 W/mK). Further work such as the addition of foaming agent to the MkGPs might help to improve the thermal conductivity to satisfy the criteria to be insulating materials as required by RILEM. Besides, the existence of the crystalline phases in geopolymer affects the thermal conductivity as the insulating ability is based highly on the amorphous nature of the material [17].

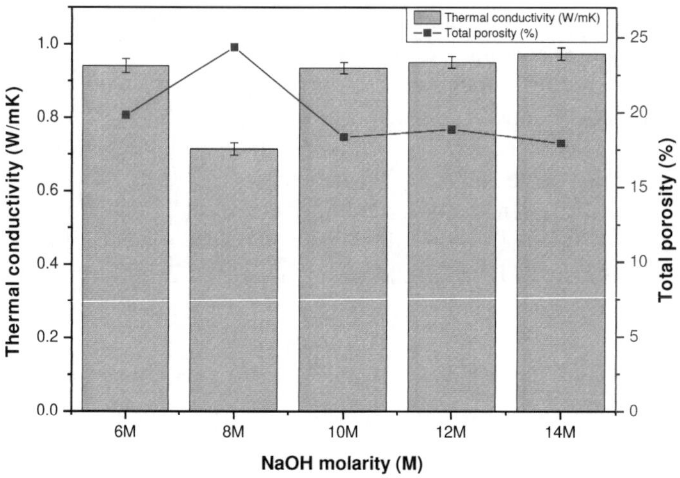

Figure 3: Thermal conductivity and total porosity of MkGPs at varying NaOH molarity.

3.4. Microstructural analysis

From the microstructural image in Figure 4, all samples showed the existence of pores that contributed to the porous structure. The MkGPs at low NaOH concentration (6M and 8M) has more pores compared to the MkGPs at high NaOH molarity (10M, 12M and 14M). Unreacted particles were observed in the samples at 6M and 8M due to the low molarity of NaOH solution to be reacted with metakaolin particles. At 10M, the unreacted particles still exist but become less due to the progress of geopolymerisation reaction. The MkGPs with high molarity (12M and 14M) has a large portion of unreacted particles.

The existence of more pores at 6M and 8M has shown to have high porosity values which correlate well with the lower densities (Figure 1) compared to MkGPs at high NaOH molarity (10M, 12M and 14M). The size of the pores also became smaller at high molarity which is showed by the dense matrix. The existence of pores reduced the density of the geopolymers and subsequently decreased the compressive strength and improved the insulating properties.

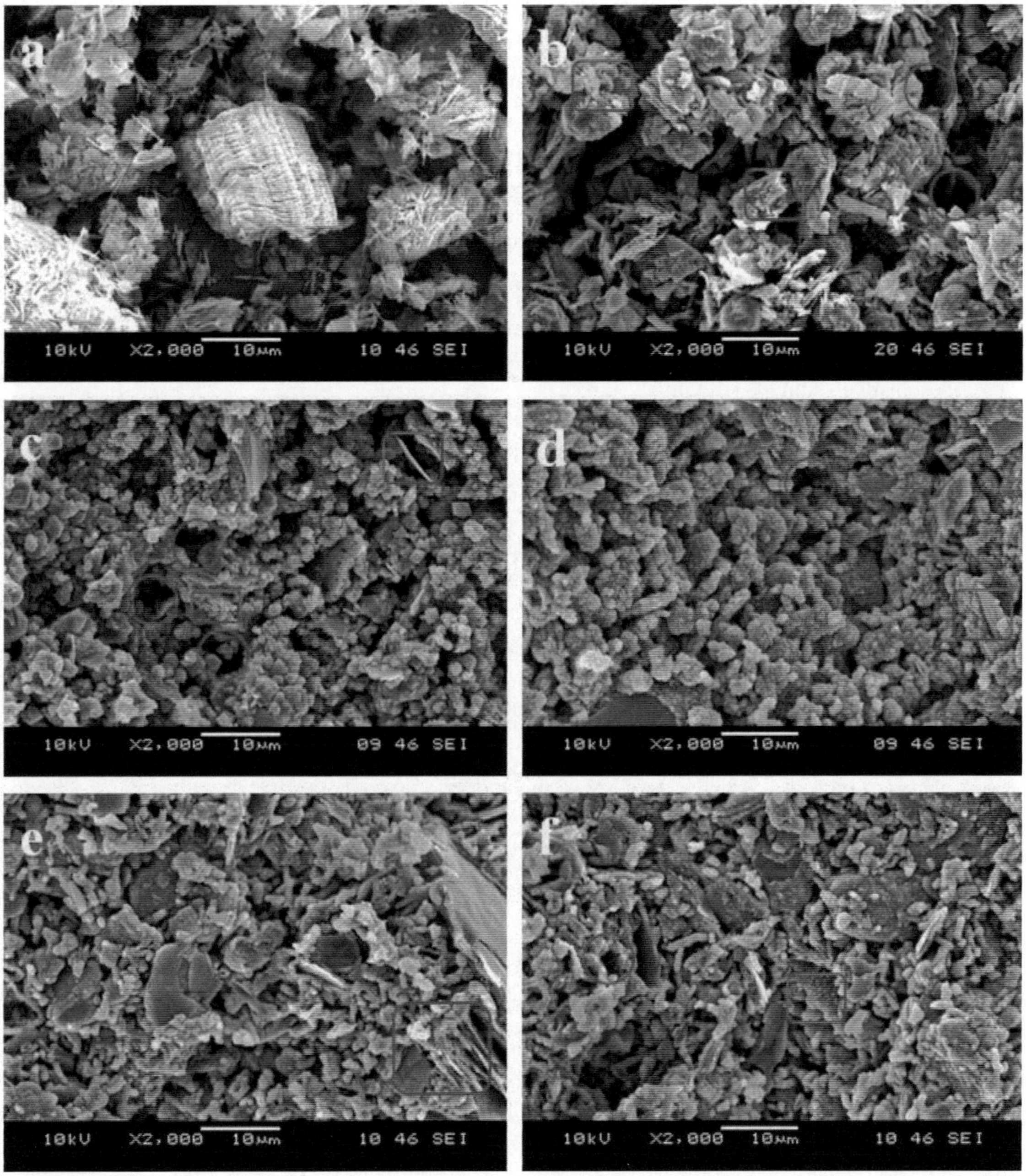

Figure 4: Microstructural images of (a) metakaolin and MkGPs with different NaOH molarity (b) 6M, (c) 8M, (d) 10M, (e) 12M and (f) 14M. (○-pores and □-unreacted metakaolin particles).

4. Conclusions

The effect of NaOH molarity on physical, mechanical and thermal properties of MkGPs has been carried out. The NaOH molarity has a significant effect on the compressive strength but a marginal effect on thermal conductivity of MkGPs. The highest compressive strength of 14.6 MPa was achieved at 10M NaOH solution. The lowest thermal conductivity of 0.71 W/mK was obtained when activated with 8M NaOH solution. The thermal conductivity was mainly affected by the bulk density and total porosity. The results obtained show that geopolymer here developed can be considered potentially useful to be used as the thermal insulating material.

Acknowledgments
The authors wish to deliver their gratitude to the funding provided by Fundamental Research Grant Scheme (FRGS/1/2015/TK05/UNIMAP/02/2) sponsored by Ministry of Education Malaysia.

References
[1] Davidovits J 1991 *J Therm Anal* **37** 1633-1656
[2] Bakharev T 2006 *Cement Concrete Res* **36** 1134-1147
[3] Silva P D and Sagoe-Crenstil K 2008 *Cement Concrete Res* **38** 870-876
[4] Duxson P, Fernandez-Jimenez A , Provis J L, Lukey G C, Palomo A, and Deventer J S J V 2007 *J Mater Sci* **42** 2917-2933
[5] Hardjito D and Rangan B V 2005 Aust J Struct Eng **6** 1-9
[6] Kovalchuk G and Krivenko P V 2009 *Producing fire- and heat-resistant geopolymers, Geopolymers: Structure,processing, properties and industrial applications* ed Woodhead Publishing Limited (UK) pp 227-263
[7] Jaarsveld J G S Vand Deventer J S J V 1996 *Miner. Eng* **10** 659-669
[8] Aly A, Vance E R, Perera D S, Hanna J V, Griffith C S, Davis J, and Durce D, 2008 *J Nucl Mater* **378** 172-179
[9] Brooks R, Bahadory M, Toviac F, and Rostami H 2010 *Int J Sustain Eng* **3** 211–218
[10] Sagoe-Crentsil K and Weng L 2007 *J Mater Sci* **42** 3007-3014
[11] Bell A T 1999 *Colloids Surf A* **158** 221-234
[12] Atiş C D, Görür E B, Karahan O, Bilim C, İlkentapar S, and Luga E 2015 *Constr Build Mater* **96** 673-678
[13] Provis J L, Yong, C Z, Duxson P, and Deventer J S J V 2009 *Colloids Surf A* **336** 7-63
[14] Demirboga R 2007 *Build Environ* **42** 2467–2471
[15] Kamseu E, Nait-Ali B, Bignozzi M C, Leonelli C, Rossignol S, and Smith D S 2012 *J Eur Cer Soc* **32** 1593–1603
[16] Feng J, Zhang R, Gong L, Li Y, Cao W, and Cheng X, 2015 *Mater Des* **65** 529–533
[17] Kamseu E, Ceron B, Tobias H, Leonelli E, Bignozzi M C, Muscio A, and Libbra A 2012 *J Therm Anal Calorim* **108** 1189-1199
[18] Duxson P, Lukey G C, and Deventer J S J V 2006 Ind Eng Chem Res **45** 7781-7788
[19] Kamseu E, Gouloure Z M N M, Ali B N, Zekeng S, Melo U C, Rossignol S and Leonellic C 2015 *Energ Buildings* **88** 45–56
[20] Wang H, Porter W D, Bottner H, Konig J, Chen L, Bai S, Tritt T M, Mayolet A, Senawiratne J, Smith C, Harris F, Gilbert P, Sharp J, Lo J, Kleinke H, and Kiss L 2013 J Electron Mater **42** 224-240
[21] A.C. Institute, *Building Code Requirements for Structural Concrete (ACI 318M-11) and Commentary*. 2011: Farmington Hills, Michigan
[22] Strozi-Cilla M, Morellia M R, and Colombo P 2014 *Cerams Int* **40** 13585–13590
[23] Wang H, Li H, and Yan F 2005 *Colloids Surf A* **268** 1-6
[24] Görhan G and Kürklü G 2014 *Composites Part B* **58** 371-377
[25] Gouloure Z N M N, Nait-Ali B, Zekeng S, Kamseu E, Melo U C, Smith D, and Leonelli C 2015 *J Build Eng* **3** 58–69

IOP Conf. Series: Materials Science and Engineering **343** (2018) 012016 doi:10.1088/1757-899X/343/1/012016

Temperature Effects on Stainless Steel 316L Corrosion in the Environment of Sulphuric Acid (H_2SO_4)

I.G. Ayu Arwati [1,2*], Edy Herianto Majlan[1], Wan Ramli Wan Daud[1], Loh Kee

Shyuan [1], Khuzaimah Binti Arifin[1], Teuku Husaini[1], Sagir Alfa [2], and Fakhruddien Ashidiq [2]

[1]Fuel Cell Institute, National University of Malaysia , 43600 UKM Bangi, Selangor Darul Ehsan, Malaysia.

[2] Mechanical Engineering Program, Engineering Faculty, Mercubuana, 1160, Indonesia

arwatiayu@yahoo.com

Abstract. In its application, metal is always in contact with its environment whether air, vapor, water, and other chemicals. During contact, chemical interactions emerge between metals and their respective environments such that the metal surface corrodes. This study aims to determine the corrosion rate of 316L stainless steel sulphuric acid environment (H_2SO_4) with weight loss and electrochemical methods. The corrosion rate (CR) is value of 316L stainless steel by weight loss method with sulfuric acid (H_2SO_4) with concentration of 0.5 M. The result obtained in conjunction with the increase of temperature the rate of erosion obtained appears to be larger, with a consecutive 3 hour the temperature of 50°C is 0.27 mg/cm^{2}h, temperature 70°C 0.38 mg/cm^{2}h, and temperature 90 °C 0.52 mg/cm^{2}h. With the electrochemical method, the current value increases by using a C350 potentiostal tool. The higher the current, the longer the time the corrosion rate increases, where the current is at 90 °C with a 10-minute treatment time of 0.0014736 A. The 316L stainless steel in surface metal morphology is shown by using a Scanning Electron Microscope (SEM).

1. Introduction

Metal is generally used as a support tool given its several advantages compared with other materials [1]. Certain types of metal are used as design/construction materials to create equipment, such as bipolar stack PEM fuel cells. One type of stainless steel used is 316L stainless steel the type of steel that is easy to rust. Corrosion is a metal material damage caused by a reaction between metals and their respective environments, thereby resulting in metal oxide, metal sulphide, or other reactions, such as calcification. From a chemical point of view, corrosion is essentially a metal reaction to ions on the metal surface that directly contacts aqueous and oxygen environments [1]. In its application, metal is always in contact with its environment whether air, vapour, water, and other chemicals. When metals come in contact with their respective environments, interactions between them happen. As a result, the metal surfaces connecting form anode and cathode regions. In the anode region, metals release electrons to the cathodes captured, thereby forming a hydroxyl group, which then reacts with the metal ions that

IOP Conf. Series: Materials Science and Engineering **343** (2018) 012016 doi:10.1088/1757-899X/343/1/012016

have released the electrons [2].

Stainless steel is a corrosion-resistant alloy steel, which is widely used in chemical, fertiliser, food and beverage industries and is related to seawater and all industries that require corrosion resistance [3]. Corrosion is the decline in the quality of metals due to the electrochemical reaction with their respective environments that are directly related to open air, which is generally referred to as atmospheric corrosion. Almost all corrosion products are caused by the atmospheric environment. One reason is that most metals are always associated with open air where humidity and pollutant contents may affect metal leakages. Factors such as temperature, humidity and chemical content in the air determine the degree of corrosion. Meanwhile, metal composition, metallurgical structure, and metallurgical processes also accelerate corrosion. Corrosion is a difficult metal problem, dangerous and very expensive if not treated. Generally, corrosion is a metal deterioration due a reaction between metals and the surrounding environments. The important types of corrosion are uniform, galvanic, crack, current-assisted, intragranular and high temperature corrosions. Corrosion results from a low pH effect of water in which H^+ ions react with electrons in cathodes. The factors that can cause corrosive metals are, pH, oxygen content, air, metal and temperature [4] [3]. Corrosion prevention in metal can be done in several ways, such as cathodic protection, coating and can use also use inhibitor. Inhibitor use green is the new trend in corrosion control [6][8]. Corrosion can be defined as an electrochemical event between metals or metal alloys and their respective environments, with the following components as the conditions of occurrence: (1) anode,as the cover of oxidation reaction., (2) cathode as closed the occurrence of reduction reaction, (3) media electrolyte as the conductor of electric current and (4) the emergence of electric current between the anode and the cathode.factors that can affect the corrosion process, among others. (I) Temperature—The rise in temperature results in an increase in the rate of corrosion reaction. This scenario happens because the high temperature of the kinetic energy of the reacting particles increases such that it exceeds the magnitude of the price of activation energy. Consequently, the rate of reaction corrosion also accelerates, and vice versa. (II) Fluid flow velocity or speed stirring—The corrosion rate tends to increase as the rate or velocity of fluid flow increases. This scenario happens because the contact between reactants and metals is significant such that the number of metal ions increases and Abu-Dalo [7] more loose so that the metal will experience the fragility (corrosion). (III) Concentration of corrosive material—This activity is related to the pH or acidity and alkalinity of a solution. The acidic solution is highly corrosive to the metal where the latter in the acid solution medium will rapidly corrode due to the anode reaction. Meanwhile, an alkaline solution may cause corrosion in the cathode reaction because such a reaction is always in conjunction with the anode reaction. (IV) Oxygen—Oxygen in the air can be in contact with moist metal surfaces. So the chances of becoming more corrosion. In water (open environment), the presence of oxygen causes corrosion.

Corrosion velocity can be calculated by weight gain union time unity can also be calculated by the thickness of oxidation formed time unity. Corrosion designation is generally made by adding or removing the weight graph as a function of time. The calculation of the corrosion rate (CR) uses the equation below for Abu-Dalo [7]:

$$CR = \frac{m1 - m2}{A\,t}$$

where m1 and m2 are the weight losses (mg) before and after the immersions in the test solutions, respectively, A is the surface area of the specimens (cm^2) and t is the exposure time (hour).

2. Methods

In this study, SS 316L metal, with 1 × 1 cm^2 metal size in 0.5 M sulfuric acid environment, was tested at room temperature, 50 °C, 70 °C and 90 °C using immersion times 1, 2 and 3 hours. Before the metal test was cleaned using sandpaper from grade 800 to grade 1200, the metal was then cleaned with aquades DM and alcohol. For testing, the corrosion rate was determined in this study using weight loss and electrochemical methods with potentiostate. The morphology and surface of 316L stainless steel specimens were examined after polishing and after expore to 0.5 M sulphuric acid using scaning electron microscopy (SEM).

3. Results and Discussion

To see the corrosion rate was carried out by preparation of 316L stainless steel cutting with size 1 × 1 cm^2, then immersion with sulfuric acid (H_2SO_4) with 0.5. Figure 1 shows that the 316L stainless steel metal in the sulfuric acid environment (H_2SO_4) with a concentration of 0.5 M at a temperature of 50°C, 70°C, and 90°C, at a temperature of 50°C at a resistance of 1hour 0.24mg, while in the immersion for 3 hours there was a reduction of 0.66. With the rise in a temperature rate of 70 °C in 1 hour, a weight loss of 0.66 mg and M concentration for 1, 2 and 3 hours with the variation of temperatures 50 °C, 70 °C and 90°C occurs. The weight loss (1.14 mg) in the 3-hour immersion was significant. If the temperature rate increased to 90 °C in a 1-hour immersion with a weight loss of 0.88 mg, then int 3 hours, a high weight loss of 1.14 emerged. From the data obtained it can be concluded that the longer time and temperature the higher the heavy metal 316L stainless steel means that the metal is clearly corrosive.

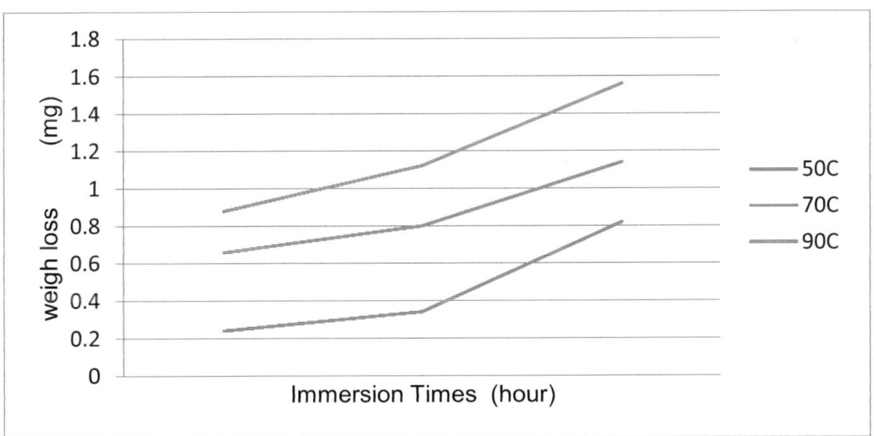

Figure 1. Graph of 316L stainless steel weight loss by immersion in acid sulphate (H_2SO_4) at concentration 0.5 M in 1, 2 and 3 hours with temperature variations of 50 °C, 70 °C and 90 °C.

From the data of corrosion rate calculation to immersion with sulfuric acid 0.5 M for 1 hour at temperature 50°C, is 0.24 mg/cm^2h. Figure 2 illustrates the highly detailed CR for 2- and 3-hour smearing. Evidently, the temperature increased, and the length of immersion times reduced the corrosion rate. After calculating a rust rate of 0.5 M in a 3-hour immersion at 90 °C, 316L stainless steel has a 0.52 mg/cm^2h rust CR content. From the data of corrosion rate

IOP Conf. Series: Materials Science and Engineering **343** (2018) 012016 doi:10.1088/1757-899X/343/1/012016

calculation to immersion with sulfuric acid 0.5 M for 1 hour at temperature 50°C is 0.24 mg/cm²h. Figure 2 exhibits the highly detailed CR for 2- and 3-hour smearing. Apparently, the temperature increased, and the length of immersion times reduced the corrosion rate. 316L stainless steel after calculating corrosion rate (CR) content of 0.5 M in a 3 hour immersion at 90°C has a 0.52 mg/cm²h corrosion. Corrosion rate decreases with increasing time. The higher the temperature and the soaking time the corrosion rate decreases. With temperature rise and immersion period in the yield at 50°C by immersing the corrosion rate of 3 hours 0.18 mg/cm²h, at 70°C temperature with 3 hours immersion 0.38 mg/cm²h, at 90°C with 3 hours immersion of 0. 52 mg/cm²h.

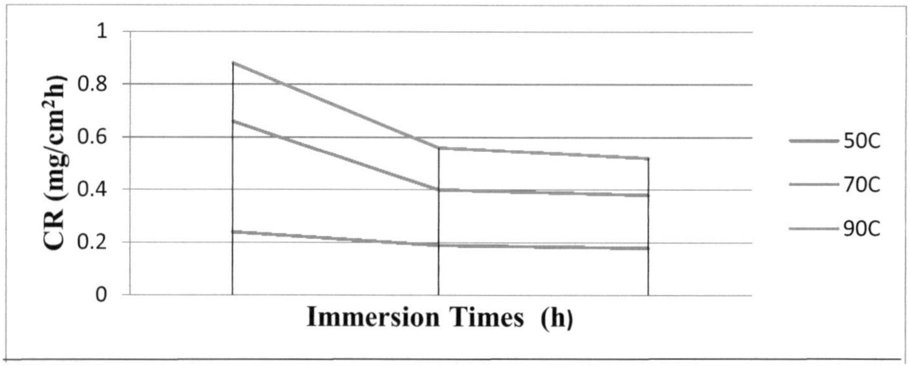

Figure 2. Graph of the average corrosion rate calculation of each 316L stainless steel test sample

Figure 3 depicts the analysis results and the usage of the potentiostat tool to determine the relationship of wear (I) vs time (s) for 30 seconds. Evidently, at a temperature rate of 50 °C, the current decreased from 0.0012675 A to 0.0012453 A; at temperature rate 70 °C, the current decreased from 0.0011972 A to 0.0011744 A and at a temperature rate of 90 °C, the current decreased from 0.0014458 A to 0.0014114. Hence, within 30 seconds, the 316L stainless steel metal environment sulphuric acid at a concentration of 0.5 M visible current decreased, meaning on the surface of this 316L stainless steel formed a passive layer means the surface of the metal has not been corroded. (For additional details, see the figure graph below.) Based on Figure 3, the current current graph with regard to the third time temperature variation is 50°C, 70°C, 90°C can be seen that at 50°C the starting point of the current value is 0.0012675 A, at 70°C temperature of the starting point of the current value is 0.0011792 A, and at point temperature 90°C initial value current is 0.0014458 A. Therefore, the higher the temperature, the higher the current value, which leads to the onset of corrosion. This test was carried out for 30 seconds, and on the graph shows that graphic upwards and downwards unstable on the surface of the 316L stainless steel metal begins to damage the pitting corrosion in certain areas that have been oxidised.

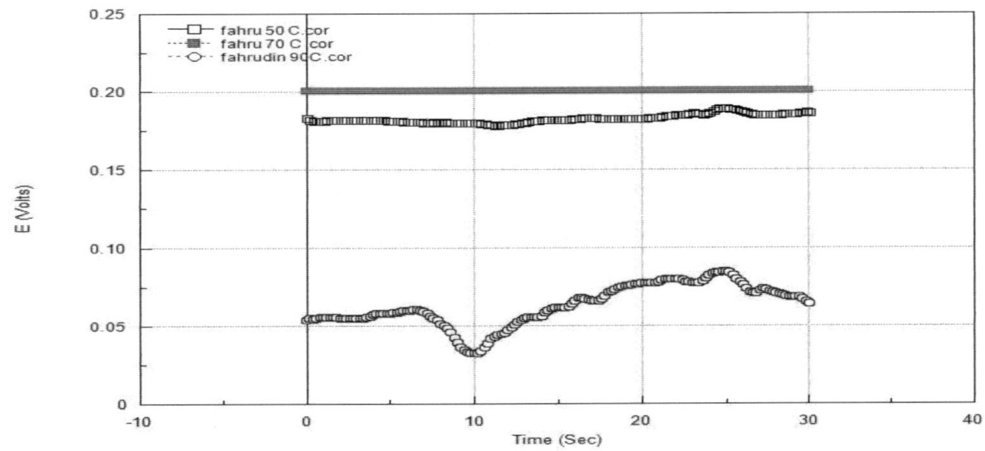

Figure 3. Graphic potential relationship (V) with time (s); 316L stainless steel in medium sulphuric acid (H2SO4) with temperature variations of 50 °C, 70 °C and 90 °C.

To view the surface morphology of AISI 316L stainless steel test material before immersion and after immersion is done using SEM photo. From the pictures shown, the immersion test material shown in Figure 4 does not suffer from smoother corrosion damage from the test material after immersion in photo Figure 5 of the metal surface seen in some locals experiencing damage to the passive layers in the form of pitting (corrosion pitting) in certain areas that have undergone previous oxidation may occur as the locals are initiated by metal defects or defects, so that the first region is subject to damage due to corrosive sulphuric acid environment, while in certain local local passive properties 316L stainless steel metal still protects wells metal.

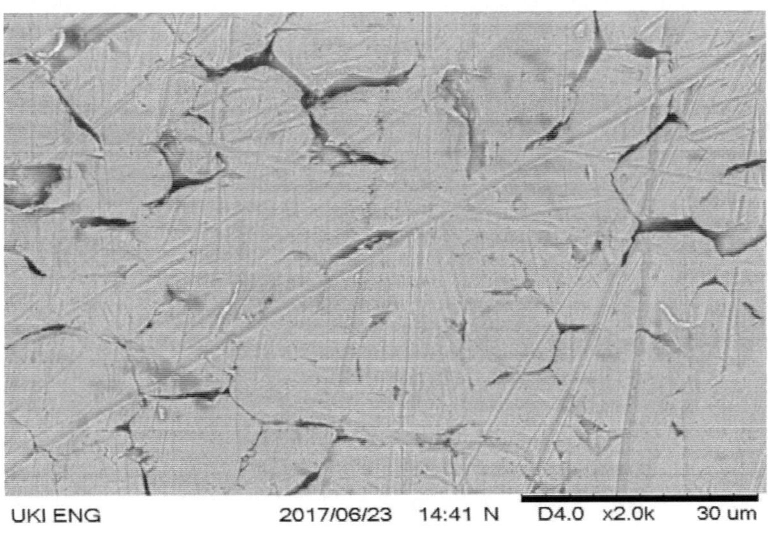

Figure 4. Photo SEM magnification 5000×, 316L stainless steel metal without immersion

Figure 5. Photos of SEM magnification 5000×, 316L stainless steel metal corrosion pitting with the immersion 0.5 H2SO4 result for 3 hours and a temperature rate of 90 °C

4. Conclusion

Corrosion analysis using the weight loss method of 316L stainless steel metal after immersion with sulphuric acid (H_2SO_4) 0.5 M concentration with variation of temperature 50°C, 70°C, and 90°C having the largest weight loss was at 90°C with immersion for 3 hours with a decrease weight of 1.56 mg. while the least weight loss occurs at 50°C by immersing for 1 hour with an average weight loss of 0.24 mg. This means that higher temperatures and duration of immersion will be found to lower the weight of the higher the start of corrosion. The corrosion analysis using the electrochemical method of 316L stainless steel metal for 30 seconds corroborated that an increase in current at 50 °C temperature of 0.00108762 A, at 70 °C temperature of 0.00115587 A and at 90°C temperature of 0.00141139 A emerged. Using the tool SEM with 5000× magnification can see the surface morphology of AISI 316L The stainless steel test material before soaking appears to be smooth surface while after 3 hours of absorption is seen on the surface of metal pitting corrosion.

Acknowledgment

The author would like to thank Mercubuana University (UMB) for facilities and support.

5. References

[1] Ornelasari, R. (2015). *Analisa laju Korosi Pada Stainless Steel 304 menggunakan Metode ASTM G31-72 pada Media Air Nira Aren*. JTM Volume 01 nomor 1 Tahun 2015.

[2] Seitovirta, M. (2013). *Handbook of Stainless Steel (Outokumpu)*. 11-20.

[3] Raharjo, R. (2015). *Tingkat Kekerasan Permukaan Stainless Steel 316L Akibat Tekanan Steelballpeening*. Proceeding Seminar Nasional Tahunan Teknik Mesin XIV(SNTTM XI

[4] R. V. M, S. K. N, and R. Balaji, "A Review on Corrosion of Metallic Bi-Polar Plates for Proton Exchange Membrane (PEM) Fuel Cells," vol. 11, no. 6, pp. 83–99, 2015.

[5] N.Guerrero, M. Cisneros, D. Gervasio, J. Francisco, and P. Robles, "Approaches to polymer electrolyte membrane fuel cells (PEMFCs) and their cost," *Renew. Sustain. Energy Rev.*, vol. 52, pp. 897–906, 2015.

[6] T.Sutharssan, D. Montalvao, Y. Kang, W. Wang, C. Pisac, and H. Elemara, "A review on prognostics and health monitoring of proton exchange membrane fuel cell," *Renew. Sustain. Energy Rev.*, vol. 75, no. November 2015, pp. 440–450, 2017.

[7] M. A. Abu-Dalo1, A. A.-R. (2012). *Exudate Gum from Acacia Trees as Green Corrosion Inhibitor.* Int. J. Electrochem. Sci, 1-5.

[8] A. S. Fouda and L. E. Lamloum, "Study on corrosion inhibition efficiency of some quinazoline derivatives on stainless steel 304 in hydrochloric acid solutions †," no. 2, pp. 112–131, 2016.

[9] Azar, V. (2010). *The Effect Of Shot Peening on FatiGue And Corrosion Behavior Of 316L Stainless Steel in Ringer's Solution.* Departement Of Material Science And Engineering, Shiraz University, Shiraz, Iran.

[10] Fontana, Mars G. (1987).*Corrosion Engineering (Third Edition).* Singapore: McGraw-Hill Book Company.

IOP Conf. Series: Materials Science and Engineering **343** (2018) 012017 doi:10.1088/1757-899X/343/1/012017

Integrated Speed Limiter and Fatigue Analyzer System

Hadi Pranoto[1,2], A.M.Leman[2], Abdi Wahab[3] and Darwin Sebayang[1]

[1]Faculty of Mechanical Engineering, University Mercu Buana Jakarta, Indonesia
[2]Faculty of Engineering Technology, University Tun Hussein Onn Malaysia 86400 Parit Raja, Batu Pahat, Johor , Malaysia
[3]Faculty of Computer Science, University Mercu Buana Jakarta, Indonesia

hadi.pranoto@mercubuana.ac.id

Abstract. The traffic accident increase in line with the growth of the vehicle, so the safety system must be developed to decrease the accident. This paper will purpose the integrated between speed limiter and fatigue analyser to improve the safety for vehicle, and also to analyse if there is an accident. The device and the software or application are developed and then integrated into one system. The testing held to prove the integrated between device and the application, and it show the system can work well. The next improvement for this system can be developing the server to collect data from internet, so the driver and the vehicle owner can monitor the system by internet.

1. Introduction

The traffic accident in Indonesia based on the report of Indonesia Police Department close to 31.234 fatalities cause by traffic accident. The average annual vehicle growth from 1996 to 2006 of 20%, the data is shown in Table 1.

Table 1. Number of Vehicle growth in Indonesia [1]

Year	Total registered vehicle	Annual growth of total vehicle (%)
1996	14530095	
1997	16821076	16
1998	17644885	5
1999	18224149	3
2000	18975344	4
2001	21201272	12
2002	24671330	16
2003	32774929	33
2004	41986814	23
2005	47654826	13
2006	50102492	5

This vehicle growth in Table 1 will also increase the percentage of traffic accident. The accident can be caused by many problems, such as fatigue, engine problems, over speed attitude and external cause [2]. From [3], the development of speed limiter has been developed, and tested in bus. The results show that the speed limiter can increase the safety for driver in bus and also in tank truck [4]. A proposed product titled on "Proposed simple electro-mechanical automotive speed control system" is a speed limiter which integrated RFD (Radio Frequency Decoder), the working principle of the proposals of this tool is RFD as a switch module speed limiter installed in the sign board on the highway that each vehicle must reduce the speed, RFD is connected to the engine cut-off system on the machine [8]. Then several patents related to the tools to prevent the fatigue of drivers among others, as in Volvo vehicles, podusen Automotive is developing a system called driver alert control, which can alert a driver drowsiness before he or she dozes of at the wheel [9]. Similar systems can also be found in other vehicles, such as the Driver Monitoring System in Toyota [10] and the Attention Assist System in the Mercedes-Benz [11]. To decimate accident rate that causes the driver fatigue factor, there has been increasing interest in the development Fatigue Warning System (FWSs) over the last two decades [12].

To improve the safety system, we propose a system by integrating the speed limiter and the fatigue analyser. The system will detect the limit speed from the vehicle and also the fatigue from driver.

2. Methodology

To achieve the purpose of this research, we propose a methodology in Figure 1, and the methodology consists of the steps how we develop the system.

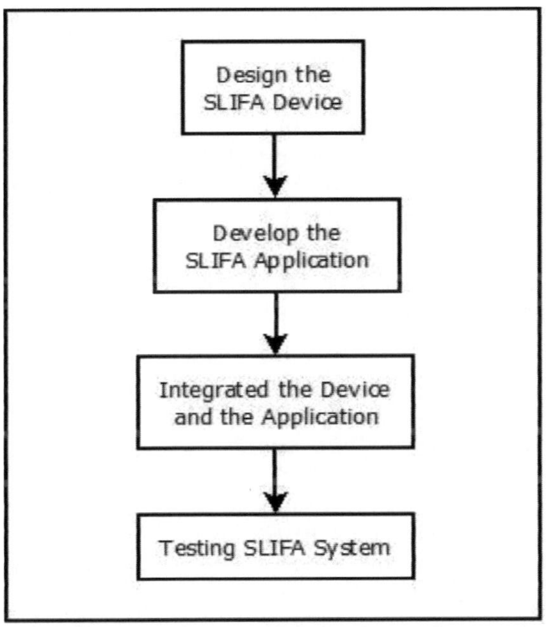

Figure. 1. The Proposed Methodology

In the first step of methodology, we design the devices for SLIFA. In this step, the speed limiter system [1] will be collaborated with the fatigue analyser system. After designing the SLIFA system, the next step is development of the SLIFA Application. This step needs the concept of SCADA and the software engineering. The next step is integrating the device and the application into one system. The final step is Testing the SLIFA System in a test bed environment. We will describe all of the step more clearly in the subsection below.

3. Design The SLIFA Device

The SLIFA device compose from the speed limiter system, and the fatigue analyser system. The SLIFA device will be bundle into one board, so the implementation of the SLIFA system will give a convenience to the user. The architecture of SLIFA system will be describe in Figure 2.

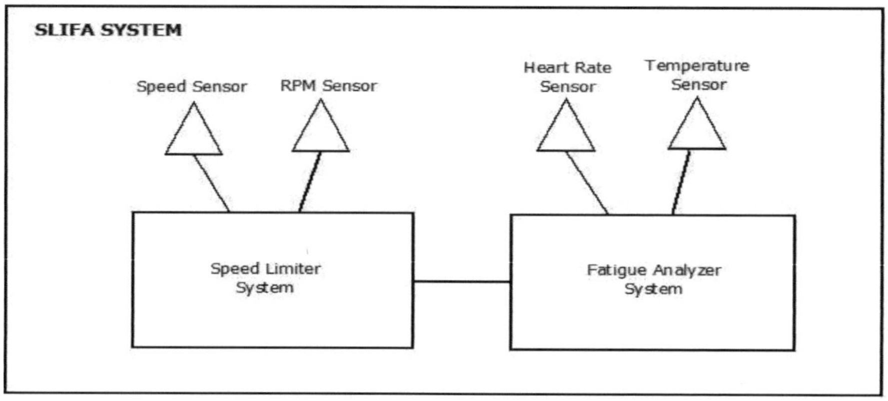

Figure. 2. Architecture of SLIFA System

Based on [1], the speed limiter that has been developed will read the speed of the vehicle and also improved with the engine rotation (RPM). These two data will be used as parameters from the speed limiter device. The speed limiter consists of sensor to read these two data from the vehicle, and save the data to the storage devices.
The fatigue analyser system will be equipped with two sensor, the first sensor is temperature sensor, and the second sensor is heart rate sensor. The temperature sensor will sense the temperature from the driver, meanwhile the heart rate sensor will take the heart rate value from the driver.
The main device for SLIFA system is an Arduino UNO, and some the sensors need by the system. All the sensors connected to Arduino board, and the Arduino manage all of the sensor for getting data are needed by SLIFA system. The sensor will be facing the object to be measured, such as the body temperature of driver, and also the heart rate of driver, the sensor will be implemented closed to the object.

3.1. Develop the SLIFA Application

This subsection will describe the development of SLIFA Application. This application will monitor data from SLIFA device and show them graphically. Also this application can read the data saved in SD Card SLIFA, the purpose to read captured data from SD Card is to analyse if there is an accident happened to the driver and the vehicle. The data could show the condition from driver when the accident happened, whether the driver in a good condition or in a tired condition.
To develop the SLIFA application, the first step is analyse the need of the software. The SLIFA application was not need much requirement, because the software only real time monitoring from the SLIFA device, especially monitoring data from the sensors. Figure 3 will show the use case diagram for this SLIFA application.

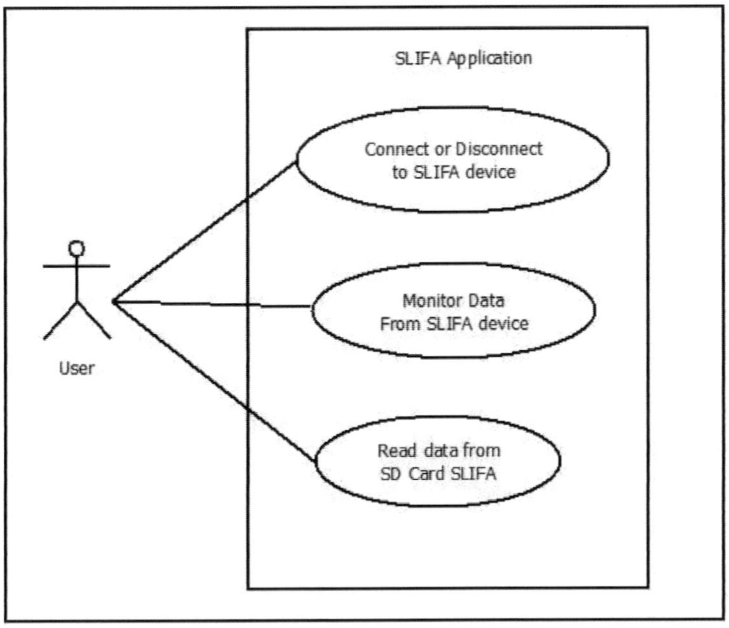

Figure 3. Use Case Diagram of SLIFA Application

Basically, from Figure 3, the user only has three features with the application. The first is connected or disconnect with the SLIFA device. This is the important thing must the user done before using the other features in the application, because without connected to the SLIFA device the application couldn't be used well. The second step is monitoring the data from SLIFA device, if the device can be connected with the application. And the last feature is reading data from SD Card SLIFA, and making the report from the data.
The development of this application uses some software development kit. Below are the requirements to develop the SLIFA application [5].

1. Programming Language. Java Development Kit version 1.7 was used to develop this application.
2. IDE for designing the application use Netbeans 7.0
3. IDE for Arduino use Arduino IDE

One thing we consider when developing this application was the way for connecting between the SLIFA device and this application. The serial connection is one of the way to communicate between Arduino device and the application [6,7]. So, we also adding the serial programming module into this application to handle the connection between both of them.

3.2. Integrated the Device and The Application

After the SLIFA device and the SLIFA application have already developed, the next step is to integrate both of them into one bundle system. The mechanism to integrate both of them physically using the Universal Serial Bus (USB) cable, and programmatically using serial programming. The SLIFA device connected as client device, while the PC or Computer installed with SLIFA application act as a server. The architecture to integrated these two devices describe in Figure 4.

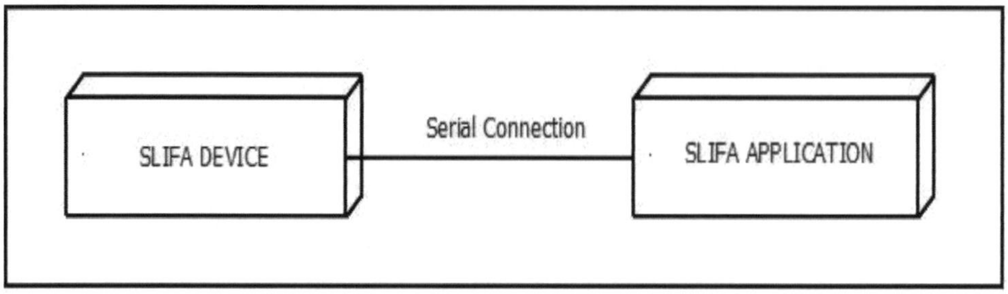

Figure 4. Architecture of Integration for SLIFA System [5]

The implementation of integration for SLIFA system is shown in Figure 5. The Notebook installed with SLIFA application and the SLIFA device. And Figure 6 will be shown a flow how to integrate and operate the SLIFA System.

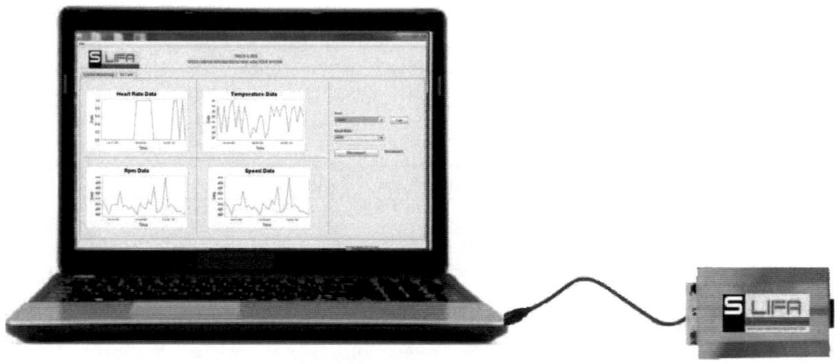

Figure 5. Architecture of Integration for SLIFA System

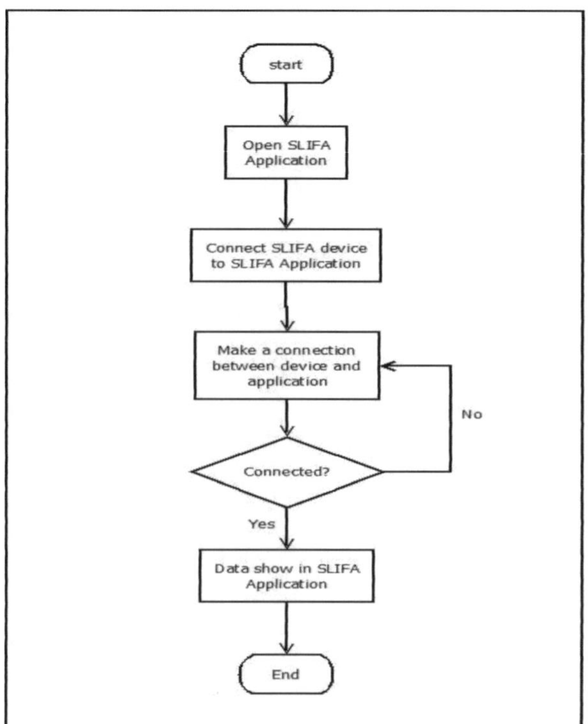

Figure 6. Flow of Integration for SLIFA System

4. Testing The SLIFA system

The testing for SLIFA system will be done in a test bed condition for the first time. The sensors will be simulated with some potensio meter, especially for speed and RPM sensor, for the last two sensors we use our body to connect the sensors. The purpose of the testing is for reaching the data into the SLIFA application, because it show the good integration between SLIFA device and SLIFA application. Figure 7 show SLIFA application visualize the sensor data in dynamic chart.

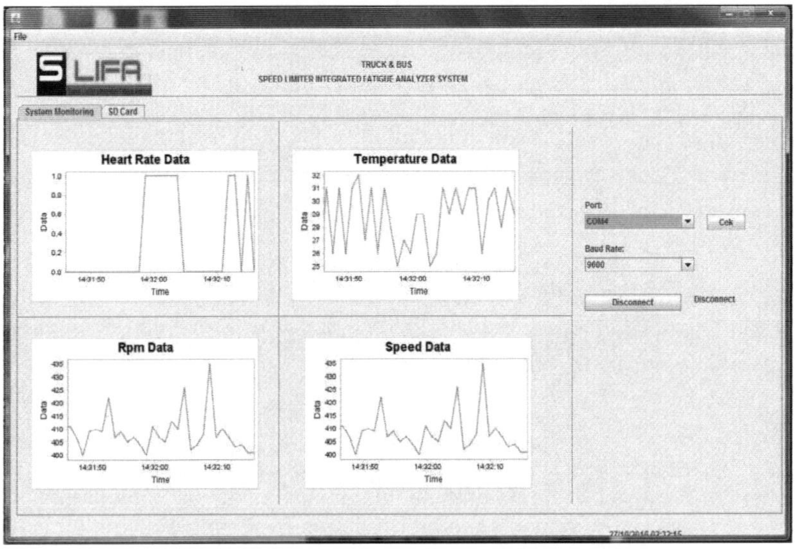

Figure 7. Testing Data from SLIFA System

5. Conclusion

Integrating SLIFA device and SLIFA Application has done the development of Integrated Speed Limiter and Fatigue Analyzer System. The Speed Limiter device develops from [1], and improves with Fatigue Analyzer in this paper. This paper describes the development of entire SLIFA system, and testing the SLIFA device with the SLIFA application. The testing shows that the device and the application can be integrated well. For the improvement of this research, it can be develop a server that can collect the data from the SLIFA system. So the owners of vehicle can monitor the vehicle and also the driver by cloud.

6. References

[1] T. Tjahjono. The Effect of traffic and road condition to the fatality rates on rural road in eastern Indonesia. Journal of The Eastern Asia Society for Transportation Studies, 8 (2010).

[2] B.S Dhillon. Safety and Human Error In Engineering System, (2013).

[3] H. Pranoto, A.M. Leman, D. Sebayang, D. Feriyanto. Improvement of Transportation Safety on Bus by Installing A Speed Limiter in Conventional Engine to Reduce Speed Rate. *MATEC Web Conference* 78, 01053 (2016).

[4] H. Pranoto, A.M. Leman, I. Baba, D. Feriyanto, G.W. Putra. Improving Road Safety of Tank Truck In Indonesia by Speed Limiter Installation.

[5] P.Deitel, H.Deitel. Java : How to Program, 9th Edition.Prentice Hall, (2011).

[6] M.Schmidt. Arduino A Quick-Strat Guide. The Pragmatic Programmer, (2015)

[7] Brian Evans. Beginning Arduino Programming, (2011).

[8] Ahmed Farouk *et al.,* Proposed simple electro-mechanical automotive speed control system (2015)

IOP Conf. Series: Materials Science and Engineering **343** (2018) 012017 doi:10.1088/1757-899X/343/1/012017

[9] Arellano, J. Volvo's Driver Alert Control Keeps Car Drivers Awake,(2009). http:// blogs.automotive.com/volvos-driver-alert-control-keeps-car-drivers-awake- 9436.html.
[10] Williamson, A.M., Feyer, A.-M., Friswell, R. The impact of work practices on fatigue in long distance truck drivers. Accid. Anal. Prev. 28 (6), (1996).
[11] Philips, T., 2008. Mercedes-Benz to Introduce Attention Assist into Series Produc- tion in Spring, (2009). http://www.emercedesbenz.com/Aug08/11_001331_Mercedes_ Benz_To_Introduce_Attention_Assist_Into_Series_Production_In_Spring_2009. html.
[12] Hartley, L., Horberry, T., Mabbott, N., Krueger, G.P. Review of Fatigue Detec- tion and Prediction Technologies. National Road Transport Commission, Mel- bourne, Australia. Technical Report, (2000).

IOP Conf. Series: Materials Science and Engineering **343** (2018) 012018 doi:10.1088/1757-899X/343/1/012018

Design and Analysis of Windmill Simulation and Pole by Solidwork Program

Tatang Mulyana[1], Darwin Sebayang[2], Akmal Muamar. D. R[1], Jauharah H. D. A[1], and Yahya Shomit, M[1].
[1]Universitas Telkom, Jakarta.
[2]Universitas Mercu Buana, Jakarta

tatang21april@gmail.com

Abstract. The Indonesian state of archipelago has great wind energy potential. For micro-scale power generation, the energy obtained from the windmill can be connected directly to the electrical load and can be used without problems. However, for macro-scale power generation, problems will arise such as the design of vane shapes, there should be a simulation and an accurate experiment to produce blades with a special shape that can capture wind energy. In addition, daily and yearly wind and wind rate calculations are also required to ensure the best latitude and longitude positions for building windmills. This paper presents a solution to solve the problem of how to produce a windmill which in the builder is very practical and very mobile can be moved its location. Before a windmill prototype is built it should have obtained the best windmill design result. Therefore, the simulation of the designed windmill is of crucial importance. Solid simulation express is a tool that serves to generate simulation of a design. Some factors that can affect a design result include the power part and the rest part of the part, material selection, the load is given, the security of the design power made, and changes in shape due to treat the load given to the design made. In this paper, static and thermal simulations of windmills have been designed. Based on the simulation result on the designed windmill, it shows that the design has been made very satisfactory so that it can be done prototyping fabrication process.
Keywords: design, analysis, simulation, windmill, pole, solidwork.

1. Introduction

As designers or engineers are required to analyse the manual calculation area of the hook is quickly damaged, the frame time timely, and of course the efficiency of materials to reduce cost. Solidworks is a design program that is widely used to work on product design, machine design, mould design, construction design, or other engineering purposes. It is equipped with tools used to simulate the design results such as stress, strain, and the influence of temperature, wind, etc. The solidworks is a parametric feature-based modelling program, meaning that all objects and relationships between geometrics can be modified again even though the geometric is finished without the need to start over again. With this method is very easy in the design process of a product or design. Structural analysis for assemblies with Solidwork Simulation enables engineers to efficiently analyse product behaviour under mechanical conditions taking into account the physical relationships between the parts of the structure to ensure high quality, performance, and safety. Every product is made of several parts or bodies. Defining how these parts interact and connect to each other is crucial for an accurate structural simulation during the product development process. Strategies and approaches related to the design method can be seen from the Montgomery literature [1] while some papers that have applied it can be

IOP Conf. Series: Materials Science and Engineering **343** (2018) 012018 doi:10.1088/1757-899X/343/1/012018

found in Haris [2, 3]. Paper by Tatang [4] presents digital control design for temporary boiler for concept and theory of automation applied to production system can be found in Groover literature [5].

2. Methodology

To create a solid 3D model we have to sketch it first. 3D model of the component then assembled into an assembled image with assembly menu [6]. After the image component or and assembly so then made drawings work using drawing facility.

Some of the benefits of creating engineering drawings using solidworks are as follows:

1. This software is quite easy to operate
2. Can help reduce errors in designing
3. Can simulate motion of design result
4. Can analyse the stress, load, temperature influence, weather, and so the design results easily without using any other software
5. Can create program for manufacturing process with CNC or industrial robot with the help of other software like mastercam, robotcam, delcam, etc.
6. Production costs to be incurred to be reduced due to a planned process

Part is a 3D image modelling drawing space of a drawing object. The basis for creating parts is a sketch image. Automatically after completion in making sketch it will switch to futures menu. Basically the picture part on solidoworks can be divided into 3 kinds namely:

a) Solid, i.e. a solid image object Suitable for design with very detailed components of machinery, furniture, construction, moulding, etc.
b) Surface, which is not a dense image object in the form of leather Matches are used in complex product design / complex such as car design, body motor design, bicycle design, etc.
c) Sheetmetal, the image object in the form of metal plate bending Suitable for designing the hook with the casing with metal plate material such as electronics casing, engine case, etc.

3. Result and Discussion

The windmill design is shown in Figure 1 and Figure 2. The designed windmill has three propellers and the pole can be up and down. Figure 1 presents some parts of the windmill and its reflections and dimensions. Figure 2 presents a windmill that has three flexi with its poles up and down.

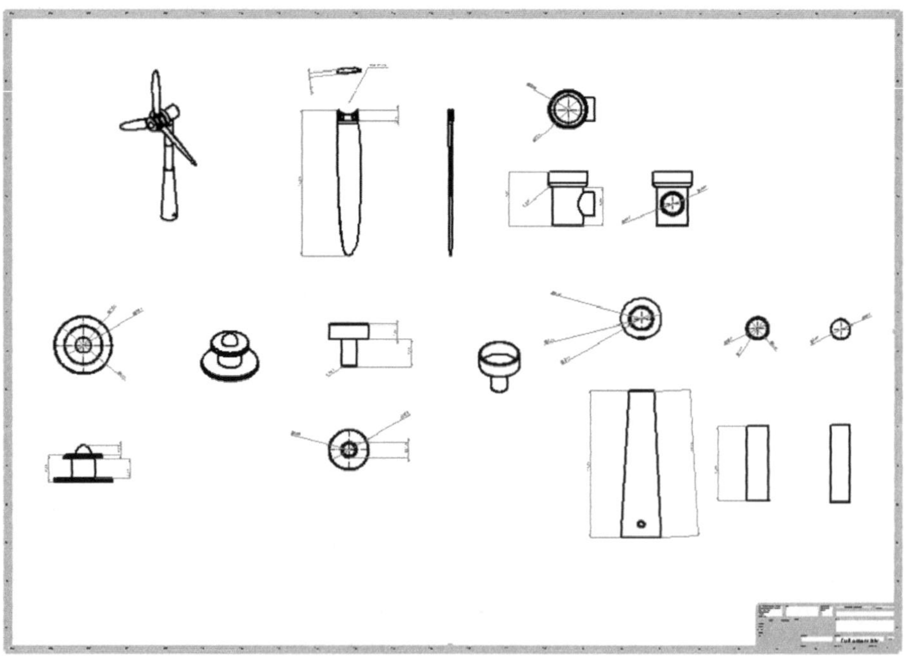

Figure 1. Design of parts of windmill.

Figure 2. Designed full assembly of windmill.

Table 1 presents part name and volumetric properties for some parts of the designed windmill. Some of these parts, including three flexi, two cut-extrude, and three boss-extrude. Figure 3 shows static simulation for some major windmill components, i.e. three flexi units, two cut-extrude units, and three boss-extrude units. Red colour indicates component get maximum force load and blue colour indicates component obtain minimum force load.

Table 1. Part name and volumetric properties of static simulation

Part Name	Volumetric Properties			
	Mass (kg)	Volume (m^3)	Density (kg/m^3)	Weight (N)
Flexi-1	0.815344	0.000103735	7859.9	7.99037
Flexi-2	0.815343	0.000103735	7859.89	7.99037
Flexi-3	0.815216	0.000103735	7858.67	7.98912
Cut-Extrude-1	12.605	0.00159557	7900	123.529
Cut-Extrude-2	0.00158839	2.01062e-007	7900	0.0155662
Boss-Extrude-1	0.384936	4.87261e-005	7900	3.77237
Boss-Extrude-2	0.00819013	1.03673e-006	7900	0.0802633
Boss-Extrude-3	3.13588	0.000396947	7900	30.7316

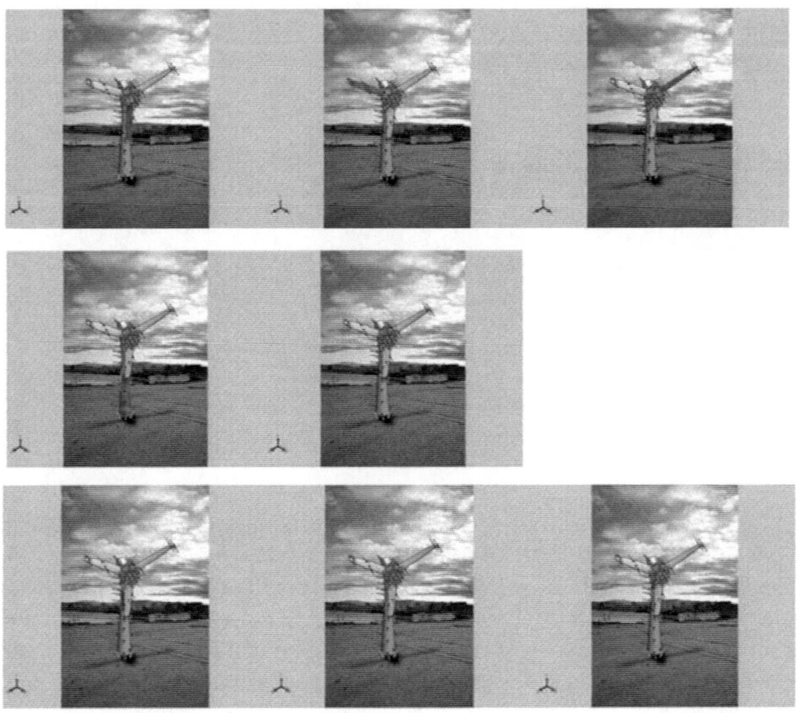

Figure 3. Results Static Simulation of flexi, cut-extrude, and boss-extrude parts assembly.

IOP Conf. Series: Materials Science and Engineering **343** (2018) 012018 doi:10.1088/1757-899X/343/1/012018

Name	Type	Min	Max
Displacement1	URES: Resultant Displacement	0.000e+000mm Node: 66068	5.445e+001mm Node: 12

Name	Type
Displacement1{1}	Deformed shape

Figure 4. Static simulation results of full assembly for displacement.

Figure 4 shows the static simulation performed on the full assembly of windmill. Can be interpreted stress on all surfaces windmill occurs in all directions. The narrower the surface area but the fixed force, the greater the voltage. The occurrence of blue colour on the surface of the windmill is due to receive a static minimum load permanently, and the windmill is not changed shape. Figure 5 shows different material properties, incorporated into full windmill assembly, i.e. 201 annealed stainless steel (SS), AISI 1020, and chrome stainless steel. Figure 6 shows the results of simulation of load and fixtures where the resultant force is 2040.38N.

IOP Conf. Series: Materials Science and Engineering **343** (2018) 012018 doi:10.1088/1757-899X/343/1/012018

Material Properties

Model Reference	Properties	Components
	Name: 201 Annealed Stainless Steel (SS) Model type: Linear Elastic Isotropic Default failure criterion: Unknown Yield strength: 292 N/mm^2 Tensile strength: 685 N/mm^2 Elastic modulus: 207000 N/mm^2 Poisson's ratio: 0.27 Mass density: 7.86 g/cm^3 Thermal expansion coefficient: 1.7e-005 /Kelvin	SolidBody 1(Flex1)(Baling-baling-6), SolidBody 1(Flex1)(Baling-baling-7), SolidBody 1(Flex1)(Baling-baling-8), SolidBody 1(Shell3)(part atas-2/cap-1), SolidBody 1(Cut-Extrude4)(part atas-2/housing-1)
Curve Data:N/A		
	Name: AISI 1020 Model type: Linear Elastic Isotropic Default failure criterion: Unknown Yield strength: 351.571 N/mm^2 Tensile strength: 420.507 N/mm^2 Elastic modulus: 200000 N/mm^2 Poisson's ratio: 0.29 Mass density: 7.9 g/cm^3 Shear modulus: 77000 N/mm^2 Thermal expansion coefficient: 1.5e-005 /Kelvin	SolidBody 1(Cut-Extrude3[1])(Part1-1), SolidBody 2(Cut-Extrude3[2])(Part1-1), SolidBody 1(Boss-Extrude3)(Part2-1), SolidBody 2(Boss-Extrude2)(Part2-1), SolidBody 1(Boss-Extrude2)(Part3-1)
Curve Data:N/A		
	Name: Chrome Stainless Steel Model type: Linear Elastic Isotropic Default failure criterion: Unknown Yield strength: 172.339 N/mm^2 Tensile strength: 413.613 N/mm^2 Elastic modulus: 200000 N/mm^2 Poisson's ratio: 0.28 Mass density: 7.8 g/cm^3 Shear modulus: 77000 N/mm^2 Thermal expansion coefficient: 1.1e-005 /Kelvin	SolidBody 1(Shell2)(part atas-2/1stcap-1)
Curve Data:N/A		

Figure 5. Static simulation results of full assembly windmill for different material properties.

Loads and Fixtures

Fixture name	Fixture Image	Fixture Details
Fixed-1		Entities: 1 face(s) Type: Fixed Geometry

Resultant Forces

Components	X	Y	Z	Resultant
Reaction force(N)	1530.87	873.17	1028.19	2040.38
Reaction Moment(N.m)	0	0	0	0

Load name	Load Image	Load Details
Pressure-1		Entities: 15 face(s) Type: Normal to selected face Value: 101325 Units: N/m^2 Phase Angle: 0 Units: deg

Figure 6. Static simulation results of full assembly for load and fixtures.

Figure 7 shows thermal simulation for some major windmill components, i.e. three flexi units, two cut-extrude units, and three boss-extrude units. Blue colour indicates component obtain minimum force load. Figure 8 shows the thermal simulation performed on the full assembly of windmill. Can be interpreted stress on all surfaces windmill occurs in all directions. The narrower is the surface area but the fixed force is the greater the voltage. The occurrence of blue colour on the surface of the windmill is due to receive a thermal minimum load permanently, and the windmill is not changed shape. Figure 9 shows different material properties, incorporated into full windmill assembly, i.e. 201 annealed stainless steel (SS), AISI 1020, and chrome stainless steel. Figure 10 shows the results of simulation of load and fixtures where the resultant force is 54.45N.

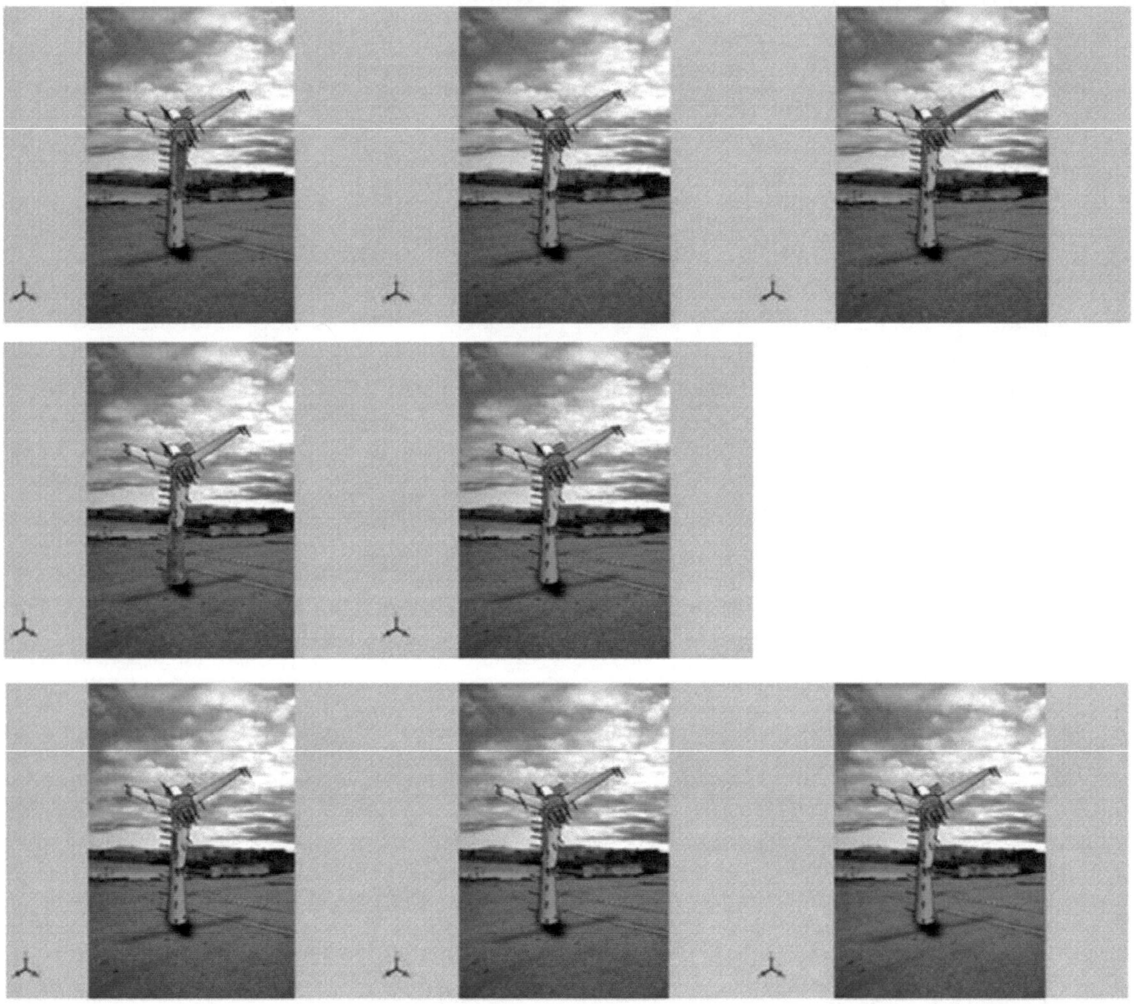

Figure 7. Thermal simulation results of three flexes, two cut-extrudes, and three boss-extrudes parts assembly.

Name	Type	Min	Max
Thermal1	TEMP: Temperature	0.000e+000Kelvin Node: 60308	3.011e+002Kelvin Node: 1

Full assembly-Thermal 2-Thermal-Thermal1

Name	Type
Displacement1	Deformed shape

Full assembly-Thermal 2-Displacement-Displacement1

Figure 8. Thermal simulation result of full assembly for displacement.

Material Properties

Model Reference	Properties	Components
	Name: 201 Annealed Stainless Steel (SS) Model type: Linear Elastic Isotropic Default failure criterion: Unknown Thermal conductivity: 0.163 W/(cm.K) Specific heat: 119.981 Cal/(kg.C) Mass density: 7.86 g/cm^3	SolidBody 1(Flex1)(Baling-baling-6), SolidBody 1(Flex1)(Baling-baling-7), SolidBody 1(Flex1)(Baling-baling-8), SolidBody 1(Shell3)(part atas-2/cap-1), SolidBody 1(Cut-Extrude4)(part atas-2/housing-1)
Curve Data:N/A		
	Name: AISI 1020 Model type: Linear Elastic Isotropic Default failure criterion: Unknown Thermal conductivity: 0.47 W/(cm.K) Specific heat: 100.382 Cal/(kg.C) Mass density: 7.9 g/cm^3	SolidBody 1(Cut-Extrude3[1])(Part1-1), SolidBody 2(Cut-Extrude3[2])(Part1-1), SolidBody 1(Boss-Extrude3)(Part2-1), SolidBody 2(Boss-Extrude2)(Part2-1), SolidBody 1(Boss-Extrude2)(Part3-1)
Curve Data:N/A		
	Name: Chrome Stainless Steel Model type: Linear Elastic Isotropic Default failure criterion: Unknown Thermal conductivity: 0.18 W/(cm.K) Specific heat: 109.943 Cal/(kg.C) Mass density: 7.8 g/cm^3	SolidBody 1(Shell2)(part atas-2/1stcap-1)
Curve Data:N/A		

Figure 9. Thermal Simulation Results of full assembly for material properties.

Load name	Load Image	Load Details
Temperature-1		Entities: 8 face(s) Temperature: 28 Celsius

Figure 10. Thermal Simulation Results of full assembly for thermal load.

4. Conclusion

From static and thermal simulations to the design of the windmill it can be concluded that the designed windmill can be forwarded to the fabrication stage. This fact is reinforced by the simulation results that show despite static or thermal loads but both parts and full assembly windmill does not change shape. Some parts that have been simulated consist of three flexions, two cut-extrude, and three boss-extrude. Several different material properties have also been incorporated for this simulation, including 201 annealed stainless steel (SS), AISI 1020, and chrome stainless steel.

IOP Conf. Series: Materials Science and Engineering **343** (2018) 012018 doi:10.1088/1757-899X/343/1/012018

5. References

[1] Montgomery D.C 2013 *Design and Analysis of Experiments* (New York: John Wesley and Sons)

[2] Haris R, Rino Andias A and Tatang M 2015 EMS-SCADA Design of AC Usage on a Building (Proceeding 8th International Seminar on Industrial Engineering and Management 2015 (ISIEM 2015)) PS45-49.

[3] Haris R and Tatang M 2015 Website Design of EMS-SCADA for AC Usage on a Building (Proceedings of the 3rd International Conference on Information and Communication Technology 2015 (ICoICT 2015)) p 17-22

[4] Tatang M, Mohamad Nor M.T, Dirman Hanafi B, and Afrodi A 2010 Digital control design for the boiler drum BDT921 (Proceeding of Conf. on Industrial and Appl. Math., Indonesia) p 257-260

[5] Groover M.P 2001 *Automation, Production System, and Computer-Integrated Manufacturing* (New Jersey: Pearson)

[6] Solidworks 2010 *An Introduction to Stress Analysis Applications with SolidWorks Simulation, Student Guide* (Solidworks Education - Concord, Massachusetts 01742 USA)

Acknowledgments

The results of the study presented in this paper are part of a study funded by a research grant from Ristekdikti in post-doctoral research scheme on fiscal year 2017 study entitled design development of power plants wind-solar hybrid system.

Design and Analysis of Solar Smartflower Simulation by Solidwork Program

Tatang Mulyana[1], Darwin Sebayang[2], Fildzah Fajrina[1], and Raihan, M. Faizal[1]
[1]Universitas Telkom, Jakarta.
[2]Universitas Mercu Buana, Jakarta

tatang21april@gmail.com

Abstract. The potential of solar energy that is so large in Indonesia can be a driving force for the use of renewable energy as a solution for energy needs. Government with the community can utilize and optimize this technology to increase the electrification ratio up to 100% in all corners of Indonesia. Because of its modular and practical nature, making this technology easy to apply. One of the latest imported products that have started to be offered and sold in Indonesia but not yet widely used for solar power generation is the kind of smartflower. Before using the product, it is of course very important and immediately to undertake an in-depth study of the utilization, use, maintenance, repair, component supply and fabrication. The best way to know the above is through a review of the design and simulation. To meet this need, this paper presents a solar-smartflower design and then simulated using the facilities available in the solidwork program. Solid simulation express is a tool that serves to create power simulation of a design part modelling. With the simulation is very helpful at all to reduce errors in making design. Accurate or not a design created is also influenced by several other factors such as material objects, the silent part of the part, and the load given. The simulation is static simulation and body battery drop test, and based on the results of this simulation is known that the design results have been very satisfactory.
Keywords: design, analysis, simulation, solar, smartflower, solidwork.

1. Introduction

The last few years are intensively promoted through various electronic media sales of solar smartflower, but not many are interested. This may be due to the very expensive price and unavailability of service. To anticipate if it will be widely used but knowledge about it is not adequate then for the solution is to design and analyse through simulation program. This paper presents some research that tries to develop the design of solar smartflower and then analyse it through simulation using solidwork before made the process of its fabrication. The smartflower is an all-in-one ground mounted solar system with a tracker that follows the sun. It name comes from its design – the solar cells are arranged on individual "petals" that open at the beginning of each day. After the sun goes down, the smartflower petals fold up and a self-cleaning process kicks in. In addition to solar cells, the smartflower system contains a dual-axis tracker that makes it possible for its "petals" to follow the sun across the sky throughout the day. Thanks to this tracking capability, the smartflower can produce significantly more electricity than a similarly-sized rooftop solar panel system – up to 51 percent more, according to smartflower website. The 12-petal, 194 square-foot structure comes with 2.5 kilowatts (kW) of electricity production, which is equivalent to a 4 kW fixed rooftop array. There are two versions of the smartflower available in the U.S., both of which is manufactured in Austria: the smartflower and the smartflower-plus. The smartflower-plus offers energy storage capabilities via an integrated battery in addition to the standard solar electricity generation that the smartflower offers.

According to press coverage about the product, the smartflower is priced at approximately $16,900. The smartflower-plus, which has includes batteries for energy storage, is not yet available in the U.S., but will be priced around $22,000 when it comes to market. Strategies and approaches related to the design method can be seen from the Montgomery literature [1] while some papers that have applied it can be found in Haris [2, 3]. Paper presents digital control design in [4], and for concept and theory of automation applied to production system can be found in Groover literature [5].

2. Methodology

Static studies provide tools for the linear stress analysis of parts and assemblies loaded by static loads and drop test studies are used to analyse the stress of moving parts or assemblies impacting an obstacle [6]. The step-by-step instructions are following given:
Step 1: Creating a Study
Step 2: Assigning Material
Step 3: Applying Fixtures
Step 4: Applying Loads
Step 5: Meshing the Assembly
Step 6: Running the Analysis
Step 7: Visualizing the Results
Step 8: Save Your Work and Exit SolidWorks

3. Result and Discussion

Solar smartflower is designed to have six main parts consist of fin diesel nine units, body battery, hinge buffer, base, hinge base, and fin with hinge 1 unit. This Smartflower can open and close focus horizontally 360 degrees, you can move up and down vertically 270 degrees vertically. Figure 1 shows parts designed of the solar smartflower, and Figure 2 shows designed full assembly of the solar smartflower.

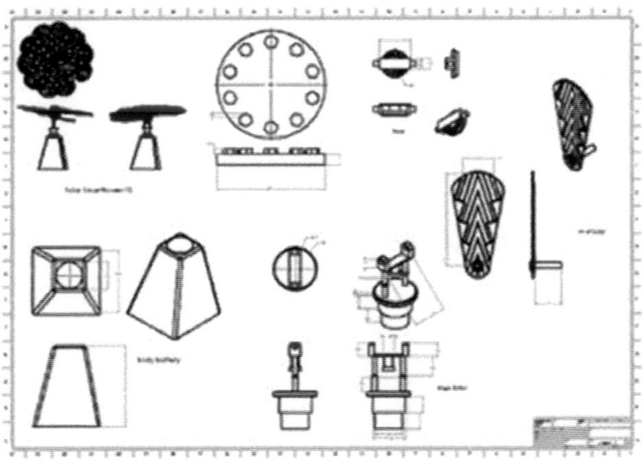

Figure 1. Design of parts of the solar smartflower.

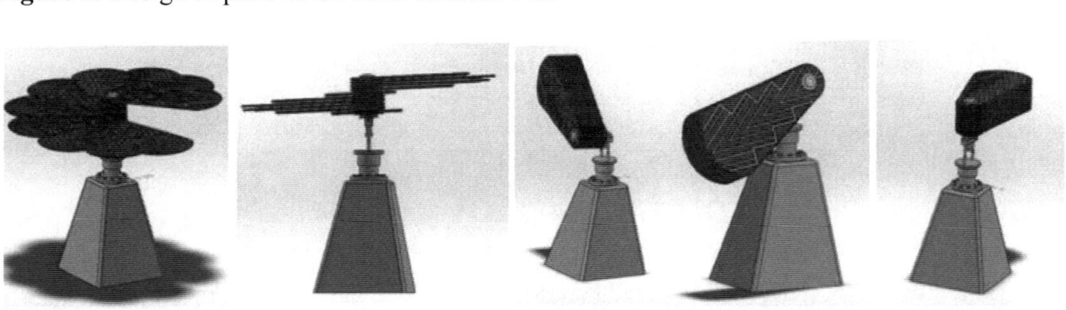

Figure 2. Designed full assembly of the solar smartflower.

Figure 3 shows the static simulation performed on the full assembly of the fan part solar smartflower. Can be interpreted stress on all surfaces occurs in all directions. The occurrence of blue colour on the surface is due to receive a static minimum load permanently, and the part is not changed shape. Figure 4 shows silicon material properties, and Figure 5 shows load and fixtures.

Model name: Fan
Current Configuration: Default

Shell Bodies			
Document Name and Reference	**Formulation**	**Volumetric Properties**	**Document Path/Date Modified**
ala panel solar_1-1-surface1 	Thin	**Thickness:20 mm** **Weight:52.1739 N** **Volume:0.00228492 m^3** **Mass:5.32386 kg** Density:2330kg/m^3	**C:\Users\hp\Desktop\Solid \Fan.SLDPRT** Aug 11 22:07:18 2017

Figure 3. The static simulation result of fan part of the solar smartflower.

Material Properties

Model Reference	**Properties**	**Components**
	**Name:** Silicon **Model type:** Linear Elastic Isotropic **Default failure criterion:** Unknown **Yield strength:** 1.2e+008 N/m^2 **Elastic modulus:** 1.124e+011 N/m^2 **Poisson's ratio:** 0.28 **Mass density:** 2330 kg/m^3 **Shear modulus:** 4.9e+010 N/m^2	SurfaceBody 1(ala panel solar_1-1-surface1)(Fan)
Curve Data:N/A		

Figure 4. Material properties simulation result of the fan part solar smartflower.

Stress simulation result of the fan part solar smartflower designed is shown in Figure 6. Figure 7 and Figure 8 show displacement and strain simulation results of the fan part solar smartflower designed. If seen in the simulation results contained in Figure 6, shows the blue colour, meaning that the material used can withstand the stress load given, as well as the load displacement and strain. This can be seen in the simulation results shown in Figure 7 and Figure 8.

IOP Conf. Series: Materials Science and Engineering **343** (2018) 012019 doi:10.1088/1757-899X/343/1/012019

Loads and Fixtures

Fixture name	Fixture Image	Fixture Details
Fixed-1	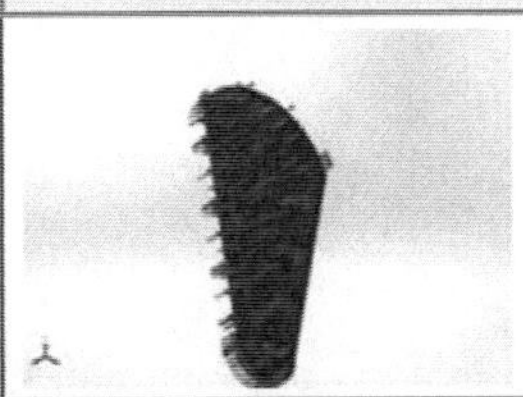	Entities: 1 face(s) Type: Fixed Geometry

Resultant Forces

Components	X	Y	Z	Resultant
Reaction force(N)	-0.0913269	-0.1022	479.983	479.984
Reaction Moment(N.m)	-3.37168	-0.00342959	-0.000377144	3.37168

Load name	Load Image	Load Details
Force-1		Entities: 16 face(s) Type: Apply normal force Value: 30 N

Figure 5. Static simulation result of the fan part solar smartflower for load and fixtures.

Name	Type	Min	Max
Stress1	VON: von Mises Stress	0 N/m^2 Node: 65309	2.22317e+006 N/m^2 Node: 11137

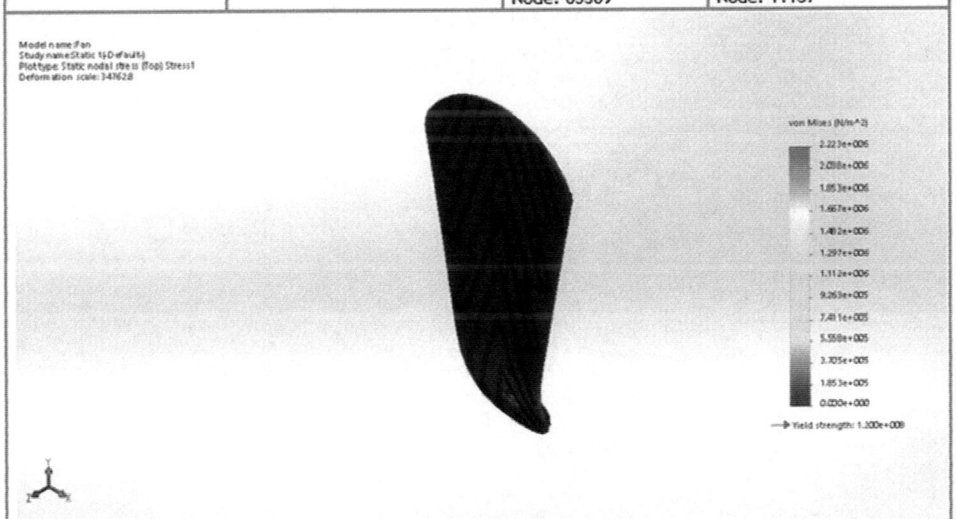

Fan-Static 1-Stress-Stress1

Figure 6. Stress simulation result of the fan part solar smartflower designed.

Figure 9 shows body battery of the solar smartflower designed, and the material properties is shown in Figure 10. Figure 11 shows stress simulation of the body battery solar smartflower designed while Figure 12 and Figure 13 show displacement and strain simulation results of the body battery solar smartflower designed.

IOP Conf. Series: Materials Science and Engineering **343** (2018) 012019 doi:10.1088/1757-899X/343/1/012019

Name	Type	Min	Max
Displacement1	URES: Resultant Displacement	0 mm Node: 176	0.00113608 mm Node: 13509

Figure 7. Displacement simulation results of the fan part solar smartflower designed.

Name	Type	Min	Max
Strain1	ESTRN: Equivalent Strain	0 Element: 126349	1.65338e-005 Element: 89022

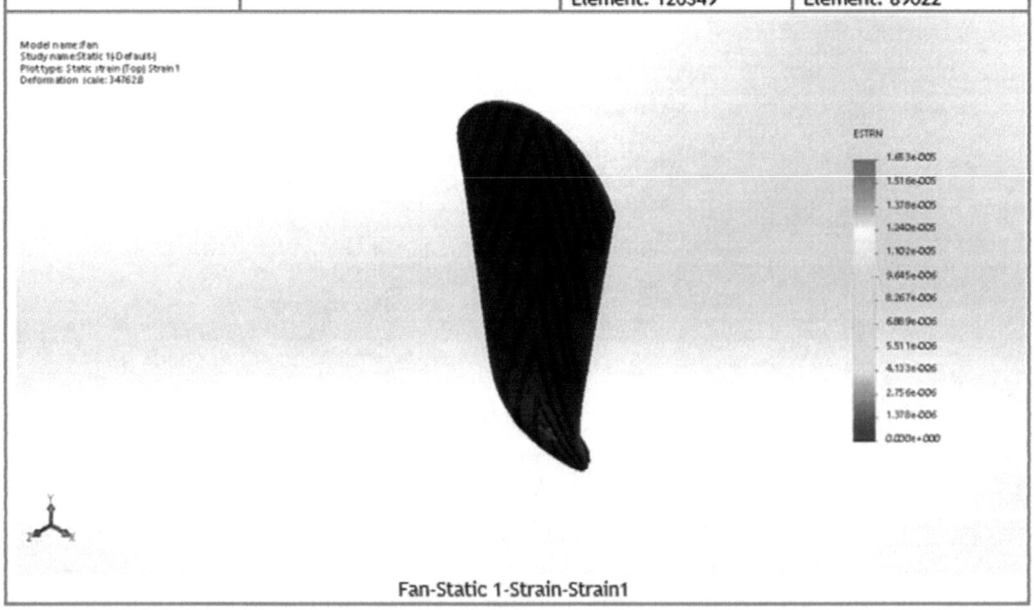

Figure 8. Strain simulation results of the fan part solar smartflower designed.

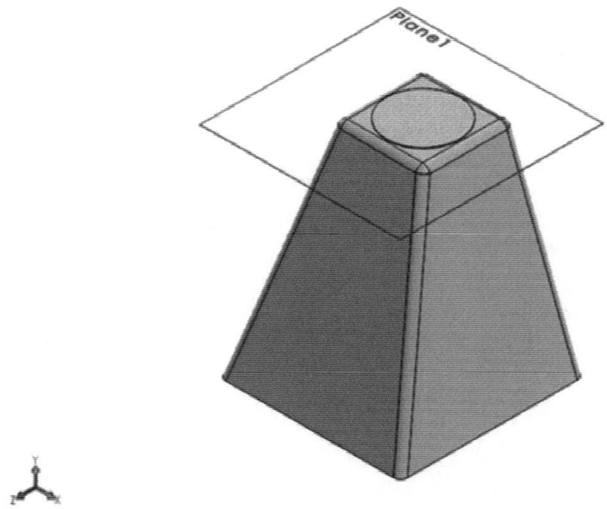

Model name: Body Battery
Current Configuration: Default

Solid Bodies

Document Name and Reference	Treated As	Volumetric Properties	Document Path/Date Modified
Boss-Extrude2 	Solid Body	Mass:85.6293 kg Volume:0.0106677 m^3 Density:8027 kg/m^3 Weight:839.168 N	C:\Users\hp\Desktop\Solid \Body Battery.SLDPRT Sep 13 01:18:31 2017

Figure 9. Body battery of the solar smartflower designed.

Material Properties

Model Reference	Properties	Components
	Name: AISI Type 316L stainless steel Model type: Linear Elastic Isotropic Default failure criterion: Unknown Yield strength: 1.7e+008 N/m^2 Tensile strength: 4.85e+008 N/m^2 Elastic modulus: 2e+011 N/m^2 Poisson's ratio: 0.265 Mass density: 8027 kg/m^3 Shear modulus: 8.2e+010 N/m^2 Thermal expansion coefficient: 1.7e-005 /Kelvin	SolidBody 1(Boss-Extrude2)(Body Battery)
Curve Data:N/A		

Figure 10. Body battery material properties of the solar smartflower designed.

Name	Type	Min	Max
Stress1	VON: von Mises Stress	2.83574e+007 N/m^2 Node: 6989	1.15494e+009 N/m^2 Node: 9108

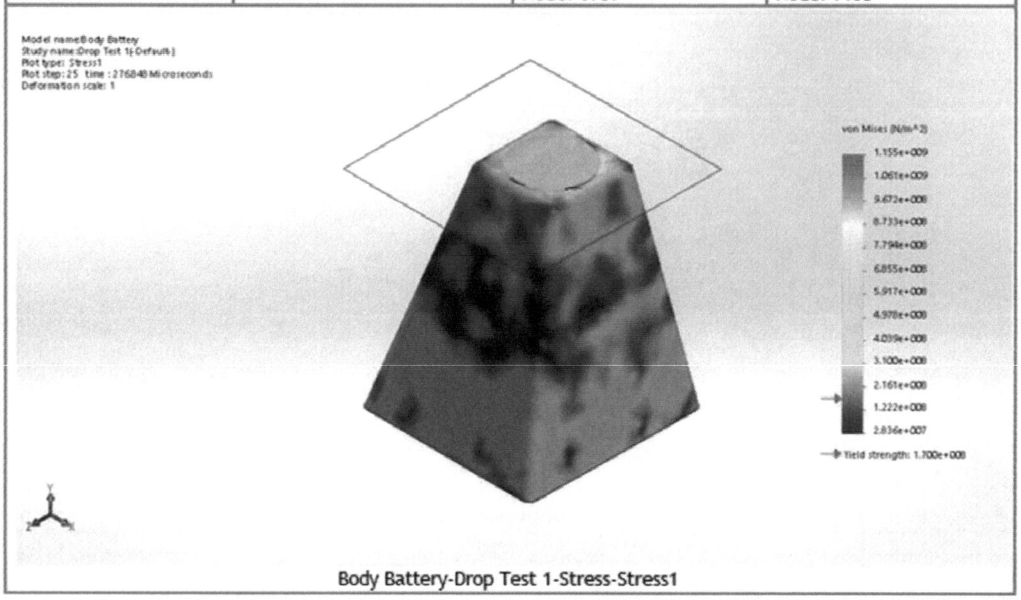

Figure 11. Stress simulation of the body battery solar smartflower designed.

Name	Type	Min	Max
Displacement1	URES: Resultant Displacement	1.74442 mm Node: 24	2.38006 mm Node: 673

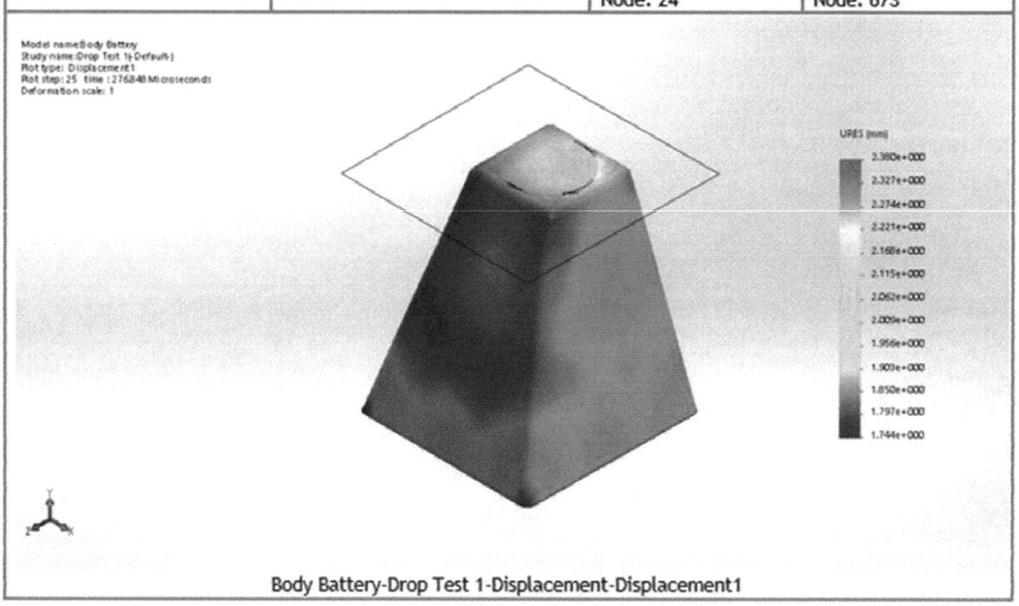

Figure 12. Displacement simulation of the body battery solar smartflower designed.

Name	Type	Min	Max
Strain1	ESTRN: Equivalent Strain	0.000110565 Element: 2135	0.00415178 Element: 2454

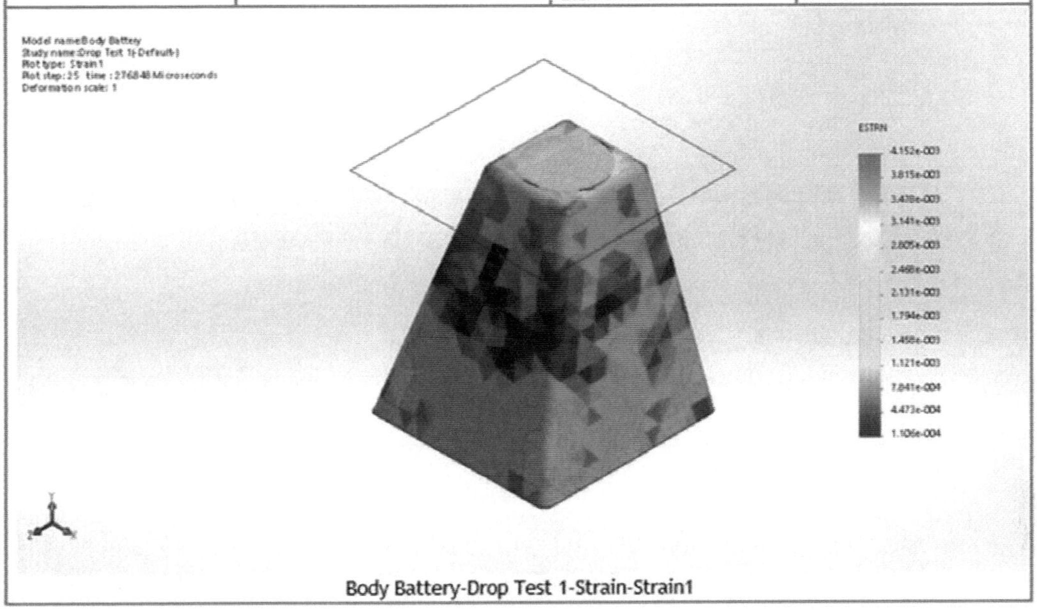

Figure 13. Strain simulation of the body battery solar smartflower designed.

4. Conclusion

Based on fan static and body battery drop test simulations it can be concluded that the solar smartflower designed can be forwarded to the fabrication stage. This fact is reinforced by the simulation results that show despite the static or drop test loads but parts of the solar smartflower designed does not change shape.

5. References

[1] Montgomery D.C 2013 *Design and Analysis of Experiments* (New York: John Wesley and Sons)

[2] Haris R, Rino Andias A and Tatang M 2015 EMS-SCADA Design of AC Usage on a Building (Proceeding 8th International Seminar on Industrial Engineering and Management 2015 (ISIEM 2015)) PS45-49.

[3] Haris R and Tatang M 2015 Website Design of EMS-SCADA for AC Usage on a Building (Proceedings of the 3rd International Conference on Information and Communication Technology 2015 (ICoICT 2015)) p 17-22

[4] Tatang M, Mohamad Nor M.T, Dirman Hanafi B, and Afrodi A 2010 Digital control design for the boiler drum BDT921 (Proceeding of Conf. on Industrial and Appl. Math., Indonesia) p 257-260

[5] Groover M.P 2001 *Automation, Production System, and Computer-Integrated Manufacturing* (New Jersey: Pearson)

[6] Solidworks 2010 *An Introduction to Stress Analysis Applications with SolidWorks Simulation, Student Guide* (Solidworks Education - Concord, Massachusetts 01742 USA)

Acknowledgments

The results of the study presented in this paper are part of a study funded by a research grant from ristekdikti in post-doctoral research scheme on fiscal year 2017 study entitled design development of power plants wind-solar hybrid system.

IOP Conf. Series: Materials Science and Engineering **343** (2018) 012020 doi:10.1088/1757-899X/343/1/012020

Comparative Investigation on Tool Wear during End Milling of AISI H13 Steel with Different Tool Path Strategies

Erry Yulian T Adesta[1], Muhammad Riza[2], Avicena[1]

[1] Department of Manufacturing and Materials Engineering, International Islamic University Malaysia, Malaysia
[2] Jurusan Teknik Mesin, Universitas Bandar Lampung, Indonesia

E-mail: eadesta@iium.edu.my

Abstract. Tool wear prediction plays a significant role in machining industry for proper planning and control machining parameters and optimization of cutting conditions. This paper aims to investigate the effect of tool path strategies that are contour-in and zigzag tool path strategies applied on tool wear during pocket milling process. The experiments were carried out on CNC vertical machining centre by involving PVD coated carbide inserts. Cutting speed, feed rate and depth of cut were set to vary. In an experiment with three factors at three levels, Response Surface Method (RSM) design of experiment with a standard called Central Composite Design (CCD) was employed. Results obtained indicate that tool wear increases significantly at higher range of feed per tooth compared to cutting speed and depth of cut. This result of this experimental work is then proven statistically by developing empirical model. The prediction model for the response variable of tool wear for contour-in strategy developed in this research shows a good agreement with experimental work.

1. Introduction

Tool wear prediction plays a significant role in machining industry for proper planning and control machining parameters and optimization of cutting conditions. Some researches related to tool wear has investigated during a few decades. In this paper, the recent researches in this area are reviewed.

Optimum cutting speed range is analyzed in order to minimize the tool wear when milling AISI H13. The analysis of the tool wear mechanisms show that, as the number of cutting speed range increases, on the tool flank face, oxidation wear influences more greatly, while the effect of adhesive wear decreases [1]. The most factors dominating factor effect over tool wear is cutting speed followed by feed rate and depth of cut. Abrasion was the principal wear mechanism observed at higher conditions. The feed rate is most significant parameter influences on surface roughness followed by depth of cut and cutting speed [2]. It was found that by milling of AISI H13 under different cutting speeds, as the cutting speed increases, the influences of fracture and chipping resulting from mechanical load on tool wear were reduced. At the same time the influences of adhesion, oxidation, and thermal crack accelerated by high cutting temperature became greater [3]. The tool wear under different tool path strategies that are straight-in entry and the rolling-in was investigated under milling process. It was found that the tool wear is lower for the experiment using the when roll-in strategy applied. This strategy proposes the entry of the cutter rotating on a pivot point. The cutting tool enters gradually until the nominal width of cut is reached. This technique aims to ensure the smallest chip thickness possible when exiting the cut [4]. Research on high-speed and ultra-high-speed face milling of AISI H13 steel (46–47 HRC) was conducted in order to acquire a thorough understanding of the tool wear evolution process and tool wear mechanisms in high-speed hard milling. The analysis of the tool wear mechanisms show that, as the number of cutting speed range increases, on the tool flank face, oxidation wear influences more significantly, while the effect of adhesive wear decreases [5].

Many researches have been carried out to investigate tool wear during machining process specifically in milling operation as mentioned previously. However, research related to tool path application and tool wear during milling operation have not yet established. In addition to that, this research is valuable for the die and mould maker to select proper cutting condition to avoid increase in wear during milling operation.

2. Experimental Setup

The workpiece material used in the present works was a block of H13 steel which hardened and tempered to 48 HRC. The volume of the workpiece used was 130×100×30mm. Cutting tool 20 mm in diameter was utilized in pocketing process. The cutting tool accommodates 2 indexable inserts in every single cutting process. Coated carbide mill inserts CoroMill 490 were employed in the current study with various machining parameters including cutting speeds were 150, 200 and 250 m/min with feed per tooth were 0.05, 0.10 and 0.15 mm/tooth and depths of cut involved were 0.10, 0.15 and 0.20 mm. CoroMill 490 facilitates four-edge inserts which every cutting edge is only applied in different cutting parameter. Pocket was cut using a new cutting edge. The CoroMill 490 insert tools were securely fastened to the tool holder. The experimental works were performed on CNC vertical machining center Mazak Nexus 410A-II CNC Mill without any cutting fluid. These experiments adapted 3 different tool path strategies which are contour-in, zigzag and inclined tool path strategy as seen in Figure 1. Pocket millings were performed by varying the cutting parameters. Response Surface Method (RSM) design of experiment with a standard called Central Composite Design (CCD) was employed. There are 20 runs for every tool path strategies as output from Design of Experiment software were tested. All the process and parameters were programmed in CNC programming. When the machining process completed wear of inserts were measured by optical microscope HISOMET II.

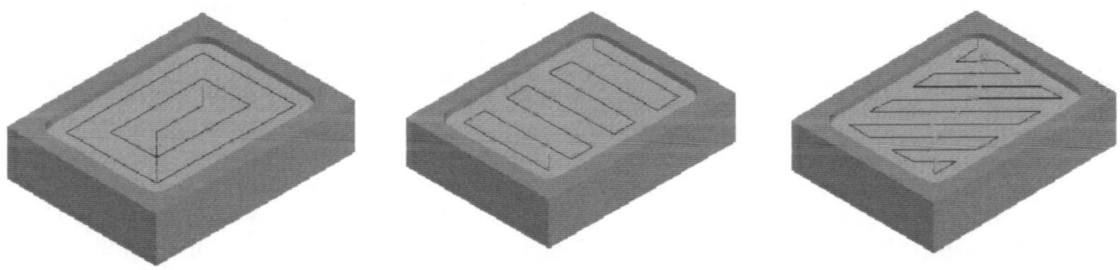

(a) Contour-in Strategy (b) Zigzag Strategy (c) Inclined Strategy

Figure 1. Tool Path Strategies Applied in Experiments

3. Results and Discussion

Experiment results reveals that the number of path segments including the finishing cut in order to make pocket for contour-in, zigzag and inclined tool paths are 11, 13 and 20, respectively. Besides that, some turning points are involved as the transition point before the cutting tool changes its direction to cut the material with different direction. The number of turning points are also different depends on the movement of tool path. The number of turning points involved in during pocketing by applying contour-in, zigzag and inclined are 9, 13 and 20 points, correspondingly. Figure 2 schematically shows the numbers of path segments, finishing cut and turning points for three tool path strategies.

IOP Conf. Series: Materials Science and Engineering **343** (2018) 012020 doi:10.1088/1757-899X/343/1/012020

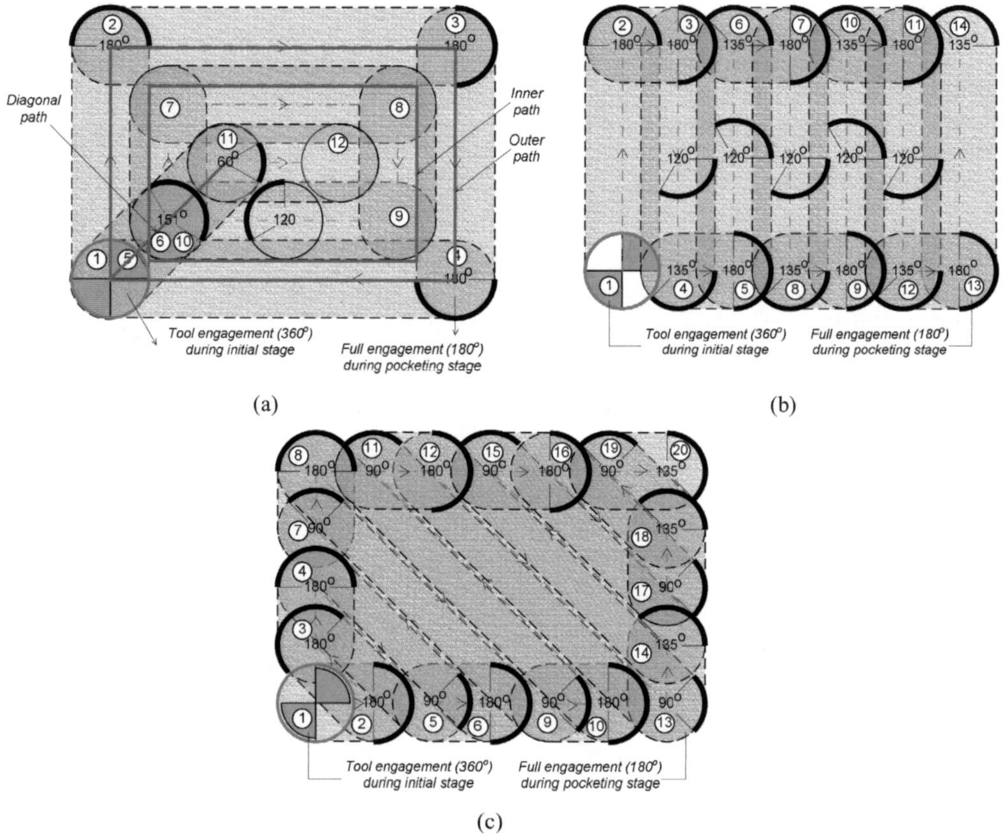

Figure 2. The Number of Tool Path Segments and Turning Points Applied for (a) Contour-in (b) Zigzag and (c) Inclined Tool Path Strategy

3.1 Effect of Tool Path Strategies on Tool Wear

The effect of utilizing three different tool paths strategies to tool wear development during end milling of AISI H13 is depicted in Figure 3. Contour-in is claimed as tool path strategy with the lowest tool wear development tailed by zigzag and inclined strategy. The poor tool wear caused by the number of tool engagement conditions. It is believed that the high impact forces produced when the cutting tool frequently engaged into the corner regions and turning points with combination of high cutting temperature as effect of high cutting force were the reason for poor tool wear found. The number of turning corner regions and turning points are dominant in inclined while they are medium in zigzag and small in inclined strategy as mentioned earlier in Figure 2. It should be noted that in this research, up and down milling are performed simultaneously in every segment of tool path strategy.

The results of tool wear measurements in average obtained from the research for each strategy include contour-in 0.04 mm, zigzag 0.05 mm and inclined 0.08 mm. The result is obtained from experimental work reveals that contour-in and zigzag strategy produce tool wear 49% and 31% lower than inclined strategy. The tool wear obtained are shown in Figure 4 under optical microscope (20X magnification) for three tool path strategies.

IOP Conf. Series: Materials Science and Engineering **343** (2018) 012020 doi:10.1088/1757-899X/343/1/012020

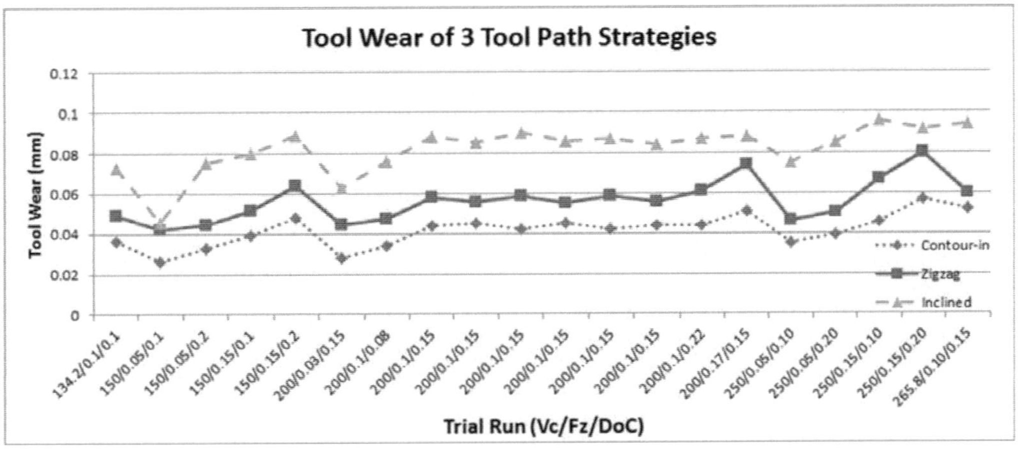

Figure 3. Comparisons of Tool Path Strategies for Tool Wear

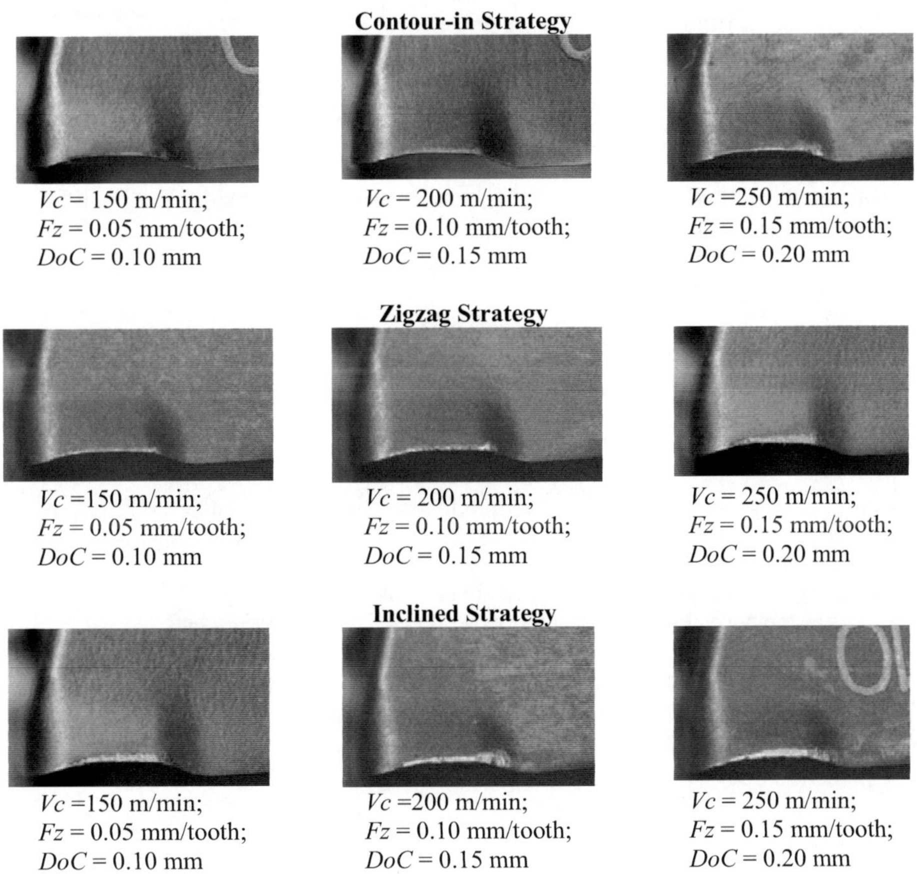

Figure 4. Tool Wear Development for (a) Contour-in, (b) Zigzag, (c) Inclined strategy

IOP Conf. Series: Materials Science and Engineering **343** (2018) 012020 doi:10.1088/1757-899X/343/1/012020

3.2 Empirical Modeling of Tool Wear

Analysis of tool wear was carried out using the analysis of variance (ANOVA) only for Contour-in strategy. This is because contour-in gives the best result among three strategies. Backward elimination procedure is selected to automatically eliminate the non-significant term to the model which resulting ANOVA table for Response Surface Reduced Quadratic Model for all strategies tested.

Table 1 shows the ANOVA model for tool wear model under contour-in strategy. The Model with F-value of 65.09 indicated that the model is significant with probability F < 0.0001. There is only 0.01% a chance that the F-Model is wrong which due to noise. The values of Prob > F is less than 0.05 indicating the significance of model terms. In this case, two level interaction of cutting speed with feed per tooth (AB), two level interaction of cutting speed with depth of cut (AC) and second order effect of cutting speed (A^2) are not the significant factor to the tool wear. Which means, main effect of cutting speed (A), main effect of feed per tooth (B), main effect of (depth of cut), two level interaction of feed per tooth with depth of cut (BC), second order effect of feed per tooth (B^2) and second order effect of depth of cut (C^2) are significant model terms. Apparently, the most significant effect to the tool wear is feed per tooth with F-value of 226.63.

Table 1. ANOVA Model for Tool Wear Model under Contour-in Strategy

Source	**Sum of Squares**	**df**	**Mean Square**	**F Value**	**p-value Prob > F**	**Significance**
Model	1.145E-003	6	1.908E-004	65.09	< 0.0001	significant
A-Cutting Speed	2.364E-004	1	2.364E-004	80.64	< 0.0001	
B-Feed per Tooth	6.643E-004	1	6.643E-004	226.63	< 0.0001	
C-Depth of Cut	1.701E-004	1	1.701E-004	58.03	< 0.0001	
BC	1.012E-005	1	1.012E-005	3.45	0.0859	
B^2	2.195E-005	1	2.195E-005	7.49	0.0170	
C^2	2.978E-005	1	2.978E-005	10.16	0.0071	
Residual	3.811E-005	13	2.931E-006			
Lack of Fit	2.877E-005	8	3.597E-006	1.93	0.2436	not significant
Pure Error	9.333E-006	5	1.867E-006			
Cor Total	1.183E-003	19				
$R^2 = 0.9678$	Adj. $R^2 = 0.9529$		Pred. $R^2 = 0.9139$		Adeq. Precision = 31.603	

The model is good as the Lack of Fit is not significant with 24.36% relative to the pure error. The R^2 is 0.97 which is high and close to 1. The $R^2_{predicted}$ of 0.9139 is reasonable agreement with the R^2_{adj} of 0.9529. Ratio greater than 4 for adequate precision is desirable. In this model, the value of adequate precision is 31.603, which beyond the limit of minimum value acceptable ratio.

Figure 5 shows the normal probability plot of the studentized residuals which is used to check for normality of residuals. The plot revealed that the points on the plot lie reasonably close to the straight line which indicated that the errors are distributed normally. This support the conclusion that cutting speed (A), feed rate (B) and depth of cut (C) are the significant effects and the underlying assumptions of the analysis are satisfied for all strategies employed.

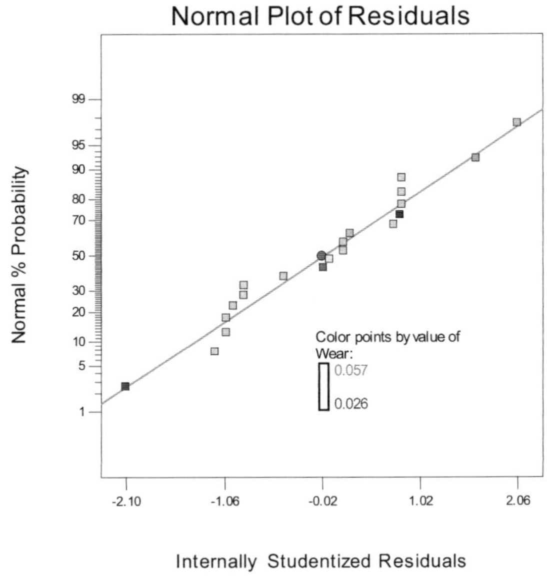

Figure 5. Normal Probability of Residuals for Tool Wear Data under Contour-in Strategy

Perturbation plot from Figure 6 generated by Design Expert software reveals the effect of input parameters in coded form on the output response (tool wear). It can be seen that the three input parameters, cutting speed, (A), feed rate, (B) and depth of cut (C) have quadratic comparable effect on tool wear. The figure also shows that at lower range of input cutting parameters (cutting speed, feed per tooth and depth of cut), the tool wear is lower compared to the higher cutting range. Feed rate affecting the tool wear less at lower range (-1). However, at higher range of machining (+1), feed per tooth reduces the tool wear the most. The tool wear increases significantly at higher range of feed per tooth compared to cutting speed (A) and depth of cut (C). It is because when the feed rate is increased, the chip thickness is increased. Due to the combined effects of radial depth of cut (a_e) and feed rate (a_f) resulted in the higher value of chips thickness as expressed in Equation (4.8) on page 148. As chip thickness increase, the cutting force and energy needed to remove the material from the workpiece is increased. The higher feed rate leads to the higher cutting force and cutting temperature. In turn, tool wear is occurred rapidly. At the higher cutting parameters range depth of cut affect the tool wear the least. This is because the higher depth of cut leads to the bigger contact area between cutting tool and workpiece. However, the factor of feed per tooth is higher than depth of cut in order to influence the tool wear. Such conditions are found in zigzag strategy where the feed rate affecting tool wear much more than cutting speed and depth of cut.

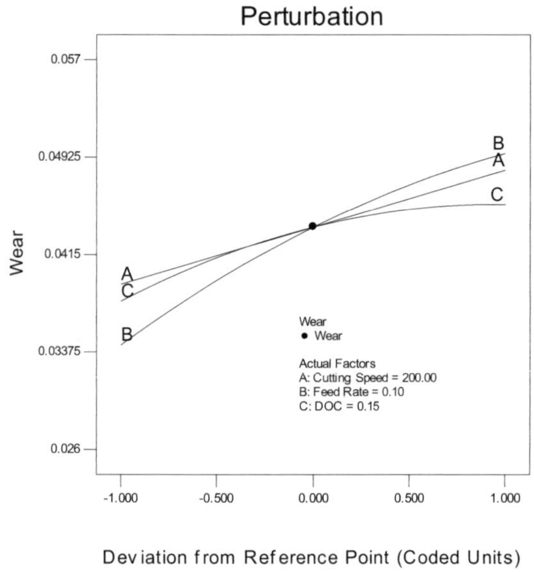

Figure 6. Perturbation Plot of Tool Wear under Contour-in Strategy (Vc = 200 m/min, Fz = 0.10 mm/tooth and DoC = 0.15 mm)

The three dimensional graph for tool wear under contour-in strategy is shown in Figure 7. The graph has a curvilinear profile which is quadratic model fitted. It can be seen that as the cutting speed increase, the tool wear is increased. Same pattern is observed for feed rate.

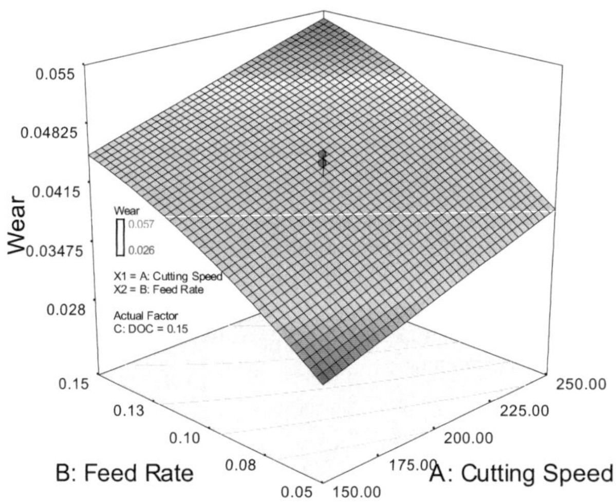

Figure 7. Contour Graph of Predicted Tool Wear under Contour-in Strategy

The model for the response variable of tool wear for contour-in strategy can be concluded as in Equation (1). The Equation (1) describes the quadratic surfaces as in Figure 7.

$$Wear_{cont.} = 0.044 + (4.541\times10^{-3}) * A + (7.612 \times 10^{-3}) * B + (3.852 \times 10^{-3}) * C + (1.125\times10^{-3}) * B * C - (1.752\times10^{-3}) * B^2 - (2.040\times10^{-3}) * C^2 \quad (1)$$

4. Conclusion

1. Contour-in is claimed as tool path strategy with the lowest tool wear development compared to zigzag and inclined.
2. Analysis of tool wear was carried out using the analysis of variance (ANOVA) for Contour-in strategy revealed that the most significant effect to the tool wear is feed per tooth, cutting speed and depth of cut subsequently.
3. At lower range of input cutting parameters (cutting speed, feed per tooth and depth of cut), the tool wear is lower compared to the higher cutting range. Feed rate affecting the tool wear less at lower range. However, at higher range of machining, feed per tooth reduces the tool wear the most.

5. References

[1] Cui X, Zhao J, & Tian X 2013b *Int. Journal of Advanced Manufacturing Tech.* **64** 1737. https://doi.org/10.1007/s00170-012-4137-9
[2] Suresh R & Basavarajappa S 2014 *Procedia Materials Sci.* 5 1450. https://doi.org/10.1016/j.mspro.2014.07.464
[3] Cui X, Zhao J & Tian X 2013a *Int. Journal of Advanced Manufacturing Tech.* **64** 1737. https://doi.org/10.1007/s00170-012-4137-9
[4] Perez H, Diez E, Perez J & Vizan A 2013 *Procedia Eng.* **63**, 573. https://doi.org/10.1016/j.proeng.2013.08.193
[5] Cui X, Zhao J, Jia C and Zhou Y 2012 *Int. Journal Advanced Manufacturing Tech.* **61** 1. DOI: 10.1007/s00170-011-3684-9.

IOP Conf. Series: Materials Science and Engineering **343** (2018) 012021 doi:10.1088/1757-899X/343/1/012021

Measurement value analysis overall equipment effectiveness (OEE) packaging process in line 2 (Case Study of PT. MBI Tbk)

Erry Rimawan, Muhammad Kholil, Hendri

Industrial Engineering Department, Faculty of Engineering, Mercu Buana University, Jakarta

Jl. Meruya Selatan Kembangan, Kebun Jeruk, Jakarta Barat, Jakarta

erry_rimawan@mercubuana.ac.id

Abstract. PT. MBI Tbk is engaged in the manufacture of beverage industry, where the company's production is based on the magnitude of customer demand that is marketing offices that had been scattered in various regions of Indonesia. In the packaging process steps in PT.MBI through the line 3 lines including racking, canning line, bottling line. In the canning process to existing packing on Line 2 (canning line), there are some machines that are used continuously, among other Depalletizer machine, filler machine, can seamer machine, pasteurizer machine, machine FLD, Wrap Around engine, engine Shrink Wrap. Due to the large demand from customers that is relentless, therefore the calculation of overall equipment effectiveness (OEE) as a whole on line 2 (canning line) is needed in order to make improvements continuously (Continuous Improvement) at line 2 (canning line). This study aims to determine the value of overall equipment effectiveness (OEE) and Losses of the most influential of the big six OEE Losses focused on equipment or machinery as a whole into a single unit that is on the line 2, which will then be known root cause of the losses that occur from the research over the field. From the calculation of overall equipment effectiveness (OEE), there are two ratios are still poor and under world-class standards, while the ratio of the availability of 88.85% of the world-class standards by 90% and the performance ratio of 78.51% of the standard world class by 95%, whereas for quality ratio has entered the world-class standard that is equal to 99.90%. Thus the value of OEE on Line 2 line is below world class standards. In this study there were only five losses, which can be identified, and while the losses were very influential, namely the Speed Reduced Losses, losses, these losses accounted for the largest percentage of the value of the rate of 19.12%, of the results of this study losses occurred due to poor surveillance systems (less good) that causes the employee or operator does not perform the work in accordance with a predetermined.

1. Introduction

At this time the repair or maintenance problems that exist on the average of every manufacturing occurs due to the absence or lack of effective system or method that is capable of measuring the performance of existing equipment and can provide a solution to the root of the problems encountered. Thus the selection of performance measurement methods is very important for the company to achieve company goals. As a performance measurement method that is

IOP Conf. Series: Materials Science and Engineering **343** (2018) 012021 doi:10.1088/1757-899X/343/1/012021

widely used by companies, particularly been applied by many Japanese firms are able to overcome the problem of equipment is the method of Overall Equipment Effectiveness (OEE). This method is part of a maintenance system which has been widely adopted by Japanese companies, namely Total Productive Maintenance (TPM) [3].

PT. MBI Tbk is engaged in the manufacture of beverage industry, where the company where the production is based on the magnitude of the demand of costumer who is marketing offices that had been scattered in various regions of Indonesia. At this stage of the packaging process in PT.MBI through with them racking line 3 line (Line1), a canning line (Line 2), the bottling line (Line 3). As for the canning process to existing packing line 2 takes place on a continuous basis is relentless, Line 2 is thus need special attention and can not be separated from the problem machine or equipment effectiveness as a whole into one (Line 2), therefore the absence of repair or maintenance efforts as well as the good methods of canning production process on Line 2 is not running as optimally as possible and desired product quality is achieved even less, so it can lead to failure to achieve the desired productivity and profitability.

2. Problem Identification

The problem that arises is on the packaging line at line 2, that is the constraint that is often experienced by the packaging during the production process in line 2 takes place, and less effective and less efficient its existing machines on line 2 of the current production process. So from these events will be calculated as well as the effectiveness of the error rate on the second line with OEE method. The objectives are: 1) Getting OEE values on the overall machine or equipment lines on existing packaging process on line 2. 2) Knowing the losses and the root causes that lead to the low value of OEE. 3) Provide the suggestions solving problems that occur.

Authors restrict the problem to the following: 1) The study was conducted on lines Line 2. 2) The data were taken in the period August - October 2013 per week. 3) This study focuses on the value of the OEE calculation and analysis of the measurement results, and determines the amount of loss on machinery or equipment (six big losses). The word is taken from the Greek language maintenance terrain means caring, nurturing, and maintaining. Maintenance is a combination of various measures taken to keep an item in, or fix it to an acceptable condition [1].

According to Corder [1], the purpose of maintenance is as follows: 1) Extend the useful life of the asset (ie any part of a workplace, building, and contents). 2) To ensure optimum availability of equipment installed for production (or service) and get a return on investment (return on investment) is maximum. 3) Ensure the operational readiness of all equipment needed in emergency activities at any time, for example: a backup unit, the unit firefighters and rescuers, and so on. 4) To ensure the safety of those who use the facility. This type of maintenance until now divided into three ways: planned maintenance (planned maintenance), unplanned maintenance (unplanned maintenance) and maintenance of self (Autonomous Maintenance).

2.1 Maintenance planned (planned maintenance)

Planned maintenance is maintenance that was organized and carried out with the thought that will come, controlling and recording of pre-defined according to plan. As according to Corder [1] planned maintenance is divided into two main activities namely preventive maintenance (Preventive Maintenance) and corrective maintenance (corrective maintenance). Preventive maintenance is a periodic inspection to detect conditions that may lead to the production of stalled or reduced engine combined with maintenance functions to eliminate, control, these conditions and restore the machine to its original state or in other words self detection and treatment of abnormal engine condition before the condition the cause disability or loss. According to Corder [1], the corrective maintenance is carried out repeatedly or maintenance done to fix a part (including tuning and repair) which has been stalled to meet an acceptable condition.

2.2 Unplanned maintenance

There is only one form of unplanned maintenance is maintenance emergency (emergency / breakdown maintenance). Corder [1] defines as the maintenance actions which need to be implemented immediately to prevent serious consequences, such as loss of production, major damage to the equipment, or for workplace safety.

2.3 Autonomous Maintenance

Autonomous maintenance is an activity that can improve the productivity and efficiency of machinery / equipment through the activities carried out by operators to maintain machine / equipment that they handle themselves.

2.4 Total Productive Maintenance (TPM)

Total Productive Maintenance (TPM) has the following purpose: 1) Total are indicates that the TPM consider various aspects and involving all personnel are, starting from the top level to the bottom. 2) Productive is focuses on the efforts to try to do maintenance to keep production running conditions and minimize the problems that occur are produced when maintenance is performed. 3) Maintenance means to maintain and keep the equipment independently done by production operators to keep equipment good condition and well maintained street cleaning, lubrication and watched him perform. In general, the notion of Total Productive Maintenance is a way to plan all maintenance activities include inspection, minor repairs to the stage overhaul implemented by all relevant personnel department through small group activities. In general, the notion of Total Productive Maintenance is a way to plan all maintenance activities include inspection, minor repairs to the stage overhaul implemented by all relevant personnel department through small group activities. In the measuring instruments used TPM is Overall Equipment Effectiveness (OEE) or the effectiveness of the machine as a whole, where the calculation of OEE based on loss of engine stopped due to the damage, the engine must be slowed, and the resulting product defects.

The ideal value of OEE is a world class company standards are as follows:

Table 1: World Class OEE Factor

OEE Factor	*World Class*
Avaibility	>90.0%
Performance Rate	>95.0%
Quality Rate	>99.9%
OEE	>85.0%

So that the success of a program is the TPM achievement OEE value up to> 85.0% [2].

The shape of the measurement of the effectiveness of a machine or machine OEE has three parameters which measure the variables related artifacts in the formation of each of these parameters which include, among others: 1) Availability is a comparison between actual operating time with a loading (plane operating time). These parameters take the level of readiness of existing tools, and used to operate. Value availability can be calculated by the following formula:

$$Avaibility\ Rate = \frac{Operating\ Time}{Loading\ Time}\ x\ 100$$
$$= \frac{Loading\ Time - Down\ Time}{Loading\ Time} \times 100$$

2) Performance in determining the performance of a device or machine results will show how far the level of success of maintenance program which has been implemented in the company. Efficiency describes the performance of the engine operating conditions where a machine can be operated under the actual capacity of the machine. Performance or efficiency performance values can be calculated as follows:

$$Performance\ Rate = \frac{Processed\ amount \times Theoritical\ Cycle\ Time}{Operating\ Time} \times 100\%$$
$$Performance\ Rate = \frac{(Finish\ Good + Reject)}{Standar\ Speed \times (Loading\ Time - Down\ Time)} \times 100\%$$

3) Quality Rate (Product Quality) determination of the value of quality products is measured by the ability of a machine to produce a qualified production quality that has been standardized by the company. The formula used for the measurement of this ratio is:

$$Quality\ Rate = \frac{Processed\ Amount - Defect\ Amount}{Processed\ Amont} \times 100\%$$

One of the objectives of the TPM program and measuring instruments used (OEE) is to reduce or eliminate the six big losses. The six big losses (six big losses) are as follows:

IOP Conf. Series: Materials Science and Engineering **343** (2018) 012021 doi:10.1088/1757-899X/343/1/012021

A. *Downtime Losses*

1) (*Equipment Failure Losses / Breakdown*). The formula is as follows:

$$Breakdowns\ losses = \frac{Total\ Breakdowns\ Time}{Loading\ Time} x100\%$$

2) *(Setup and adjustment).*The formula is as follows:

$$Setup\ and\ Adjustment\ Losses = \frac{Setup\ time}{Loading\ Time} x100\%$$

B. (*Speed Losses*)

3) *(Idling and Minor Stoppages)*. The formula is as follows:

$$Idling\ and\ minor\ stoppages = \frac{non\ productive}{Loading\ Time} x\ 100\%$$

4) *(Reduced speed)*. The formula is as follows:

$$Reduced\ Speed = \frac{OT - (Ideal\ cycletime\ x\ Processed\ amount)}{Loading\ Time} \times 100\%$$

Description: OT = *Operation Time*

C. (*Defect Losses*)

5) *(rework losses / Defect Losses)*. The formula is as follows:

$$Defect\ Losses = \frac{Ideal\ Cycle\ Time\ x\ Total\ Reject}{Loading\ Time} x100\%$$

6) *(Reduced yield)*. The formula is as follows:

$$Yield/Scrap\ Loss = \frac{Ideal\ Cycle\ Time\ x\ scrap}{Loading\ Time} x100\%$$

3. Methodology

On the stage of this methodology is described that phase will be conducted in the study, ranging from problem identification to final conclusions through the flow chart below.

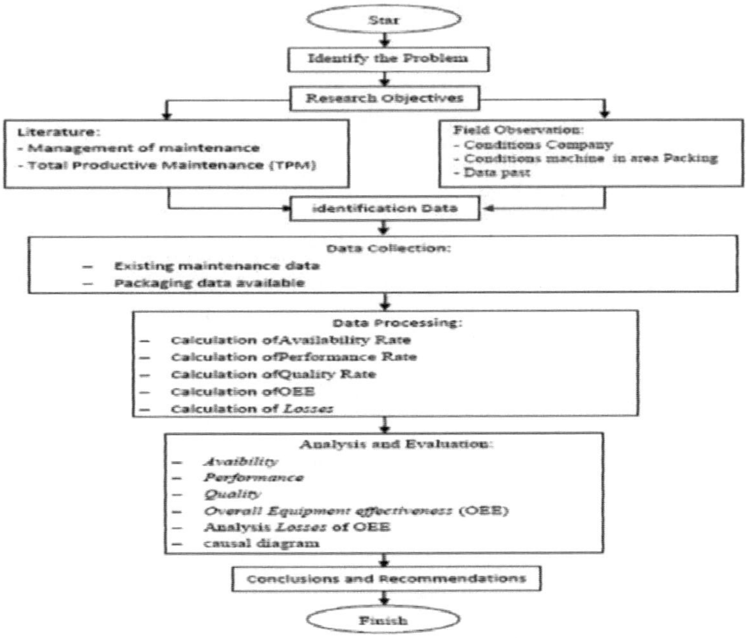

Figure 1. Flowchart of Research

IOP Conf. Series: Materials Science and Engineering **343** (2018) 012021 doi:10.1088/1757-899X/343/1/012021

3.1 Data collection and processing

Here is a picture sequence of the machine on line 2:

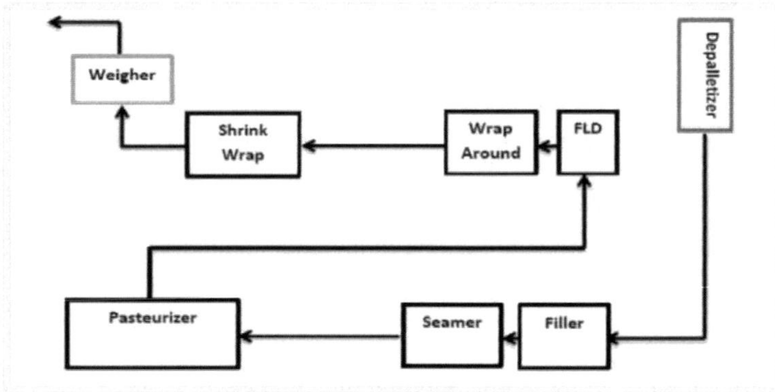

Figure 2: The sequence of line 2

From the above process, the sequence data obtained as follows:

A. The production data in Line 2

Table 2. Data Production Line 2

Mount	*Week*	*Good Product (can)*	*Reject Product (can)*	*Rework*	*Total Product (can)*
August	32	2405294	3001	0	2408295
	33	2477269	2906	0	2480175
	34	3338046	2168	0	3340214
	35	2843586	2001	0	2845587
September	36	2988309	1812	0	2990121
	37	3119730	2196	0	3121926
	38	2372857	1552	0	2374409
	39	2487262	5027	0	2492289
October	40	3132287	6096	0	3138383
	41	2950005	3481	0	2953486
	42	2553302	1446	0	2554748
	43	3055888	2740	0	3058628
	44	2300529	1710	0	2302239

B. Data *Available Time*

Table 3. Data Available Time

Mount	Week	Total Available Time (Minutes)	Total Available Time (Hour)
August	32	10080	168
	33	10080	168
	34	10080	168
	35	10080	168
September	36	10080	168
	37	10080	168
	38	10080	168
	39	10080	168
October	40	10080	168
	41	10080	168
	42	10080	168
	43	10080	168
	44	10080	168

C. Data *Planned Down Time*

Table 4. Data Planned Down Time

Mount	Week	Planed Downtime (minutes)	Planned Downtime (Hours)
August	32	30	0,50
	33	40	0,67
	34	0	0,00
	35	10	0,17
September	36	225	3,75
	37	125	2,08
	38	260	4,33
	39	107	1,78
October	40	155	2,58
	41	180	3,00
	42	440	7,33
	43	110	1,83
	44	115	1,92

Data *Down Time*

Table 5. Data Down Time

Mount	Week	Cange Over (minutes)	Breakdown (Minutes)	Stop Line (Minutes)	Downtime (Minutes)	Downtime (Hours)
August	32	747	456	130	1333	22.22
	33	305	625	125	1055	17.58
	34	730	456	75	1261	21.02
	35	507	623	80	1210	20.17
September	36	380	470	155	1005	16.75
	37	405	210	70	685	11.42
	38	325	367	80	772	12.87
	39	998	705	190	1893	31.55
October	40	710	453	167	1330	22.17
	41	295	480	315	1090	18.17
	42	140	360	65	565	9.42
	43	600	360	140	1100	18.33
	44	515	540	60	1115	18.58

With the the same calculation, the following overall availability calculations table:

Table 6. Calculation Availability

Mount	Week	Total Available Time (Minutes)	Planned Downtime (Minutes)	Loading Time (Minutes)	Downtime (Minutes)	Operating Time (Minutes)	Avaibility Rate (%)
Aug	32	10080	30	10050	1333	8717	86.74%
	33	10080	40	10040	1055	8985	89.49%
	34	10080	0	10080	1261	8819	87.49%
	35	10080	10	10070	1210	8860	87.98%
Sep	36	10080	225	9855	1005	8850	89.80%
	37	10080	125	9955	685	9270	93.12%
	38	10080	260	9820	772	9048	92.14%
	39	10080	107	9973	1893	8080	81.02%
Oct	40	10080	155	9925	1330	8595	86.60%
	41	10080	180	9900	1090	8810	88.99%
	42	10080	440	9640	565	9075	94.14%
	43	10080	110	9970	1100	8870	88.97%
	44	10080	115	9965	1115	8850	88.81%
Total		131040	1797	129243	14414	114829	88.85%

1) Example of calculation of Availability on line 2 for week 32:

Loading Time = *(Total available time – Planned downtime)*
= (10080 minutes – 30 minutes)
= 10050 minutes.

Availability $= \frac{Loading\ Time - Down\ Time}{Loading\ Time} \times 100$

$$= \frac{10050 \text{ minutes} - 1333 \text{ minutes}}{10050 \text{ minutes}} \times 100 = 86{,}74\ \%$$

2) Example of calculation of *Performance* on line 2 for week 32:
Processed Amount = (Finish Good + Reject)
= (2405294 + 3001)
= 2408295
Standard Speed used on Line 2 in the yield of products is 400 cpm.so

$$Performance\ Rate = \frac{(\text{Finish Good} + \text{Reject})}{\text{Standar Speed} \times (\text{Loading Time} - \text{Down Time})} \times 100\%$$

$$= \frac{(2408295)}{400 \times (10050 - 1333)} \times 100\%$$

= 69,07%

With the the same calculation, the following table thoroughly Performance calculation:

Table 7. Calculation Performance

Mount	Week	Total Product (can)	Standar Speed (bpm)	Loading Time (Minutes)	Downtime (Minutes)	Performance Rate (%)
Aug	32	2408295	400	10050	1333	69.07%
	33	2480175	400	10040	1055	69.01%
	34	3340214	400	10080	1261	94.69%
	35	2845587	400	10070	1210	80.29%
Sep	36	2990121	400	9855	1005	84.47%
	37	3121926	400	9955	685	84.19%
	38	2374409	400	9820	772	65.61%
	39	2492289	400	9973	1893	77.11%
Oct	40	3138383	400	9925	1330	91.29%
	41	2953486	400	9900	1090	83.81%
	42	2554748	400	9640	565	70.38%
	43	3058628	400	9970	1100	86.21%
	44	2302239	400	9965	1115	65.04%
Total		36060500	400	129243	14414	78.51%

3) Example of calculation of *Quality* on line 2 for week 32:

$$Quality\ Rate = \frac{\text{Processed Amount} - \text{Defect Amount}}{\text{Processed Amont}} \times 100\%$$

$$= \frac{2408295 - 2195}{2408295} \times 100\%$$

= 99,88%

With the the same calculation, the following overall Quality calculations table:

Table 8. Calculation Quality

Mount	Week	Good Product (can)	Total Product (can)	Quality Rate (%)
Aug	32	2405294	2408295	99.88%
	33	2477269	2480175	99.88%
	34	3338046	3340214	99.94%
	35	2843586	2845587	99.93%
Sep	36	2988309	2990121	99.94%
	37	3119730	3121926	99.93%
	38	2372857	2374409	99.93%
	39	2487262	2492289	99.80%
Oct	40	3132287	3138383	99.81%
	41	2950005	2953486	99.88%
	42	2553302	2554748	99.94%
	43	3055888	3058628	99.91%
	44	2300529	2302239	99.93%
Total		36024364	36060500	99.90%

4) Example of calculation of OEE on line 2 for week 32:
OEE = (*Avaibility* × *Performance* × *Quality)*
= (86,74% × 69,07% × 99,88%)
=59,84%

With the same calculation, the following overall OEE calculation table:

Table 9. Calculation OEE

Mount	Week	Avaibility Rate (%)	Performance Rate (%)	Quality Rate (%)	OEE (%)
Aug	32	86.74%	69.07%	99.88%	59.83%
	33	89.49%	69.01%	99.88%	61.68%
	34	87.49%	94.69%	99.94%	82.79%
	35	87.98%	80.29%	99.93%	70.60%
Sep	36	89.80%	84.47%	99.94%	75.81%
	37	93.12%	84.19%	99.93%	78.35%
	38	92.14%	65.61%	99.93%	60.41%
	39	81.02%	77.11%	99.80%	62.35%
Oct	40	86.60%	91.29%	99.81%	78.90%
	41	88.99%	83.81%	99.88%	74.50%
	42	94.14%	70.38%	99.94%	66.22%
	43	88.97%	86.21%	99.91%	76.63%
	44	88.81%	65.04%	99.93%	57.72%
Total		88.85%	78.51%	99.90%	69.68%

5) Example of calculation of *Breakdown Losses* on line 2 for week 32:

$$Breakdowns\ losses = \frac{Total\ Breakdowns\ Time}{Loading\ Time} x100\%$$

$$Breakdowns\ losses = \frac{456}{10050} x100\%$$

Breakdown Losses = 4,54 %

With the the same calculation, the following overall Breakdown Losses calculation table:

Table 10. Calculation Breakdown Losses

Mount	Week	Breakdown (Minutes)	Loading Time (Minutes)	Equpment Failure Losses (Minutes)
Aug	32	456	10050	4.54%
	33	625	10040	6.23%
	34	456	10080	4.52%
	35	623	10070	6.19%
Sep	36	470	9855	4.77%
	37	210	9955	2.11%
	38	367	9820	3.74%
	39	705	9973	7.07%
Oct	40	453	9925	4.56%
	41	480	9900	4.85%
	42	360	9640	3.73%
	43	360	9970	3.61%
	44	540	9965	5.42%
Total		6105	129243	61.33%
Average		469.62	9941.77	4.72%

6) Example of calculation of *Setaup and Adjustment* on line 2 for week 32:

$$Setup\ and\ Adjustment\ Losses = \frac{Setup\ time}{Loading\ Time} x100\ \%$$

IOP Conf. Series: Materials Science and Engineering **343** (2018) 012021 doi:10.1088/1757-899X/343/1/012021

$$Setup\ and\ Adjustment\ Losses = \frac{747}{10050} x100\ \%$$

Setup and adjustment Losses = 7,43 %

With the the same calculation, the following overall *Setup and Adjustment* calculation table:

Table 11. Calculation Setup and Adjustment

Mount	Week	Cange Over (minutes)	Loading Time (Minutes)	Setup & Adjustment Losses (%)
Aug	32	747	10050	7.43%
	33	305	10040	3.04%
	34	730	10080	7.24%
	35	507	10070	5.03%
Sep	36	380	9855	3.86%
	37	405	9955	4.07%
	38	325	9820	3.31%
	39	998	9973	10.01%
Oct	40	710	9925	7.15%
	41	295	9900	2.98%
	42	140	9640	1.45%
	43	600	9970	6.02%
	44	515	9965	5.17%
Total		6657	129243	66.76%
Average		512.08	9941.77	5.15%

7) Example of calculation of *Idling & Minor Stoppages* on line 2 for week 32:
8)

$$Idling\ and\ minor\ stoppages = \frac{non\ produktive}{Loading\ Time} x\ 100\%$$

$$Idling\ and\ Minor\ Stoppages = \frac{130}{10050} x\ 100\%$$

Idling and Minor Stoppages = 1,29 %

With the the same calculation, the following overall *Idling & Minor Stoppages* calculation table:

Table 12. Calculation Idling & Minor Stoppages

Mount	Week	Stop Line (Minutes)	Loading Time (Minutes)	Idling & Minor Stoppages (%)
Aug	32	130	10050	1.29%
	33	125	10040	1.25%
	34	75	10080	0.74%
	35	80	10070	0.79%
Sep	36	155	9855	1.57%
	37	70	9955	0.70%
	38	80	9820	0.81%
	39	190	9973	1.91%
Oct	40	167	9925	1.68%
	41	315	9900	3.18%
	42	65	9640	0.67%
	43	140	9970	1.40%
	44	60	9965	0.60%
Total		1652	129243	16.62%
Average		127.08	9941.77	1.28%

9) Example of calculation of *Reduced Speed* on line 2 for week 32:
Ideal Cycle Time = 1/400 = 0,0025

$$Reduced\ Speed = \frac{OT - (Ideal\ cycletime\ x\ Processed\ amount)}{Loading\ Time} = \frac{8717 - (0{,}0025\ x\ 2408295)}{10050} \times 100\%$$

$$Reduced\ Speed = \frac{2696{,}2625}{10050} \times 100\%$$

Reduced Speed = 26,83 %

With the the same calculation, the following overall *Reduced Speed* calculation table:

Table 13. Calculation Reduced Speed

Mount	Week	Operating Time (Minutes)	Ideal Cycle Time (cpm)	Total Product (can)	Loading Time (Minutes)	Reduced speed (%)
Aug	32	8717	0.0025	2408295	10050	26.83%
	33	8985	0.0025	2480175	10040	27.73%
	34	8819	0.0025	3340214	10080	4.65%
	35	8860	0.0025	2845587	10070	17.34%
Sep	36	8850	0.0025	2990121	9855	13.95%
	37	9270	0.0025	3121926	9955	14.72%
	38	9048	0.0025	2374409	9820	31.69%
	39	8080	0.0025	2492289	9973	18.54%
Oct	40	8595	0.0025	3138383	9925	7.55%
	41	8810	0.0025	2953486	9900	14.41%
	42	9075	0.0025	2554748	9640	27.89%
	43	8870	0.0025	3058628	9970	12.27%
	44	8850	0.0025	2302239	9965	31.05%
Total		114829		36060500	129243	248.61%
Average		8833	0.0025	2773884.6	9941.7692	19.12%

10) Example of calculation of *Defect Losses* on line 2 for week 32:
Ideal Cycle Time = 1/400 = 0,0025

$$Defect\ Losses = \frac{Ideal\ Cycle\ Time\ x\ Total\ Reject}{Loading\ Time} x100\%$$

$$Defect\ Losses = \frac{0{,}0025\ x\ 3001}{10050} x100\%$$

Defect Losses = 0,07 %

With the same calculation, the following overall *Defect Losses* calculation table:

Table 14. Calculation Defect Losses

Mount	Week	Reject Product (can)	Ideal Cycle Time (cpm)	Loading Time (Minutes)	Defect Losses (%)
Aug	32	3001	0.0025	10050	0.07%
	33	2906	0.0025	10040	0.07%
	34	2168	0.0025	10080	0.05%
	35	2001	0.0025	10070	0.05%
Sep	36	1812	0.0025	9855	0.05%
	37	2196	0.0025	9955	0.06%
	38	1552	0.0025	9820	0.04%
	39	5027	0.0025	9973	0.13%
Oct	40	6096	0.0025	9925	0.15%
	41	3481	0.0025	9900	0.09%
	42	1446	0.0025	9640	0.04%
	43	2740	0.0025	9970	0.07%
	44	1710	0.0025	9965	0.04%
Total		36136		129243	0.91%
Average		2779.6923	0.0025	9941.7692	0.07%

4. Conclusion

From the research that has been done in the company, it can be concluded that: 1) From the calculation of OEE values on lines Line 2, a score of 88.85% Availability, performance score of 78.51%, and 99.90% of the value of Quality. Thus the value of Availability and Performance is still not good value because it is below the world-class standards, thus causing the OEE value is below the world-class standard that is> 85%. For the achievement gap OEE values obtained under standard is 15.31%, while for the achievement of Quality values can already be said to be good because it makes a world-class standard. 2) From the research, there are only five losses, which can cause low OEE value, while the average - average losses are as follows: a) The biggest Losses Losses are on the Speed Reduced by 19.12%, from field research during these losses tend to occur because of poor surveillance systems (less stringent) which causes the employee or operator does not perform the work in accordance with, so this lead to high losses low value of the performance. b) then the next highest losses Setup & Adjustment losses amounted to 5.15%, from field research during these losses occurred due to the time and manner set out to do the changeover is less effective, in addition to the lack of training for employees as a result of employees who did not master the job each particular regarding technical issues. c) The next Losses are potentially low value of the OEE Losses Breakdown by 4.72%, from field observations during these losses occur that the plan is still less effective preventive maintenance is done, so that even this results in lower losses availability value. d) The next Losses i.e. Idling and Minor Stop by 1.28%, is relatively small losses but also affect the low value of OEE, as for the results of field observations during these losses occurred due to poor surveillance systems (less stringent). e) And last Losses smallest i.e. Defect Losses of 0.07%. These losses have been worth very little so that it does not cause a defect losses low value of OEE. 3) The recommendation that can be achieved and the value of OEE losses above can be reduced, it can be concluded as follows: a) Perform a good surveillance system and the controllers are always held there every shift field. b) Perform a good evaluation of the time and the right way to do change over. c) Fix preventive maintenance plan.

5. References

[1] Corder. A, "*Teknik Manajemen Pemeliharaan," Erlangga*,Jakarta,1992
[2] Dal. B, "Overall Equipment Effectiveness as Measure of Operational Improvement, a Practical Analisis: *Intenational Journal of Operation and Production Management*, Vol.20 No. 12, 2000
[3] Hadi. M. D. K., "*Analisa dan Pengendalian TPM untuk Meningkatkan Produktivitas dengan Menggunakan Metode OEE di PT Dynaplast V,"* Undergraduate Thesis, Jakarta, 2012

Implementation of Inventory Management System (IMS) case study on XYZ online store business unit

R Bagus Yosan[1], Muhammad Kholil[2], Bhetriza Hanum[3]
[1,2,3] Faculty Industrial Engineering, Universitas Mercu Buana

bagus.yosan@mercubuana.ac.id

Abstract XYZ baby and child shop is a new player in the business of selling toys and baby gears. Consumers target for XYZ toy shop is segmentation of middle and upper financial ability buyers. XYZ toy store sells imported toys and baby products with high quality products, using international grade production materials and ISO production certificates: 9002. With this conceptual, establishment XYZ business line is as a provider (seller) toys and baby equipment premium brand; then the scope of the product sold must meet all expectations of market demands. Meet the expectations of market demand is not easy; because by fulfilling all expectations, then the impact is the high stock of goods that must be available. The high stock of goods will directly impact on the needs of a large storage area of goods and reduce money.
Based on the previous description, the formulation of the problem in this research is: "What inventory management strategy can be applied to improve the productivity of Lean Manufacturing business unit of XYZ online store?"

1. Introduction

XYZ baby and child shop is a new player in the business of selling toys and baby gear. Product toy shop XYZ is a product-product with segmentation of consumer buyers with the ability of middle and upper finance. XYZ toy store sells imported toys and baby products with high quality products, using international grade production materials and ISO production certificates: 9002.

The low level of brand image values encourages management to take action. The real action is to focus on the market penetration of the existence of XYZ toy store. The XYZ toy store has to be brave to grab the potential market of child toy buyers and baby supplies that are already familiar with the brand image of a competitor's business. To emphasize the existence of XYZ store then it can start with a focus on multiply the product variants sold. The number of product variants that are sold will certainly foster interest to consumers and prospective customers to buy from the XYZ store.

With this conceptual establishment XYZ business line is as a provider (seller) toys and baby equipment premium brand; then the scope of the product sold must meet all expectations of market demand. Meet the expectations of market demand is not easy; because by fulfilling all expectations, then the impact is the high stock of goods that must be available. The high stock of goods that must be available of course directly impact on the needs of a large storage area of goods.

IOP Conf. Series: Materials Science and Engineering **343** (2018) 012022 doi:10.1088/1757-899X/343/1/012022

Figure 1. Warehouse Condition

2. Theoretical Background

According to Barry Render's Operations Management book Jay Heizer. Inventory is one of the most expensive assets in many companies, reflecting as much as 40% of the total capital invested. Operations managers around the world have long realized that good inventory management is very important. On the one hand, a company can reduce costs by lowering inventory levels at hand. On the other hand, consumers will feel dissatisfied when a product out of stock. Therefore, companies must achieve a balance between inventory investment and customer service levels.

All organizations have several types of inventory planning and control systems. A bank has a method of controlling its cash reserves. Hospitals have methods used to control blood supplies and drugs. Actually every organization feels an interest in inventory planning and control.

Based on the previous description, the formulation of the problem / state of the art in this research is: "What inventory management strategy can be applied to improve productivity of Lean Manufacturing business unit of XYZ online store?"

3. Research Method

Research will be conducted research depicted in the flow chart depicted in Figure 2 as follows:

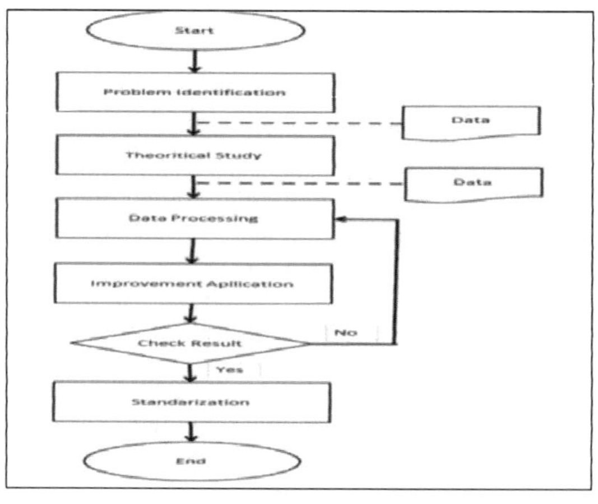

Figure 2. Research Method

4. Result and Discussion

From table 1 can be analyzed that the inventory value always experiences an increasing trend from month to month. The significant increase is in June - July where quantity increase is in 90 units of goods. The quantity of this quantity must of course be a moving asset (liquid asset) which is an embedded capital that can be done immediately when sales are made. Increased inventory with low / long selling capabilities is highly undesirable.

Table 1. Warehouse Stock for each brand Jan – Aug 2017

	Month							
Brand	Jan	Feb	Mar	Apr	May	Jun	Jul	Aug
Babyelle	24	35	33	33	28	47	36	18
Barbie	4	4	4	4	3	2	1	1
Ching ching	26	31	37	32	34	39	45	43
Crayola	1	0	6	6	4	8	6	4
Grow n Up	1	34	47	42	88	63	38	7
Haenim	15	3	6	0	0	0	0	20
JOIE	15	24	19	20	17	38	51	44
Labeille	69	92	137	115	112	67	117	105
Learning R	10	5	15	14	11	11	8	8
Lerado	0	4	3	6	3	1	2	1
Little tikes	121	159	132	163	186	196	160	130
Melissa&Dough	0	0	0	12	9	8	76	65
Pliko	48	66	57	43	36	54	62	58
The First Year	6	6	6	6	6	6	3	3
Tommie Tippee	92	115	105	103	102	128	162	154
Vtech	98	97	91	95	89	131	122	94
Grand Total	530	675	698	694	728	799	889	755

From table 2 we can see that the demand for goods from January to August. In August Little takes brand position on ranks 1 or 2 from total sales. This data is required for EOQ calculations.

Table 2. Goods sold Jan – Aug 2017

	Month							
Brand	Jan	Feb	Mar	Apr	May	Jun	Jul	Aug
Babyelle	7	2	0	5	1	15	18	7
Ching ching	13	18	11	22	27	28	60	26
Crayola	1	0	0	2	2	2	2	1
Grow n Up	30	44	35	38	49	34	53	40
Haenim	12	3	6	0	0	0	34	8
JOIE	3	5	2	3	6	23	33	11
Labeille	42	62	66	71	73	50	99	67
Learning R	5	2	1	3	0	3	0	2
Lerado	27	1	0	9	2	0	9	7
Little tikes	10	27	32	58	65	76	93	52
Pliko	13	9	14	7	6	9	3	9
Tommie Tippee	13	10	2	4	10	20	8	10
Vtech	0	6	17	15	24	9	28	14
Barbie	0	0	0	1	1	1	0	1
Melissa&Dough	0	0	0	3	1	4	11	3
The First Year	0	0	0	0	0	3	0	1
Grand Total	176	189	186	241	267	277	451	259

4.1 Calculation of Booking Fees

A = Calculation of Booking Cost Per Order Year 2017 Brand Little Tikes

1. Phone Fee

Reservations are assumed for 10 minutes.
Phone charges = number of minutes per message x phone rates per minute
= 10 x Rp 600 = Rp 6.000 / order

2. Internet Charges
In addition to via phone booking via email is also done, ordering via email is assumed as much as 5000 KB.
Internet cost = number of KB once message x internet rate per KB
= 5000 x Rp 0.1 = Rp 500 / order

3. Loading and unloading costs
The loading and unloading costs of little tikes brands
Cost of loading and unloading per hour per employee = Rp 15.000 / Jam / Employee.
The loading and unloading duration is 1 hour per shipment and 3 staff of warehouse are done.
= Rp 15.000 x 1 x 3 = Rp 45.000 / booking

4. Shipping costs
Shipping fee is not available due to order until warehouse

5. The cost of correspondence
Reservations are assumed to require 2 sheets of paper
The cost of correspondence = the amount of paper once the message x the A4 sheet of paper tariff
= 2 x Rp 625 = Rp 1.250

6. Administration Fee
Administrative costs = office stationery costs (ATK)
Total number of ordering frequencies in one year = 100 orders from all brands
ATK cost = Rp 9.600.000
Warehouse staff administration fee = Rp 9,600,000 per year: 100 = Rp 96,000

7. Warehouse Manager Salary Cost
Warehouse Manager Salary Costs who place a product order are calculated one time salary x 10% x13 times salary / total order frequency
= Rp 25.000.000 x 10% x 13
= Rp32.500.000 / 100 = Rp 325,000

Table 3. Total Ordering Cost

Ordering Cost	1 time ordering cost (Rp)
1. Phone Cost	Rp 6.000
2. Internet Cost	Rp 500
3. Load-Unload Cost	Rp 45.000
4. Mailing Cost	Rp 1.250
5. Administration Cost	Rp 96.000
6. Warehouse Manager Fee	Rp 325.000
Total Ordering Cost	**Rp 473.750**

4.2 Calculation of storage costs
h = Storage Cost / Unit / Time Unit

1. Payroll Supervisor
Salary Supervisor and Warehouse Implementer = {(1 warehouse staff x salary / month) + warehouse head salary} x 13 times salary in one year

Payroll and Warehouse Implementer = {(1 x Rp 2.300.000) + 4,000,000} x 13
= Rp 81.900.000 per year

2. Electricity Fee
A. Lights
= Number of x wattage bulbs used (converted in kw) x number of flame clocks used daily) x electricity tariff
A. = {2 units x 20 watts (0.02 Kwh x loss RRD = 115%} x 9 Working hours x Rp 1352
= Rp 560 / day
B. = {6 units x 80 watts (0.08 Kwh) RRD loss = 115%} x 4 Working hours x Rp. 1352
= Rp 2.985 / day
The total cost of electricity for the lamp is
= Rp. 560 + Rp 2.985 = Rp 3.545 x 312 days = Rp 1.106.040 / year
B. AC
= {1 unit x 880 watts (0.88 Kwh) x loss RRD = 115%} x 9 hours work x Rp 1352
= Rp 12,314 x 312 days = Rp 3,841,968
C. Computers and Notebooks
Computer
= {1 computer unit x 100 watts (0.1 Kwh) x loss RRD = 115%} x 9 hours work x Rp 1352 = Rp 1.399 x 312 days = Rp 436. 488 / year
Notebook
= {1 unit Notebook x 50 watts (0.05Kw) x loss RRD = 115%} x 9 hours work x Rp 1352 = 700 x 312 = Rp 268,200 / year
Total cost of electricity for computers and notebooks in one year = Rp 436. 488 + Rp 268,200 = Rp 704,688 / year
Total cost of electricity is cost for lamp + charge for ac + charge for computer and notebook Rp 1,106,040 / year + Rp 3,841,968 / year + Rp 704,688 / year = Rp 5,652,696

3. Warehouse Depreciation Fee
By the straight-line method then calculate the depreciation cost of the warehouse is
= Acquisition Price: Economic Life (calculated per month, as depreciation expense is calculated per month)
= 100,000,000: (3 × 12) (number 3 = 3 years, 1 year there are 12 months So 3 x 12 = 36 months)
Acquisition cost: (3x12)
Rp 100,000,000: 36 months = Rp 2.777.777 (Depreciation value per month)
Calculate the accumulated depreciation from January 2017 to August 2017
= Depreciation per month x (number of months from January 2017 to August 2017)
= Rp 2.777.777 x 8 months = Rp 22,222,222

4. Shipping Costs Products
Assumption of car operating expenses per one day x number of cars x 26 days x 12 months
Rp 50.000 x 1 x 26 x 12 = Rp 15,600,000 / Year
Total Storage Cost
= Payroll salary + Electricity costs + Warehouse depreciation costs + Product shipping costs
= Rp 81.900.000 + Rp 5.652.696 + Rp 22.222.222 + Rp 15.600.000 = Rp 125,374,918
h = Storage cost per unit per year
H = Total storage cost per one year / Total inventory per one year

4.3 Determine the value of EOQ
EOQ method formula
EOQ = √ ((2.A.D) / h)
INFORMATION
A = COST OF BOOKING / ONE TIME MESSAGE

D = TOTAL DEMAND / ONE YEAR
h = STORAGE COST / UNIT / UNIT TIME
h = Storage cost per unit per year this cost is obtained from the formula as follows:
h = (Total Storage Cost) / (Total Inventory)
h = (Rp 125.374.918) / 155
h = Rp 808,870

EOQ = √ ((2.A.D) / h)
EOQ = √ ((2. (Rp 473,750) (413)) / 808,870)
EOQ = √ (391,317,500 / 808,870)
EOQ = √483
EOQ = 21.99 ≈ 22 units
So the amount of little tikes brand stock that can be ordered based on economical order quantity method or Economic Order Quantity (EOQ) is 22 units per order.

4.4. Annual booking fee
These are costs incurred in connection with the ordering activities of raw materials / products.
Annual booking fee = (Request year x message cost per message) / (Number of times each message)
Booking cost per year = D / EOQ x A
Annual booking fee = 413/22 x 473,750
Annual booking fee = Rp 8.893.579

4.5. Cost of storage per year
Is the cost incurred in connection with the storage of raw materials purchased.
Storage cost per year = average inventory level x storage cost per unit per year
Storage cost per year = EOQ / 2 x h
Storage cost per year = 22/2 x 808,870
Storage cost per year = Rp 8,897,570

D. Total raw material inventory cost (Total Cost)
The optimal total raw material inventory is the summation of the total cost of the message and the total cost of storing raw materials.
C = goods price = Rp 1,000,000
TC = Total message cost + Total save cost + One year fee
TC = {(D / EOQ) x A + (EOQ / 2) x h + (CD)}
TC = Rp 8.893.579 + Rp 8.897.570 + (Rp 1.000.000 * 413)
TC = Rp 430.791.149

4.6 Safety Stock
Safety Stock = Zq
Where: Z = Standard deviation
q = squared error; where q is derived from the formula
√ ((€ (x-y) ²) / n)
Where:
x = Inventory
y = Usage
n = Number of months
Z = 5% ≈1.65
€ (x-y) ² = 90.384
n = 8 months
√ ((€ (x-y) ²) / n)
√ (90384/8)

√11.298
106 units per month
So the amount of safety stock needed to avoid running out of little tikes merchandise is 106 units per month.

4.7. Determining Order Frequency
Frequency = D / EOQ
Information
D = Total product demand from one-year consumer
Q = Quantity of one year order
Frequency = D / EOQ
Frequency = 413/22
Frequency = 18.7 ≈ 19 times / one year

4.8. Specifies the ROP value
Reorder or Reorder Point When reorder or reorder point (ROP) is the time when the company must place the order back material, so acceptance of materials ordered can be timely.
ROP formula
1. Without safety stock policy
Long production turnaround = 1 year effective corporate work / order frequency
= 312 days / year: 19
= 16.42 ≈ 16 days / every order
ROP = EOQ / (Old production turnaround) x lead time
ROP = 22/16 x 3
ROP = 4.125 ≈ 4 units
So re-ordering can be done at the time of the amount
supplies remaining 4 units
2. With safety stock policy
ROP = (EOQ / (Old production turnaround) x lead time) + Safety stock
ROP = (22/16 x 3) + 106 pcs
ROP = 110 units

4.9 Warehouse Layout Improvement
Layout arrangement is also not clearly indicated allocation placement of goods stock. The initial layout placement is based on the large items allocated in the front area and the small goods at the back of the Figure 3. The result is the placement of goods of different dimension different in one area (not tidy and inefficient).

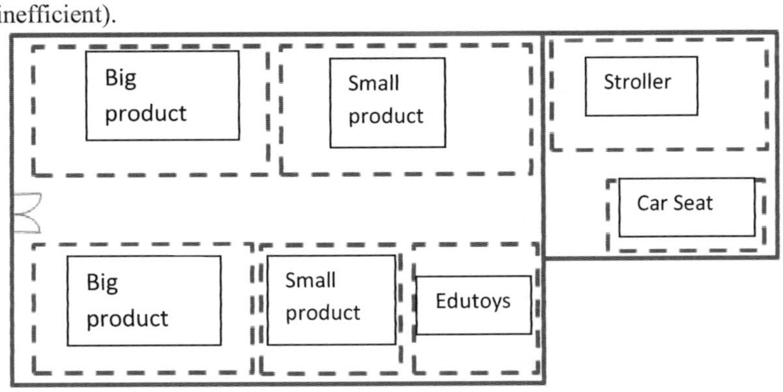

Figure 3. Warehouse Layout before improvement

Preparation of a new layout is clearly indicated allocation placement of goods stock. The layout placement is based on the type of goods. The result is the placement of items of the same or similar dimension in one area (placement is more neat and efficient). Figure 4 placement of goods with a new layout.

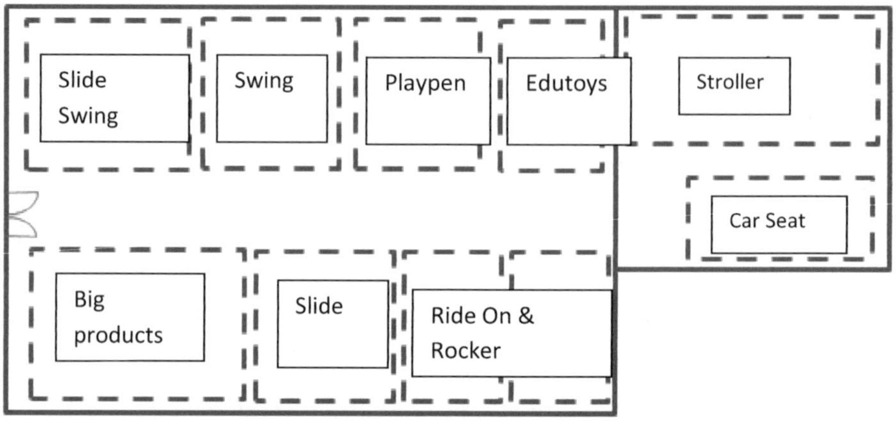

Figure 4. Warehouse Layout after improvement

Figure 5. Warehouse after improvement

5. Conclusion

The conclusions that can be drawn from this final project are:

1. From the calculation and data processing, it is found that the ROP for Little Tikes stock type is when the value of 110 units. And the order quantity is 19 times in one year.
2. Repairs Warehouse layout needs to be done and standardized and the allocation of placement in the warehouse in accordance with the type of stock items.
3. EOQ with a value of 22 units / order has answered the needs of the warehouse order, so anticipation of the warehouse can be better.

6. References

[1] R Bagus Yosan 2017 Analisis Kualitas Produk Dan Korelasinya Terhadap Tingkat Kepuasan Konsumen Dengan metoda Statistical Process Control. Jurnal Sinergi Vol 21, No 3 (2017) hal 231-237 http://dx.doi.org/10.22441/sinergi.2017.3.010

[2] W Adrianto, M Kholil, 2016, Analisis Penerapan Lean Production Process Untuk Mengurangi Lead Time Process Perawatan Engine (Studi Kasus PT. GMF Aeroasia), Jurnal Optimasi Sistem Industri Jilid 14,hal 299-309

[3] R Bagus Yosan 2016 Increasing Productivity with Objective Matrix Method Case Study on Building Maintenance Management PIO PT. XYZ, Jakarta
[4] Zulfa Fitri Ikatrinasari, 2016, Evaluasi Kapabilitas Sistem Informasi Manajemen Akuntansi Barang Milik Negara di Satuan Kerja TNI Angkatan Laut, Performa (2016) Vol. 15, No.1: 59-69
[5] R Bagus Yosan 2012 Usulan Peningkatan Poduktivitas Dengan Perbaikan Layout - ARD Analysis Pada Dies Manufacturing Division (DMD) PT. Pratama, Jakarta
[6] Arikunto, S. 2010. Prosedur penelitian : Suatu Pendekatan Praktik. (Edisi. Revisi). Jakarta
[7] David, Fred.R. Strategic 2009 Management Concepts and Cases 12th Edition. Pearson International Edition. Florence South Carolina.
[8] Hunger, J. David, Wheelen, Thomas L. 2006, Manajemen Strategis. Alih bahasa Julianto Agung. : Andi Press, Yogyakarta.
[9] Muslim, Erlinda. 2009 Diktat Manajemen Strategi Industri. Departemen Teknik Industri UI. Depok.
[10] Rutoto, Sabar. 2007. Pengantar Metedologi Penelitian. FKIP: Universitas Muria. Kudus
[11] Umar Husein (2005). Riset Pemasaran & Perilaku Konsumen, Jakarta: PT. Gramedia.
[12] Gaspersz, Vincent, 1992, Analisa Sistem Terapan Berdasarkan Pendekatan Teknik Industri Gramedia Pustaka Utama.
[13] Gaspersz, Vincent, 1998, Manajemen Produktivitas Total, Strategi Peningkatan Produktivitas Bisnis Global, Jakarta, Gramedia Pustaka Utama.

IOT for Agriculture: Food Quality and Safety

Gunawan Witjaksono[1*], Almur Abdelkreem Saeed Rabih[1], Noorhana bt Yahya[2], Sagir Alva[3]

[1]Department of Electrical and Electronic Engineering, Universiti Teknologi PETRONAS
Seri Iskandar, 32610, Perak Darul Ridzuan, Malaysia

[2]Department of Fundamental Applied Science, Universiti Teknologi PETRONAS
Seri Iskandar, 32610, Perak Darul Ridzuan, Malaysia

[3]Mechanical Engineering Department, Mercu Buana University
Jl. Meruya Selatan no.1 Jakarta Barat, Indonesia

gunawan.witjaksono@utp.edu.my

Abstract—Food is the main energy source for the living beings; as such food quality and safety have been in the highest demand throughout the human history. Internet of things (IOT) is a technology with a vision to connect anything at anytime and anywhere. Utilizing IOT in the food supply chain (FSC) is believed to enhance the quality of life by tracing and tracking the food conditions and live-sharing the obtained data with the consumers or the FSC supervisors. Currently, full application of IOT in the FSC is still in the developing stage and there is a big gap for improvements. The purpose of this paper is to explore the possibility of applying IOT for agriculture to trace and track food quality and safety. Mobile application for food freshness investigation was successfully developed and the results showed that consumer mobile camera could be used to test the freshness of food. By applying the IOT technology this information could be shared with all the consumers and also the supervisors.

1. Introduction

For the past few years, food quality and safety has been a major concern and a hot topic in all around the globe, and especially in china for many reasons [1]. For instance, the incident of 2008 *Sanlu melamine milk powder* scandal has shocked the world due to its effect on thousands of babies, which resulted in death of several of them. Another incident that stunned the Chinese community and the world happened in 2011, when the meat from the Shuanghui Group (the largest meat supplier in China) was discovered to contain a substance (*Clenbuterol hydrochloride*) that is prohibited to add to food stuffs in China [2]. Food quality and safety gains attentions of both government agencies and consumers due to the major loss to the state revenue, and also the consumers health risks that caused by breaching the food standards. Therefore, there is an imperative need to develop technologies that can guarantee the safety of food along the whole food supply chains (FSCs) that include production, processing, packing, transportation, storage, distribution and sale for the consumers. Many researches have been conducted around the world with the development of some technologies like internet of things (IOT) to help the factories to resolve issues related to food quality and safety. IOT is a vision to

IOP Conf. Series: Materials Science and Engineering **343** (2018) 012023 doi:10.1088/1757-899X/343/1/012023

connect to anything at anytime and anywhere, and it is expected to drastically change the status of our life in years to come. It is defined by European commission information society as *"Things having identities and virtual personalities operating in smart spaces using intelligent interfaces to connect and communicate within social, environmental, and user contexts"* [3]. IOT technologies are believed to be able to provide potential solutions to resolve issues of traceability, visibility and controllability along the FSCs line [4]. IOT technologies have already been for many years in the pharmaceutical supply chain to enhance the quality and to keep the drugs safe [1]. IOT will play a role in resolving the issues of food quality and safety in term of managing factory staffs, food products and also in term of providing useful information to the customers to make them feel easier and comfortable to track or consume the food with great confidence [5].

Food packaging plays an important role to preserve the quality and safety along the line of the supply chain. Traditional packaging is only meant to protect the food from the environmental changes such as, temperature, humidity, light, gaseous emissions or microbial attacks. On the other hand, active packaging systems are those which have interactive communication between the packed food and the packaging environment to provide protection to extend the shelf life of the food. Active packaging protects the packed food by containing active components that release or absorb substances into/from the food to keep the conditions under control. In contrary intelligent packaging does not contain active components. However, it has intelligent systems that communicate with the supervisors or consumers to provide the current status of the food [6]. Smart packaging devices can be divided into two types. The first type is called data carriers such as barcode labels and radio frequency identification (RFID) tags, while the second type is called package indicators, which are used to monitor environmental changes such as time-temperature indicators and gas indicators. Fadable ink for time-temperature control of food freshness has been widely used in printing labels on food packages for the freshness indication. The working principle is that, as time passes, the ink will disappear and indicating that the food is no longer fresh [7]. The label ink will react to the oxygen which will make the colour to fade with time or temperature change. The rate of colour change is varied by the chemical composition used to make the ink. **Figure. 1** shows how the label ink is faded with time. Gas indicators are labels that detect the presence of a gas due to the food condition change inside the package [6]. Polyaniline (PANI) film was used in [8] as a sensing layer to check the spoilage of fish based on reaction of the film with ammonia released as the fish is spoiled. PANI films will react to the ammonia released due to the fish spoilage as a result of bacterial growth. The reaction will lead to the change of the films colour. Another smart packaging indicator was reported in [9], which is a novel sticker sensor based on methyl red immobilized on bacterial cellulose membrane for broiler chicken freshness assessment. The colour of the sticker changes from red to yellow to indicate the chicken spoilage as shown in **Figure. 2**. As the chicken spoilage progresses, the pH level increases due to increase of spoilage volatile amines, which will occupy the headspace of the food package.

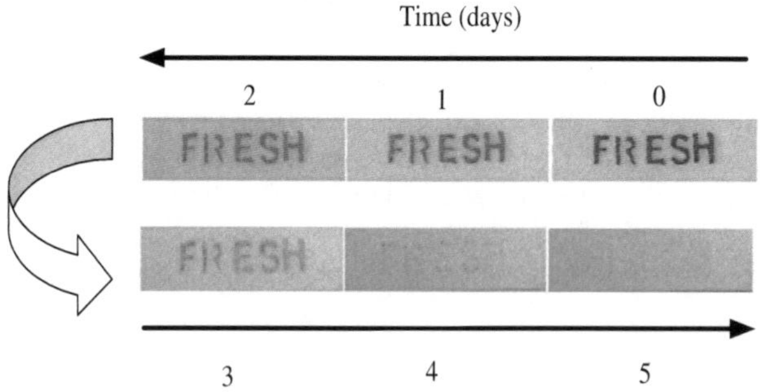

Figure. 1. Fadable ink for time – temperature control of food freshness [7]

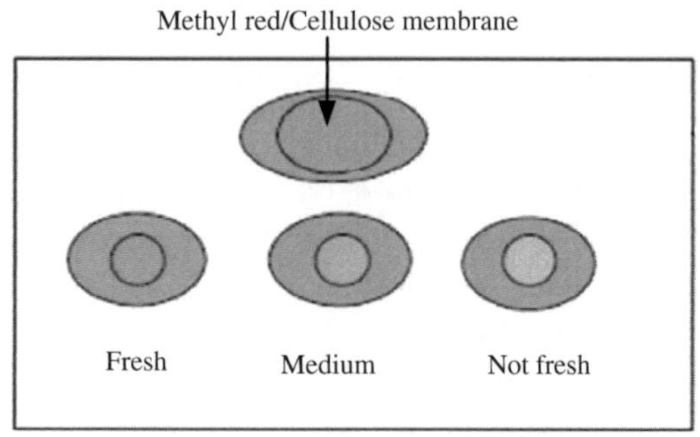

Figure. 2. Sticker sensor design based on methyl red/cellulose membrane [9]

The reaction of the volatile amine with the methyl red/cellulose membrane will cause the color of the membrane to change to yellow. The aim of this paper is to explore the possibility of using the internet of thing technology to provide a solution for the food quality and safety control along the whole food supply chain.

2. Architecture of IOT technology

IOT technologies composes of different distributed resources that can be selected depending on the need of the different users and organizations around the globe [1]. By considering the vertical view of the IOT technology which focuses on the technical implementation, the resources of IOT can be categorized under three different layers, which are considered to be the main architecture layers of the IOT [10]. These layers are the sensing layer, the network layer and application layer as shown in **Figure 3**.

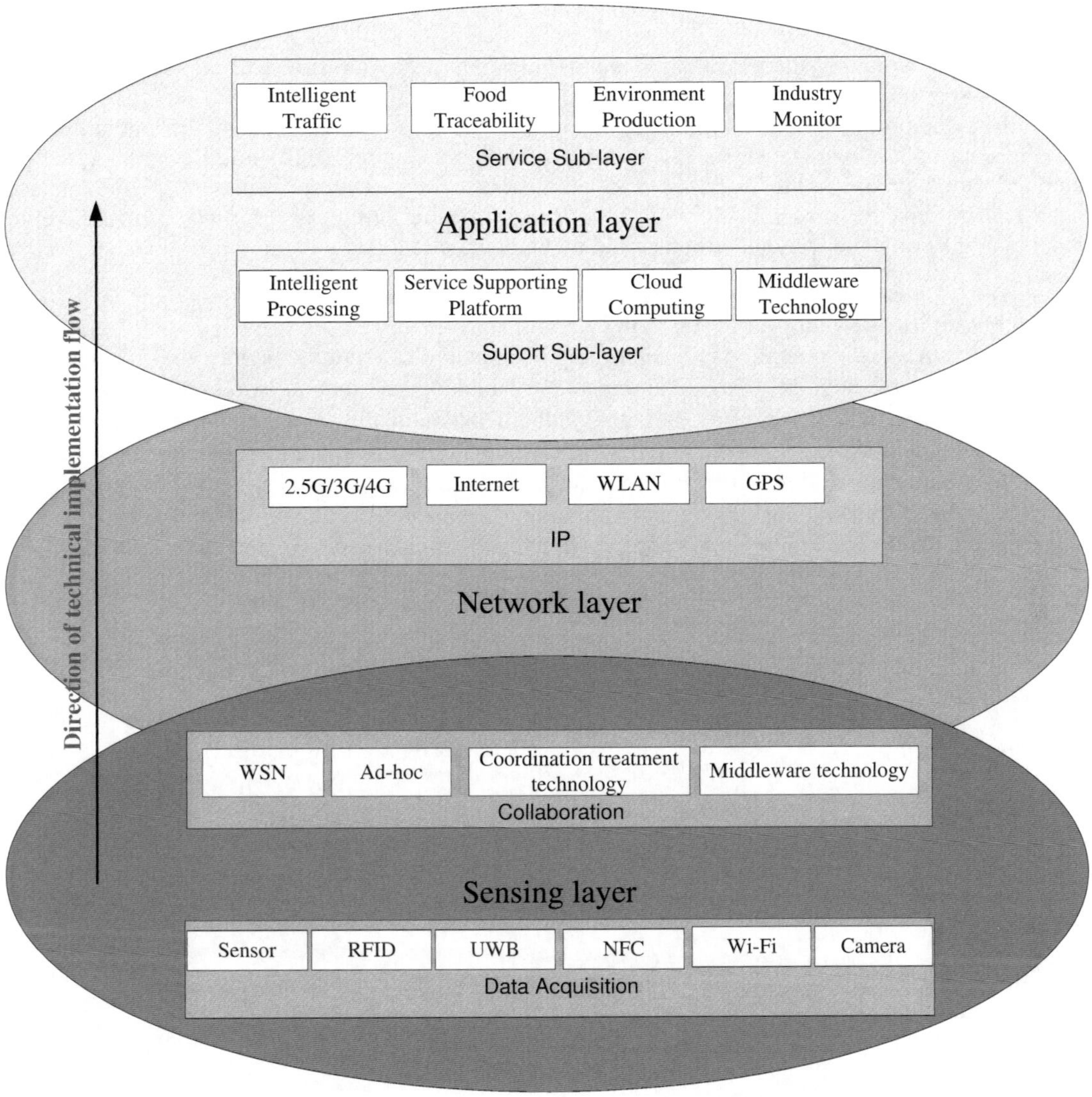

Figure. 3. Layers of the IOT architecture from the perspective of technical implementation [10]

2.1 Sensing layer

Sensing layer can be subdivided into two layers based on the functions of the units. They are data acquisition and collaboration layers. In the data acquisition layer the things from the physical world like temperature, humidity, pressure, concentration, multimedia data..etc., are sensed using devices like sensors, Radio Frequency Identification (RFID) tags, Ultra-wide Band (UWB), Near field communication (NFC), WiFi and cameras. On the other hand, collaboration layer focusses on the technologies applied in short distance data transmission, context awareness and massive information processing. For these applications technologies like Wireless sensor network (WSN), Ad-hoc network, coordination treatment technology and middleware technology.

2.2 Network layer

Network systems have been used for many years already, and it is a heterogeneous system that comprises backbone, mobile communication, wireless local area network (WLAN), satellite communication networks, Global positioning system (GPS), Bluetooth etc. The data acquired by the sensing layer is transmitted to the application layer through this network layer.

2.3 Application layer

Application layer is sub-divided into two layers. They are support sub-layer and service sub-layer. Support sub-layer process the data using intelligent processing unit, and then will deliver to the user based on the issued request. Due to the large volume of the data involved the system integrated high processing data technologies such as cloud computing, peer to peer (P2P) middleware technology. Service sub-layer provides the interface and platforms to support the individual public services. Typical applications of this sub-layer will be security monitoring, disaster monitoring, vehicle scheduling monitoring and intelligent household appliances monitoring.

3. IOT technologies application in food supply chain

As shown in **Figure 4**, the overall view of applying IOT technology in food supply chain comprises of heterogeneous devices, with ubiquitous data exchange, tracking and localization systems [4]. There is well defined connectivity throughout the supply chain between the food producers, transportation bodies and also the food consumers. For instance when the temperature exceeds the desired value inside the food truck, the system will trigger an alarm and send a signal to the refrigeration controls to increase the ventilation to re-adjust the temperature and keep it under control. Same process is applied for other parameters like, humidity, gas concentrations, and ambient pressure and so on.

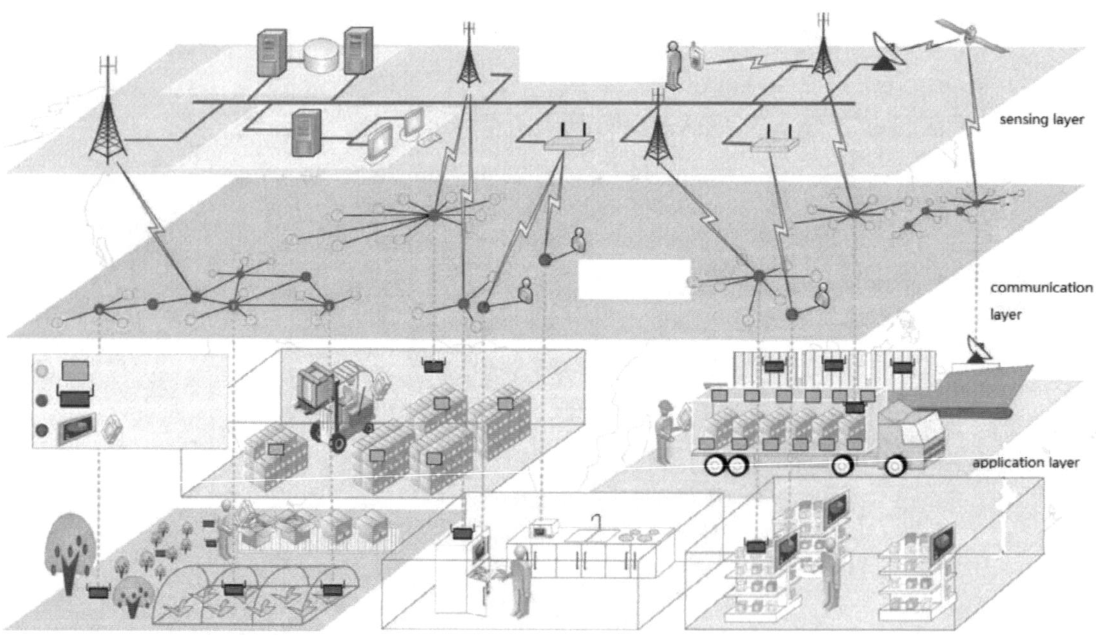

Figure. 4. Overall picture view of applying IOT in food supply chain [4]

4. Some applications of IOT sensing layer

As was mentioned before, this layer is the first layer that is used to collect the data required for the IOT technology to function. Following are two examples of applications used for food freshness level detection and food authonticity identification.

4.1 Mobile camera app development for food freshness detection

As shown in **Figure 3** one of the devices used for data acquisition is camera. For this purpose to take advantage of the mobile camera of the food consumers, a mobile app has been developed as part of the IOT technology used for the food supply chain. This app is to check the freshness of the prawn and displays the results on the screen of consumer's mobile. More information about the mobile app is presented in our previous published article in [11]. In brief, the mobile app developed is based on chromatic parameter to display the freshness level of the food directly by reading the RGB colour from the colour changing labels, and correlating it with the colour of pH paper. Then the app will display the freshness of the food by stating whether the food is "fresh, "good" or "spoiled". **Figure 5** shows the results of the app from the screen of the mobile phone.

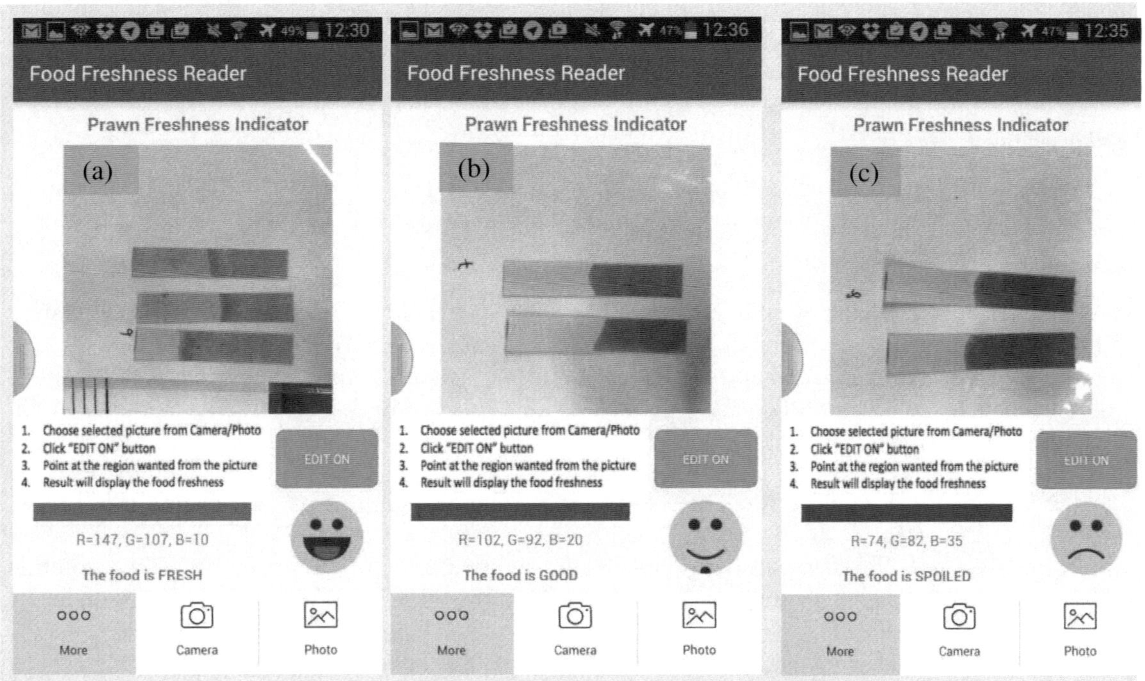

Figure. 5. Testing of Apps on pH papers of (a) pH 6, (b) pH 7 and (c) pH 8

4.2 Food type authenticity

Another application example of developing sensors for IOT sensing layer is using Fourier transform infrared spectroscopy (FTIR) to differentiate non-halal meat from halal meat which is necessary for the Muslims consumption. The data collected by FTIR sensor could be communicated directly through the communication layer to the application layer, where consumers and also supervisors can directly access the type of the meat source. FTIR was used to differentiate pig fat (lard) from the fats of cow, chicken and lamb in a broad range of mid infrared spectroscopy ranges from 400 – 4000 cm^{-1}. The full description of the results analysis are reported earlier in [12]. Results at wavenumber 2361.05 cm^{-1} is shown in **Figure 6**.

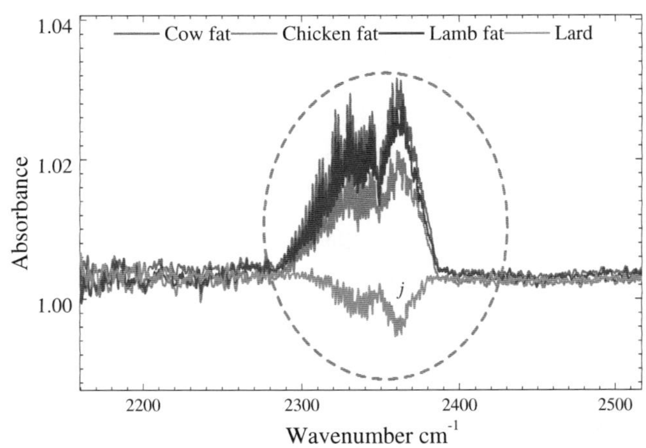

Figure. 6. FTIR peaks of lard, cow, chicken and lamb fats at 2361.05 cm^{-1}

5. Conclusion

This paper presented an idea of utilizing internet of things (IOT) for agriculture, food quality and safety. A mobile application (app) for the sensing layer of the IOT technology was developed and the app shows that the freshness of food could be investigated by studying the picture of the food and comparing it to the reference picture. The condition of the food whether it is fresh, good or spoiled could be shared with all consumers and food supervisors through the network and application layers of the IOT technology.

Acknowledgment

The authors of this paper acknowledge the financial support by UTP under STIRF-UTP, project code: 0153AA-F71 and by Petronas Research Fund (PRF) under by 0153AB-A32.

References

[1] Y. Gu, W. Han, L. Zheng, and B. Jin, "Using IoT Technologies to Resolve the Food Safety Problem – An Analysis Based on Chinese Food Standards," in *Web Information Systems and Mining: International Conference, WISM 2012, Chengdu, China, October 26-28, 2012. Proceedings*, F. L. Wang, J. Lei, Z. Gong, and X. Luo, Eds., ed Berlin, Heidelberg: Springer Berlin Heidelberg, 2012, pp. 380-392.

[2] Y. Liu, W. Han, Y. Zhang, L. Li, J. Wang, and L. Zheng, "An Internet-of-Things solution for food safety and quality control: A pilot project in China," *Journal of Industrial Information Integration,* vol. 3, pp. 1-7, 2016.

[3] Z. Pang, Q. Chen, W. Han, and L. Zheng, "Value-centric design of the internet-of-things solution for food supply chain: Value creation, sensor portfolio and information fusion," *Information Systems Frontiers,* vol. 17, pp. 289-319, 2015.

[4] X. Zhao, H. Fan, H. Zhu, and H. Fu, "The Design of the Internet of Things Solution for Food Supply Chain," presented at the 2015 International Conference on Education, Management, Information and Medicine, 2015.

[5] B. Song and Q. Xing, "On security detecting architechture of food industry based on Internet of Things," in *2011 IEEE International Conference on Automation and Logistics (ICAL)*, 2011, pp. 81-85.

[6] K. B. Biji, C. N. Ravishankar, C. O. Mohan, and T. K. S. Gopal, "Smart packaging systems for food applications: a review," *J Food Sci Technol,* vol. 52, pp. 6125–6135, 2015.

[7] Y. Galagan and W. F. Su, "Fadable ink for time–temperature control of food freshness: Novel new time–temperature indicator," *Food Research International,* vol. 41, pp. 653-657, 7// 2008.

[8] B. Kuswandi, Jayus, A. Restyana, A. Abdullah, L. Y. Heng, and M. Ahmad, "A novel colorimetric food package label for fish spoilage based on polyaniline film," *Food Control,* vol. 25, pp. 184-189, 5// 2012.

[9] B. Kuswandi, R. O. Jayus, A. Abdullah, and L. Y. Heng, "A Novel On-Package Sticker Sensor Based on Methyl Red for Real-Time Monitoring of Broiler Chicken Cut Freshness," *Packag. Technol. Sci,* vol. 27, pp. 69–81, 2014.

[10]L. Zheng, H. Zhang, W. Han, X. Zhou, J. He, Z. Zhang*, et al.*, "Technologies, applications, and governance in the internet of things," *Internet of things-Global technological and societal trends. From smart environments and spaces to green ICT,* 2011.

[11]G. Witjaksono, N. H. F. B. M. Hussin, A. A. S. Rabih, and S. Alfa, "Real Time Chromametry Measurement for Food Quality Detection using Mobile Device," presented at the Nommensen International Conference on Technology and Engineering Nommensen HKBP University, Medan Indonesia, 2017.

[12]G. Watjaksono, I. Saputra, M. Latief, I. Jaswir, R. Akmeliawati, and A. A. S. Rabih, "Non-Halal biomarkers identification based on Fourier Transform Infrared Spectroscopy (FTIR) and Gas Chromatography-Time of Flight Mass Spectroscopy (GC-TOF MS) techniques," presented at the The International Conference on Applied Photonics and Electronics, Avillion Port Dickson, Malaysia, 2017.

Hydrocoil Turbine Performance at 3 m, 4 m, and 5 m Head Analysis Using Computational Fluid Dynamics Method

A A Luthfie[1], S E Pratiwi[1], and P Hidayatulloh[1]
[1]Universitas Mercu Buana, Jakarta, Indonesia

alief.avicenna@mercubuana.ac.id

Abstract. Indonesia is a country which has abundant renewable energy resources, comprises of water, solar, geothermal, wind, bioenergy, and ocean energy. Utilization of water energy through MHP is widely applied in remote areas in Indonesia. This utilization requires a water-converting device known as a water turbine. Rosefsky (2010) developed a water turbine known as the Hydrocoil turbine. This turbine is an axial turbine which is a modification of screw turbine. This turbine has a pitch length that decreases in the direction of the water flow and is able to work at relatively low water flow and head. The use of Hydrocoil turbine has not been widely applied in Indonesia, therefore this research is focused on analyzing the performance of Hydrocoil turbine. The analysis was performed using Computational Fluid Dynamics (CFD) method. Hydrocoil turbine performance analysis was performed at 3 m, 4 m, and 5 m head respectively as well as rotational speed variations of 100 rpm, 300 rpm, 500 rpm, 700 rpm, 900 rpm, 1,100 rpm, 1,300 rpm, 1,500 rpm, 1,700 rpm, and 1,900 rpm. Based on simulation result, the largest power generated by the turbine at 3 m head is 1,134.06 W, while at 4 m and 5 m are 1,722.39 W and 2,231.49 W respectively. It is also found that the largest turbine's efficiency at 3 m head is 93.22% while at 4 m and 5 m head are 94.6% and 89.88% respectively. The result also shows that the larger the head the greater the operational rotational speed range.

1. Introduction

Indonesia is a country which has abundant renewable energy sources, comprises of water, solar, geothermal, wind, bioenergy, and ocean energy. According to the data from *Kementrian Energi dan Sumber Daya Mineral (ESDM) Republik Indonesia* in 2016, renewable energy utilized in Indonesia is only 1.1% of total renewable energy potential (there is 810 GW of renewable energy potential in Indonesia and only 8.78 GW that were utilized) [1]. The largest utilization in renewable energy is in water sector. There is 5.25 GW or approximately 28% of total water energy potential that were utilized. Utilization of water energy through Micro Hydro Power (MHP) is widely used in remote area in Indonesia. It is because MHP is a simple system so that it can be easily installed. This utilization requires a water-converting device known as water turbine. In 2010 Rosefsky developed a water turbine known as Hydrocoil turbine [2]. This turbine is an axial turbine which is a modification of screw turbine. It has pitch length that decreases in the direction of water flow and is able to work in relatively low water flow and head [2]. This turbine is not widely used in Indonesia, so this paper will focus on the analysis of this turbine performance to get information how good the turbine is. The analysis will be conducted by the use of *Computational Fluid Dynamic* (CFD) method which is simulation of fluid using computer. The software used in this research is ANSYS CFX 15.0. The aim of this research is to get a graph that shows power generated by the turbine and the turbine's efficiency which is applied at 3 m, 4 m, and 5 m head as well as rotational speed variations of 100 rpm, 300 rpm, 500 rpm, 700 rpm, 900 rpm, 1,100 rpm, 1,300 rpm, 1,500 rpm, 1,700 rpm, and 1,900 rpm..

2. Methodology

Figure 1 shows the Micro Hydro Power system and its dimensions in this research. Figure 2 shows the geometry inserted to ANSYS CFX 15.0. The drawing of the geometry is conducted using Inventor 2013. According to figure 2, the inlet boundary of the system is the inlet of penstock, and the outlet boundary is the outlet of hydrocoil turbine.

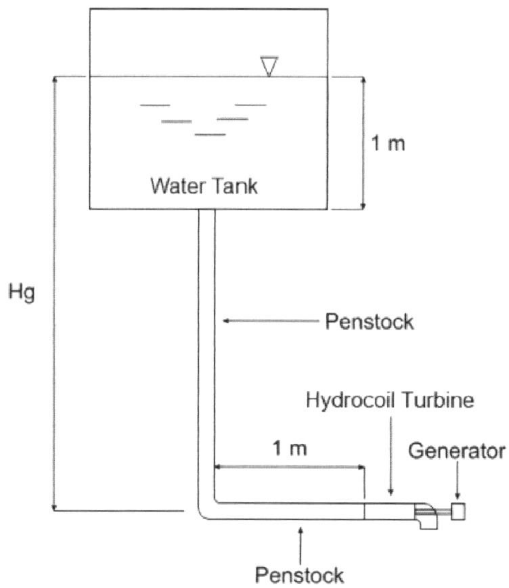

Figure 1. Schematic of Micro Hydro Power

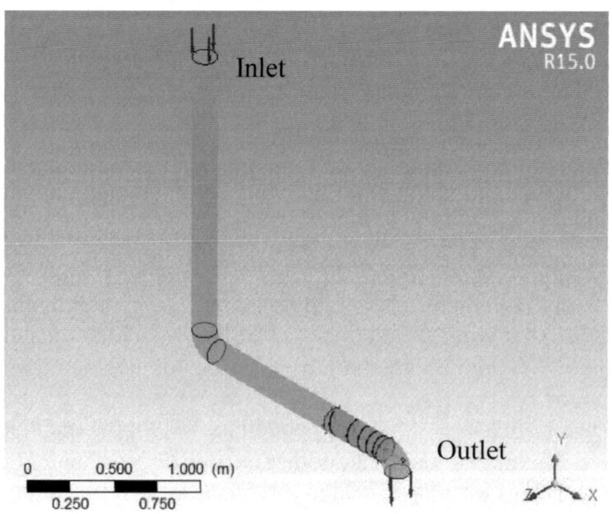

Figure 2. Geometry Inserted to ANSYS CFX 15.0

Figure 3 shows the research flowchart. According to the figure, before CFD can be done, parameters at boundary conditions of the system should be calculated.

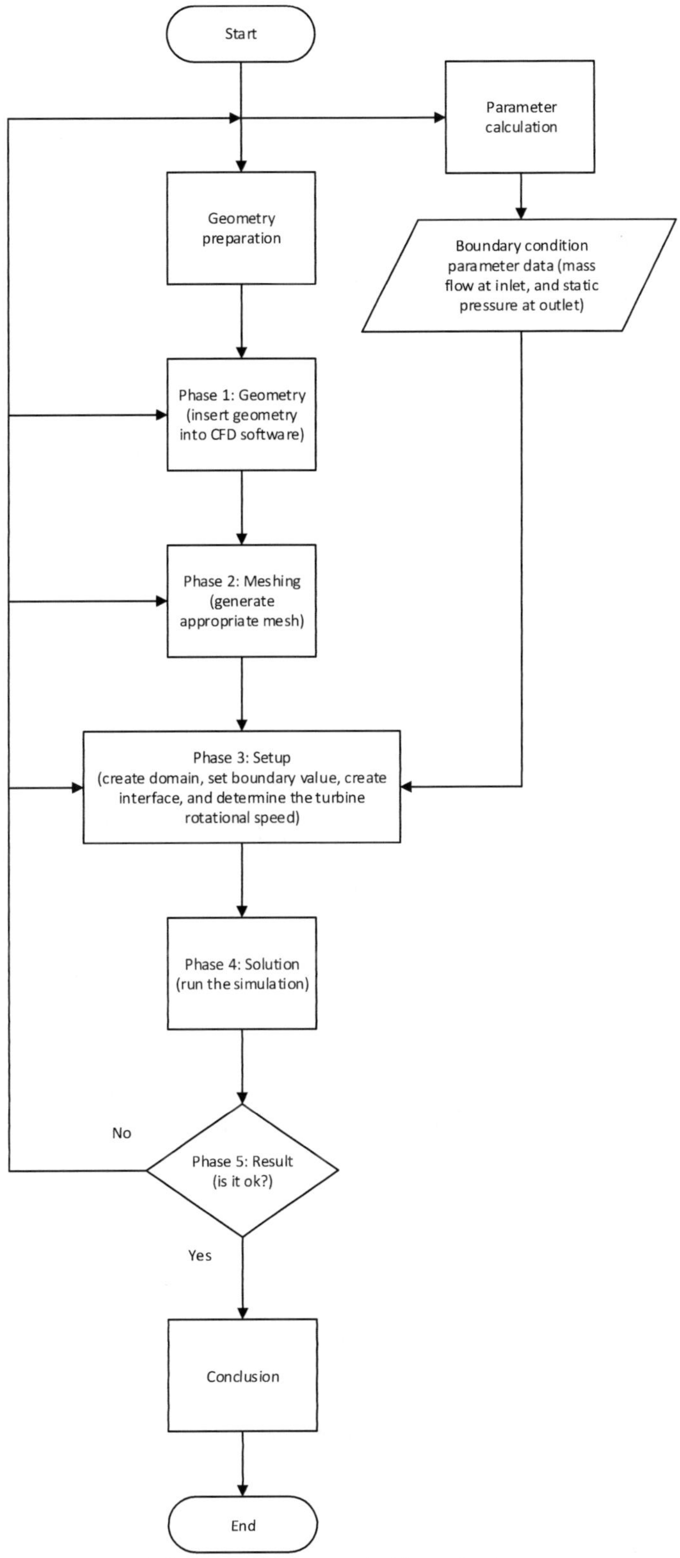

Figure 3. Research's Flowchart

2.1. *Parameter Calculation*

According to Figure 3, parameter at inlet boundary that need to be known is mass flow. Equation 1 shows the equation needed to calculate this parameter.

$$\dot{m} = \rho . Q_{opt} \tag{1}$$

At Equation 1, ρ is water density, whisch is 1.000 kg/m^3; Q_{opt} is water flow through penstock. This parameter can be calculated using Equation 2 [3].

$$Q_{opt} = \frac{2}{3} A_p \sqrt{g \frac{7H_g}{10C_L}} \tag{2}$$

At Equation 2, A_p is cross sectional area of the penstock, H_g is gross head showed in figure 1, and C_L is losses coefficient that can be calculated using Equation 3 [3].

$$C_L = f \frac{L}{D_p} + \sum K_L \tag{3}$$

At Equation 3, L is total length of the penstock, D_p is penstock diameter, K_L is minor losses coefficient due to bend (the value is 0.3 for 90° flanged elbow [4]), and f is friction losses coefficient or Darcy-Weisbach friction coefficient that can be calculated using Swamee-Jain formula showed in Equation 4.

$$f = \frac{0{,}25}{\left[log\left(\frac{\varepsilon}{3{,}7D_p} + \frac{5{,}74}{Re^{0{,}9}}\right)\right]^2} \tag{4}$$

At Equation 4, ε is pipe's roughness (0 if the pipe assumed to be smooth pipe), Re is Reynolds number that can be calculated using Equation 5. Swamee-Jain formula only valid if the Reynolds number is 5.000 to 10^8. At Equation 5, ν is kinematic viscosity of water which is 1x10^6 at 25°C.

$$Re = \frac{\rho v D}{\mu} = \frac{vD}{\nu} \tag{5}$$

Another parameter should be known at boundary is static pressure at outlet. Due to the water will fall freely after passing the turbine, so the static pressure at outlet is 0 Pa relative to atmospheric pressure.

2.2. *Hydrocoil Turbine*

Hydrocoil turbine used in this research is showed in figure 4 [2]. According to figure 4, the pitch length of the turbine is decreases gradually and can be divided into 3 coils. Each coil has different length. Table 1 shows the dimension of the turbine, including the dimension of each coil.

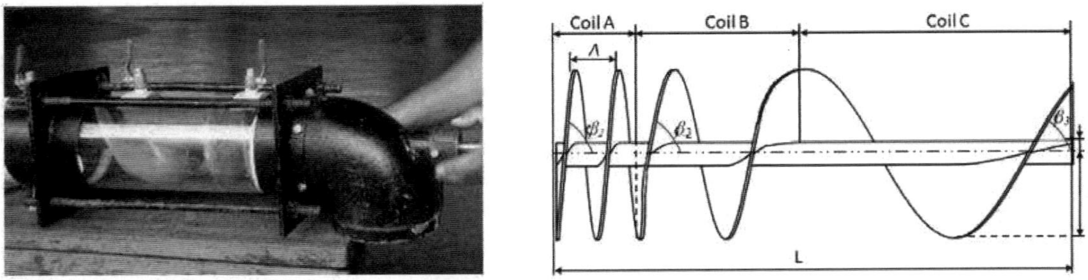

(a) (b)

Figure 4. Hydrocoil Turbine

Table 1. Dimension of Hydrocoil Turbine

No	Specification	Value
1	Length Coil A	0.085 m
2	Length Coil B	0.095 m
3	Length Coil C	0.12 m
4	β_1 in Coil A	83°
5	β_2 in Coil B	79°
6	β_3 in Coil C	71°
7	D_i ribbon drive	0.019 m
8	D_o ribbon drive	0.108 m

The performance of the turbine can be seen from turbine's power generated and turbine's efficiency. Equation 6 shows the formula to calculate optimum power that can be generated by the turbine with γ is weight of fluid per volume [3]. This is the highest power possibility that can be generated by the turbine due to losses in the system.

$$P_{opt} = \frac{76}{135}\gamma H_g A_2 \sqrt{g\frac{7H_g}{10C_L}} \quad (6)$$

3. Results and Discussions

Graph in figure 5 shows torque values of the turbine at different head and different rotational speed. According to the graph in figure 5, torque will decreases as rotational speed increases. This pattern is natural pattern of any turbine [2]. This graph also shows that the highest torque is gained at 5 m head. This is due to water velocity will increases if head applied is increases. The higher the velocity, the higher the torque gained.

Graph in figure 6 shows turbine's power. This graph shows that the power rises untill certain value at certain rotational speed and then decreaeses as the rotational speed keep increases. According to Aprilliyanto's work, this pattern also natural to turbine's performance [2]. Based on this graph, the largest power generated by the turbine at 3 m head is 1,134.06 W, while at 4 m and 5 m are 1,722.39 W and 2,231.49 W respectively. So the highest power is generated at 5 m head. This is due to the higher the torque, the higher the power generated.

IOP Conf. Series: Materials Science and Engineering **343** (2018) 012024 doi:10.1088/1757-899X/343/1/012024

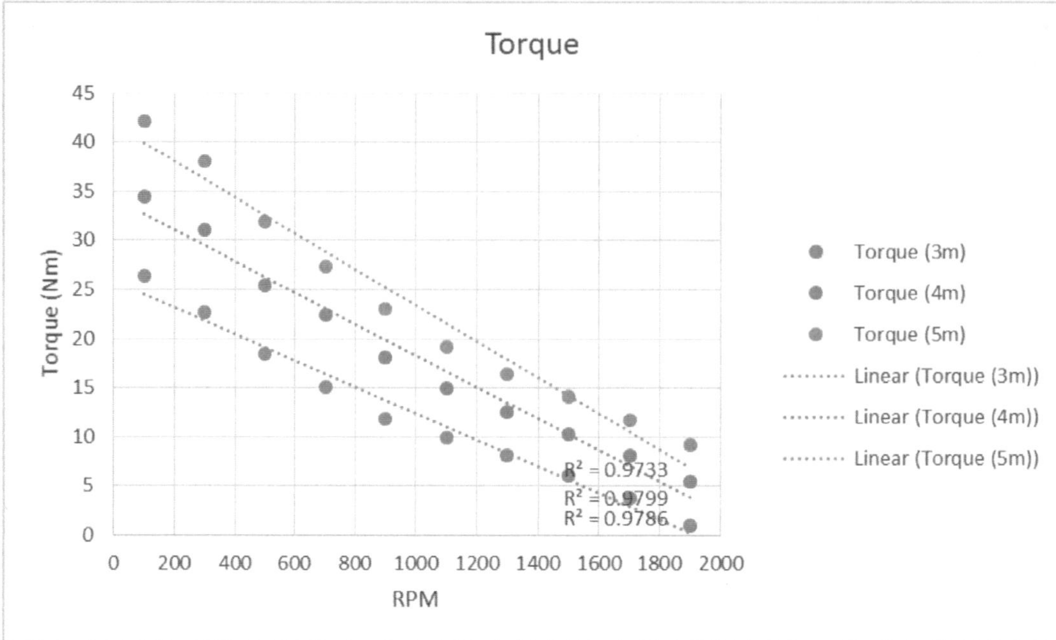

Figure 5. Turbine's Torque

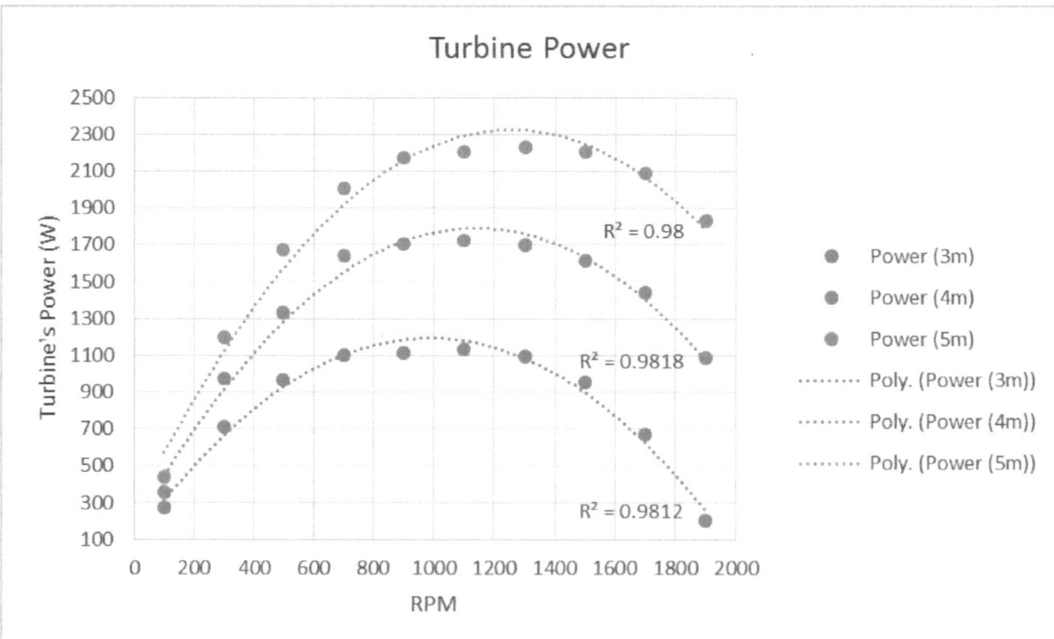

Figure 6. Turbine's Power

Graph in Figure 7 shows turbine's efficiency. The efficiency is calculated using the data given in figure 6 and the potential power generated according formula showed in Equation 6. The graph shows that the largest turbine's efficiency at 3 m head is 93.22% while at 4 m and 5 m head are 94.6% and 89.88% respectively. This result shows that higher head will not produce better performance, even though the torque and the power generated is higher. This should in consideration in designing micro hydro power using hydrocoil turbine.

The graph in figure 7 also shows that turbine's efficiency at 5 m head is the lowest untill the rotational speed reach 1.700 rpm. After that, the turbine's efficiency become the highest among all variation. This is due to the higher the head, the greater the operational rotational speed range. So 5 m head will provide greatest rotational speed range among all variation. This mean that the turbine's rotational speed can reach more than 2.000 rpm. Otherwise, 3 m head only provide rotational speed range from 0 rpm to around 1.900 rpm. This range should be a consideration in choosing appropriate generator. This is because every generator has minimum rotational speed required coupled to it.

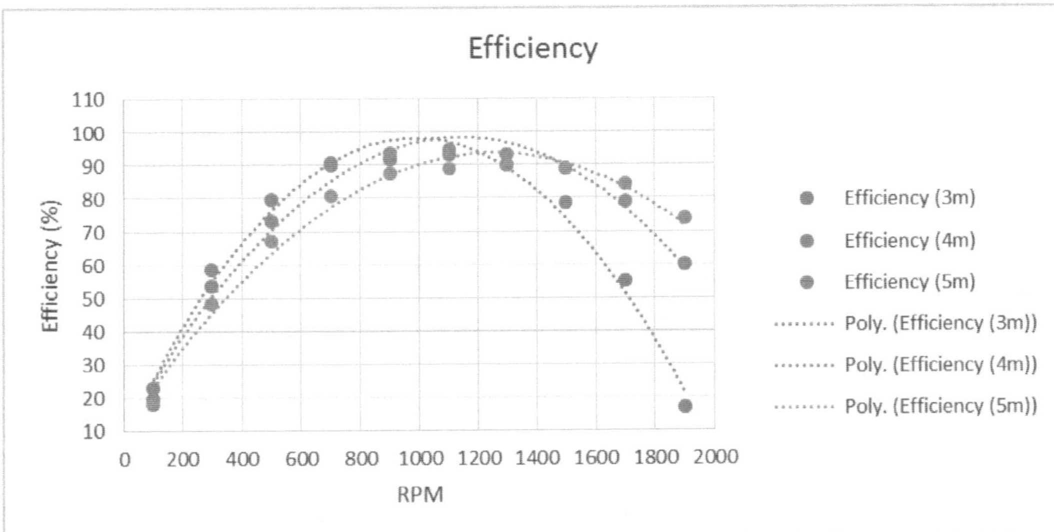

Figure 7. Turbine's Efficiency

4. Conclusions

The hydrocoil turbine power graph on head variations of 3 m, 4 m, and 5 m indicates that the largest power at 3 m head is 1,134.06 W, while the largest power at 4 m and 5 m head are 1,722.39 W and 2,231.49 W respectively. Turbine's efficiency graph shows that the largest turbine's efficiency at 3 m head is 93.22% while at 4 m and 5 m head are 94.6% and 89.88% respectively. This result shows that higher head will not produce better performance, even though the power generated is higher. This graph also shows that the larger the head, the greater the operational rotational speed range.

References

[1] Kementerian Energi dan Sumber Daya Mineral Republik indonesia 2016

[2] Aprilliyanto, A., Indarto and Prajitno 2013 *Design of A Prototype Hydrocoil Turbine Applied As Micro Hydro Solution* ASEAN Journal of System Engineering Vol. 1 p 72-76

[3] Leon, A. S. and Zhu, L. 2014 *A Dimensional Analysis for Determining Optimal Discharge and Penstock Diameter in Impulse and Reaction Turbine* Renewable Energy – Elsevier Vol. 00 p 1-14.

[4] Munson, B. R., Young, D. F. and Okiishi, T. H. 2004 Mekanika Fluida Jakarta Erlangga Publisher.

IOP Conf. Series: Materials Science and Engineering **343** (2018) 012025 doi:10.1088/1757-899X/343/1/012025

Study on aerodynamics characteristics an urban concept car for energy-efficient race

H Ambarita[1*], M R Siregar[2] and H Kawai[3]

[1]Sustainable Energy and Biomaterial Centre of Excellent, Faculty of Engineering, Universitas of Sumatera Utara, Jl. Almamater Kampus USU Medan 20155, Indonesia
[2]Sarulla Operation Ltd., The Energy Building 51st floor, SCBD Lot 11A, Jl. Jend. Sudirman Jakarta 12190, Indonesia
[3]Department of Mechanical Systems of Enginering, Muroran Institute of Technology, 27-1 Mizumoto-cho, Muroran 050-8585, Japan

himsar@usu.ac.id

Abstract. "Horas Mesin USU" is a prototype of urban concept vehicle designed by University of Sumatera Utara to participate in the energy-efficient competition. This paper deals with a numerical study on aerodynamic characteristics of the Horas Mesin USU. The numerical analyses are carried out by solving the governing equations using CFD FLUENT commercial code. The turbulent flow is closed using k-epsilon turbulence model. In the results, pathline, velocity vector and pressure distribution are plotted. By using the pressure distributions, drag and lift coefficients are calculated. In order to make a comparison, the aerodynamic characteristics of the present design are compared with commercial city car Ford-Fiesta. The averaged drag coefficients of Horas Mesin USU and Ford-Fiesta are 0.24320 and 0.29598, respectively. On the other hand, the averaged lift coefficients of the Horas Mesin USU and Ford-Fiesta are 0.03192202 and 0.09485621, respectively. This fact suggests that Ford-Fiesta has a better aerodynamic performance in comparison with Horas Mesin USU. The flow field analysis shows that there are many modifications can be proposed to improve the aerodynamic performance of the Horas Mesin USU. It is suggested to perform further analysis to improve the aerodynamic performance of Horas Mesin USU.

1. Introduction

Shell Eco-marathon is a competition that challenges students around the world to design, build and drive the most energy-efficient vehicle. There are three annual events in Asia, Americas, and Europe. In each event, student teams take to the track to see who goes further with the least amount of fuel [1]. As a note, the University of Sumatera Utara (USU) has developed a team, named as Horas team, to participate in any energy-efficient competition such as Shell Eco-marathon Asia in international level and Energy-efficient car contest (named as KMHE) in the national level of Indonesia. The Horas team consists of students from Mechanical Engineering, and they participate every year in the energy-efficient competition in international and national level. The Horas team has designed an urban concept car, and the design is named as "Horas Mesin USU", hereafter shortened as HMU. The body of HMU is designed to meet the high aerodynamic performance in order to reduce fuel consumption. One of the solution to produce an energy-efficient car is to reduce the drag coefficient of the body. Typically, in a car, 3-11% of energy from fuel is used to overcome the aerodynamic load. Thus, the study on drag coefficient in an energy-efficient car will give significant contribution to reducing fuel consumption.

Many researchers have reported studies on aerodynamic characteristics of a car in literature. Taha et al. [2] employed ANSYS Fluent to investigate aerodynamics characteristics of "Merdeka 2" a prototype of a solar vehicle that participated in the world solar challenge. The design concept of the car was based on box fish which was also claimed the concept design of Mercedes Benz minivan. Reynold-averaged Navier-Stokes with the k-epsilon model was used to model the turbulent flow. The CFD results and experimental data do not agree well. Krishnani and Pramod [3] used CFD code to study the drag reduction of a generic sports utility vehicle. Franck et al. [4] investigated flow around the Ahmed Vehicle model using numerical simulation with several turbulence models. Kim et al. [5] reported the study on salient drag reduction of a heavy vehicle using modified cab-roof fairings. The results are expected to provide useful information for the design of new cab-roof fairing models and the improvement of the aerodynamic performance of heavy vehicles, including trucks and tractor-trailers. Shim et al. [6] reported the study on the optimization of bobsleigh bumper shape to reduce aerodynamic drag. In the numerical method, k-omega shear stress transport turbulence model is used to close the three-dimensional Reynolds averaged Navier-Stokes equation. A parametric study was conducted using six parameters related to the shape of the front and rear bumpers and three parameters, i.e., the distance between the bumpers and the ground and leading angle of the front bumper as design parameters. The objective function for optimization was the drag coefficient. The results showed that aerodynamic drag of the bobsleigh was reduced by 3.08% in comparison with the reference design. Lorite-Diez et al. [7] studied the use of the adjoint sensitivity formulation to design efficient passive control strategies intending at reducing the drag coefficient of a slender blunt-based body with a straight rear cavity.

CFD has been used to explore the fluid flow characteristics in order to design and to promote modification of a racecar. Kieffer et al. [8] used CFD to study the section characteristics of Formula Mazda race car wings. STAR-CD CFD code was used to perform the simulation. The turbulent flow was modeled using k-epsilon model. The results are presented graphically, presenting pressure and velocity distributions and lift and drag coefficients for the different cases. The results were claimed to be valuable for improving the optimum handling of Formula Mazda race cars. Mariani et al. [9] presented a study that aimed to improve the external fluid-dynamics of the first prototype of the Formula-SAE (Society of Automotive Engineers) race car of the University of Perugia, Italy prepared for participation in the international competition of Varano (Parma-Italy). Two prototypes were analyzed numerically; the original prototype and redesigning prototype. The results showed that a remarkable improvement of the aerodynamics performance was obtained by the proposed modifications. Hassan et al. [10] reported a study on aerodynamic drag reduction of racing cars by using a numerical method. In the method, Favre-averaged Navier-Stokes equations closed with k-epsilon turbulence model were solved using Finite Volume Method. The results showed that the drag coefficient of the car was 0.3233 and it was evident that the drag can be reduced up to 22.13% by different rear under-body modifications and up to 9.5% by exhaust gas redirection towards the separated region at the rear of the car. Hetawal et al. [11] reported the study on rear engine Formula SAE racecar. The objective of the study was to investigate the aerodynamic characteristics of a SAE race car with a front spoiler, without the front spoiler, and with firewall vent. The study was conducted using ANSYS Fluent software with k-epsilon turbulence model. The results were graphically shown with drag coefficient and velocity contour.

The above studies were mainly focused on the improvement of high-speed racecars. Only a few studies on aerodynamics performance of energy-efficient car are found in the literature. One of the studies is the CFD analysis for Merdeka 2 solar vehicle [2]. As stated in the first paragraph the Horas team has designed and built the HMU to participate in any energy-efficient racecars. In order to improve the efficiency, the aerodynamic characteristics need to be explored. The objective of this paper is to explore the aerodynamic performance of the HMU. The performance includes pathlines, velocity vector, pressure distributions, drag coefficient and lift coefficient. The results are expected to supply the necessary information in developing high energy-efficient racecar.

2. Method

Computational Fluid Dynamics (CFD) is employed to provide aerodynamic characteristics of the HMU. The CAD model, dimension and actual picture of HMU are shown in Figure 1. In the CFD method the governing equations are converted into algebraic system equations and solved iteratively. Three-dimensional governing equations are used to solve the problem and the turbulent flow is taken into account. The continuity equation and momentum equation are written as follows.

$$\frac{\partial \rho}{\partial t}+\frac{\partial}{\partial x_i}\left(\rho u_i\right)=0 \tag{1}$$

$$\rho\frac{Du_i}{Dt}=-\frac{\partial p}{\partial x_i}+\frac{\partial}{\partial x_j}\left[\mu\left(\frac{\partial u_i}{\partial x_j}+\frac{\partial u_j}{\partial x_i}-\frac{2}{3}\delta_{ij}\frac{\partial u_i}{\partial x_i}\right)\right]+\frac{\partial}{\partial x_j}\left(-\rho\overline{u_i'u_j'}\right) \tag{2}$$

The right side of equation (2) is known as the Reynolds-stresses tensor represents the transfer of momentum due to turbulent fluctuations. This tensor can be written as

$$-\rho\overline{v_i'v_j'}=-\rho\left(\overline{v_i'v_j'}-\bar{v}_i\bar{v}_j\right) \tag{3}$$

In this study, Boussinesq hypothesis is used to relate the stresses with the mean velocity gradients. Thus, the Reynolds-stresses can be calculated using the following equation.

$$-\rho\overline{u_i'u_j'}=\mu_t\left(\frac{\partial u_i}{\partial x_j}+\frac{\partial u_j}{\partial x_i}\right)-\frac{2}{3}\left(\rho k+\mu_t\frac{\partial u_i}{\partial x_i}\right)\delta_{ij} \tag{4}$$

In order to close the above equation, additional governing equations are needed. In this study, the standard $k-\varepsilon$ turbulence model proposed by Lauder and Spalding [12] is employed. In the model two additional governing equations are proposed; they are the turbulent kinetic energy (k) equation and the turbulent dissipation rate (ε). The proposed equations are formulated in the following equations.

$$\rho\frac{Dk}{Dt}=\frac{\partial}{\partial x_i}\left[\left(\mu+\frac{\mu_t}{\sigma_k}\right)\frac{\partial k}{\partial x_i}\right]+G_k+G_b-\rho\varepsilon-Y_M \tag{5}$$

$$\rho\frac{D\varepsilon}{Dt}=\frac{\partial}{\partial x_i}\left[\left(\mu+\frac{\mu_t}{\sigma_\varepsilon}\right)\frac{\partial \varepsilon}{\partial x_i}\right]+C_{1\varepsilon}\frac{\varepsilon}{k}\left(G_k+C_{3\varepsilon}G_b\right)-C_{2\varepsilon}\rho\frac{\varepsilon^2}{k} \tag{6}$$

In the above equation, two types of viscosity are found in the equation; the actual dynamic viscosity and the turbulent viscosity μ_t. The turbulent viscosity is calculated by

$$\mu_t=\rho C_\mu\frac{k^2}{\varepsilon} \tag{7}$$

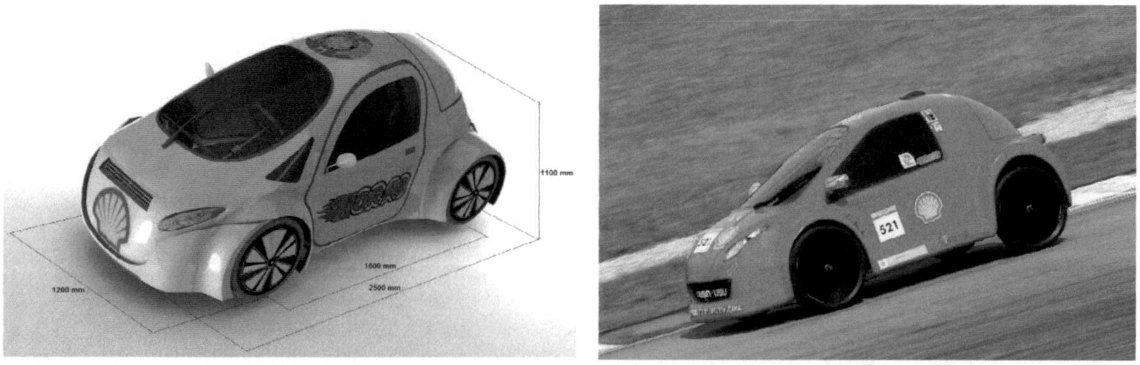

(a) CAD model (b) Actual Car

Figure 1. CAD model and Actual design of Horas Mesin USU-I

In the above equations, the parameter G_k, G_k and Y_M represent the generation of turbulent kinetic energy due to mean velocity gradient, generation of turbulent kinetic energy due to buoyancy and the contribution of fluctuating dilation in compressible turbulence to the overall dissipation rate, respectively. Furthermore, the parameter $C_{1\varepsilon}$, $C_{2\varepsilon}$, $C_{3\varepsilon}$ and C_{μ} are constants. While, σ_k and σ_ε are the turbulent Prandtl numbers for k and ε, respectively.

In this study, the governing equations will be discretized upon three-dimensional computational domain. The used mesh, computational domain, car position and the boundary conditions are shown in Figure 2. In order to save computational cost and symmetrical characteristics of the domain, only a half of the computational domain is taken into simulation. The boundary conditions are explained in Table 1.

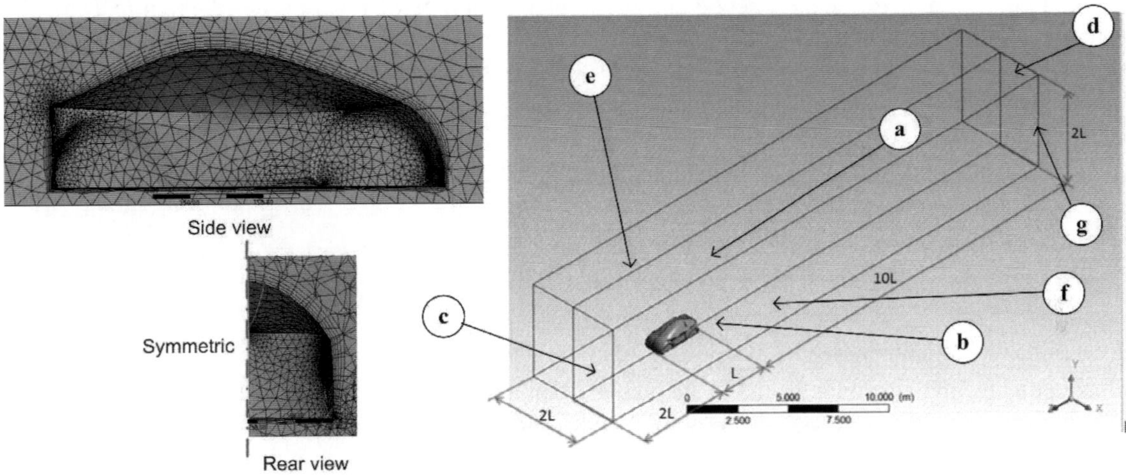

Figure 2. Computational domain, meshing and boundary conditions

Table 1. Boundary conditions

Boundary mark	Location	Boundary condition
a	Top domain	Pressure outlet (0 Pa)
b	Car surface	Wall
c	Inlet	Velocity inlet (varied from 10 to 20 m/s)
d	Outlet domain	Pressure outlet (0 Pa)
e	Symmetry Plane	Symmetry
f	Bottom/road	Wall
g	Right side	Pressure outlet (0 Pa)

The governing equations discretized using second order upwind scheme. The system equations and boundary conditions were solved iteratively using SIMPLE algorithm. The iteration will be terminated if residual in continuity and momentum are less than 10^{-6}. The aerodynamic performance will be examined using drag coefficient (C_D) and lift coefficient (C_L). These coefficients are calculated in the below equations.

$$C_D = \frac{F_D}{\frac{1}{2}\rho V^2 A_D} \tag{8}$$

Where F_D [N], ρ [kg/m^3], V [m/s] and A_D [m^2] are drag force, density, fluid inlet velocity and frontal area of drag force.

$$C_L = \frac{F_L}{\frac{1}{2}\rho V^2 A_L} \tag{9}$$

Where F_L [N] and A_L [m^2] are lift force and frontal area for lift force.

3. Results and Discussions

The simulations on the HMU have been performed on five different inlet velocities. They are 10 m/s, 12.5 m/s, 15 m/s, 17.5 m/s and 20 m/s. In the first section of the result the numerical method will be validated with previous simulation and also experimental result. The discussion will be followed by fluid flow characteristics and aerodynamic performance.

3.1. Numerical Validation

In order to validate the numerical method, a numerical validation test has been performed. The present method has been used to calculate drag coefficient of Ahmed Body. The drag coefficient of the Ahmed Body resulted from experimental and numerical from previous work are presented in Table 2. The table shows that the results of the present numerical method agree well with previous work. The discrepancy to the experimental work is only 3.93%. Thus, the present method will be used to perform the numerical analysis.

Table 2. Numerical validation

Comparison Parameter	Experimental work [4]	Simulation [4]	Present Work
Drag Coefficient	0.2300	0.2346	0.239034
Discrepancy	-	2.00%	3.93%

3.2. Pathlines and Vector velocity

Pathlines on the HMU resulted from the analysis is shown in Figure 3. The figure shows the back view and isometric view around the car's body of HMU. The pathlines are resulted from the simulation at speed of 15 m/s. The figure shows that after flowing over the car's body the fluid creates circular flow in the area behind the car. There is unbreak flow in the middle area, as shown by letter "A" in the figure. As a note, above the back wheel there is a blunt part of the body, in the figure it is shown by letter "B". Separation flow is captured in this area. This results in high drag coefficient in the body car. This suggests that HMU need improvement on the body around the rear part.

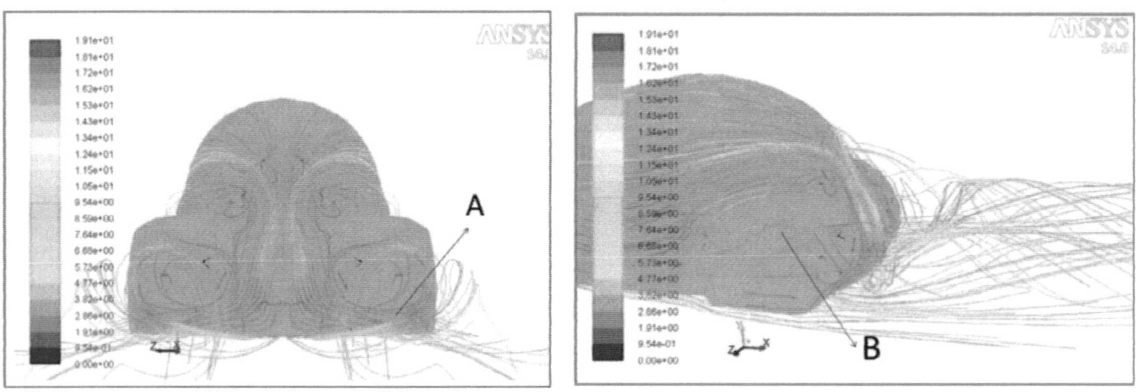

(a) Rear view (b) Isometric view

Figure 3. The pathlines of HMU

Characteristics of the velocity vector on the HMU resulted from simulation are shown in Figure 4. In the figure, four different type of velocity fields are presented. Figure 5a shows contour velocity on the symmetric plane. Figure 5b shows velocity vector after the fluid flowing from rear area. Figure 5c and Figure 5d show the contour velocity and the vector velocity after the car at a plane 0.4 m above the ground ($y = 0.4$), respectively. It can be seen that velocity vector on the vertical plane shows a good aerodynamic performance. Here, there is no weak flow is captured. On the other hand, on the horizontal plane two circular flows are captured. This fact suggest that this circular flow will increase the drag coefficient and should be modified in the next design.

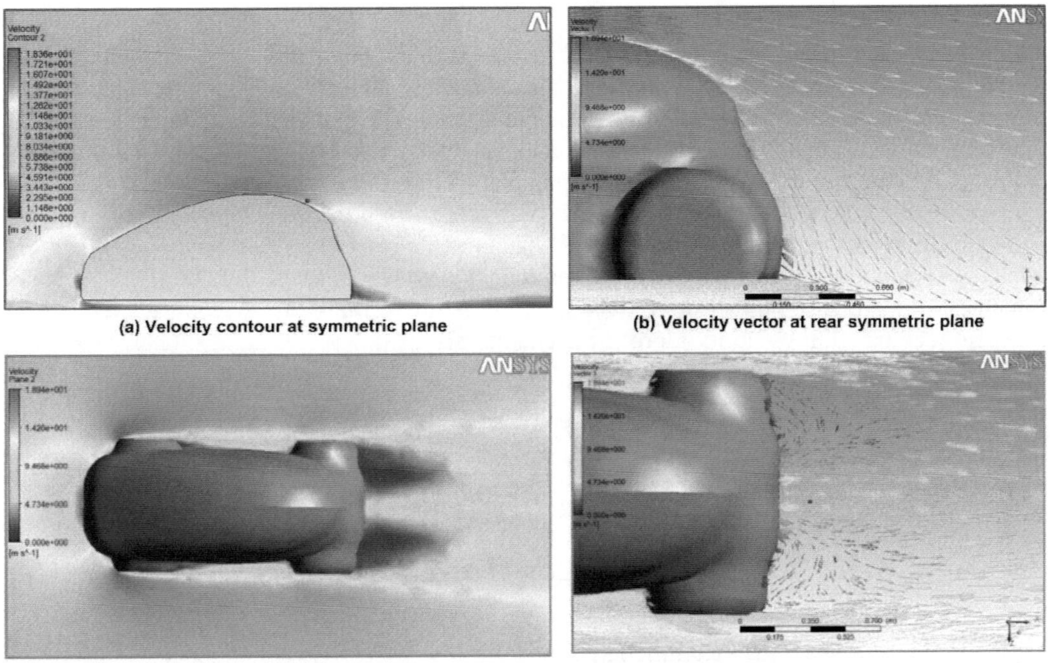

(a) Velocity contour at symmetric plane

(b) Velocity vector at rear symmetric plane

(c) Velocity contour at horizontal plane (y = 0.4m)

(d) Velocity vector at rear horizontal plane (y = 0.4m)

Figure 4. Velocity vector of the HMU

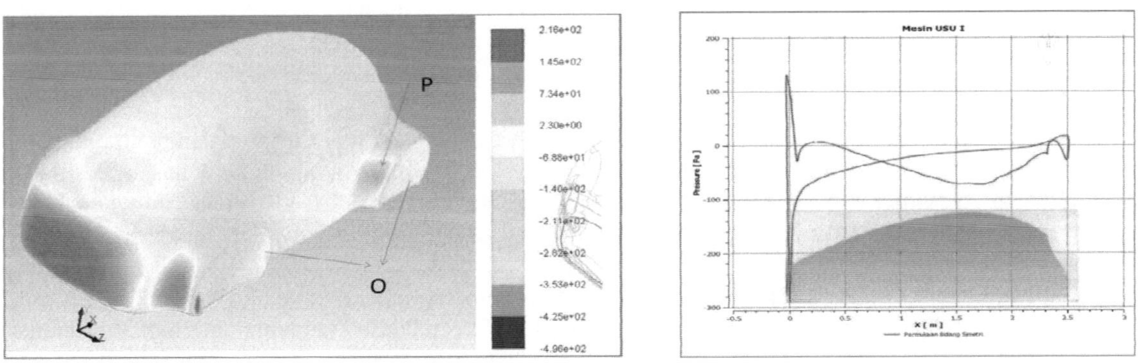

Figure 5. Pressure distributions on HMU

The objective of the present research is to explore the characteristic of the flow field in order to provide suggestion for improvement. Figure 5 shows the pressure coefficient all over the body (left figure) and pressure distribution at symmetry surface of the car (right figure). In the right figure, the value of the pressure also shown by red line. The characteristic of Pressure Coefficient distribution on the HMU reveals that there are several abrupt surfaces in on its body. In the figure, those are shown by blue colour and the value of the pressure coefficient are negative. On the other hand, the maximum pressure coefficient occurs on the front surface of the car. This is because of the velocity of the fluid on these area close to zero. The maximum pressure (shown by red colour) also occurs on the several parts of the left body such as the front of the tire covers. In the figure, these areas are shown by letter “O” and letter “P”. These points should be focused for further improvement. The pressure distribution on the symmetric surface shows that the maximum and minimum pressures are 130 Pa and -290 Pa, respectively.

3.3. Drag and Lift Coefficients

Simulation has been carried out for HMU. In order to perform comparisons, a commercial city car Ford-Fiesta that is known as aerodynamic car is also simulated. The two designs are simulated at the same condition and the velocity varies from 10 m/s to 20 m/s. The results are shown in Table 3. In the table, the drag coefficient and lift coefficient for each speed is presented. For comparison the averaged values are used. The averaged drag coefficients of Ford-Fiesta and HMU are 0.24320 and 0.29598, respectively. On the other hand, the averaged lift coefficients of Ford-Fiesta and HMU are 0.03192202 and 0.09485621, respectively. This fact reveals that Ford-Fiesta is more aerodynamics than HMU. However, the lift coefficient of HMU is better than Ford-Fiesta. Even though HMU shows less aerodynamics, however the fluid flow characteristics around the car body shows that there are many improvements can be proposed to in improve the characteristics.

Table 3. Drag and Lift coefficients

Design	Velocity (m/s)	Drag Coefficient		Lift Coefficient	
		Cd	Averaged	Cl	Averaged
Ford Fiesta	10.0	0.22980	0.24320	0.034823253	0.03192202
	12.5	0.24025		0.030879436	
	15.0	0.24184		0.028265246	
	17.5	0.25363		0.038506081	
	20.0	0.25050		0.027136104	
Horas Mesin USU	10.0	0.301946	0.29598	0.110678713	0.09485621
	12.5	0.278073		0.081272110	
	15.0	0.315233		0.079140931	
	17.5	0.285122		0.098809827	
	20.0	0.299502		0.104379470	
	20.0	0.303430		0.015054253	

4. Conclusions

In this study, fluid flow characteristics of HMU an urban concept car for energy-efficient competition has been explored using commercial code FLUENT CFD. The numerical method has been firstly validated with previous experimental and numerical works on Ahmed Body. It is shown that the present numerical method shows acceptable discrepancy. By using the validated method numerical study has been performed. The drag and lift coefficients from HMU are compared with Ford-Fiesta. The conclusions are as follows. The pathlines and vector velocity show that weak flow in HMU is still big. The pressure distribution shows that pressure coefficient of HMU. The averaged drag coefficients of Ford-Fiesta and HMU are 0.24320 and 0.29598, respectively. This fact suggests that the design of HMU is less aerodynamic than Ford-Fiesta. The flow field analysis shows that there are many improvements can be proposed to make HMU more aerodynamic. It is suggested to perform further analysis for improvement.

5. References

[1] http://www.shell.com/energy-and-innovation/shell-ecomarathon/about.html(Acessed September 25, 2017)
[2] Taha Z, Passarella, Sugiyono, Abd Rahim N, Md Sah J and Ahmad-Yazid N 2011 *Advanced Science Letter* **4**, 2807-2811
[3] Krishnani and Pramod N 2006 *CFD Study of drag reduction of a generic sport utility vehicle* (Mumbai; Mumbai University)
[4] Franck G, Nigro N, Storti M and D'Elia J 2009 *Latin American Applied Research* **39** 295-306
[5] Kim J J, Lee S, Kim M, You D and Lee S J 2017 *Journal of Wind Engineering & Industrial*

Aerodynamics **164** 138-151
[6] Shim H S, Lee Y N and Kim K Y 2017 *Journal of Wind Engineering & Industrial Aerodynamics* **164** 108-118
[7] Lorite-Diez, Jimenez-Gonzales J I, Gutierrez-Montes C and Martinez-Bazan C 2017 *Journal of Fluids and Structure* **74** 158-177
[8] Kieffer W, Moujaes S and Armbya N 2006 *Mathematical and Computer Modelling* **43** 1275-1287
[9] Mariani F, Poggiani C, Risi F and Scappaticci L 2015 *Energy Procedia* **81** 1013-1029
[10] Hassan S M R, Islam T, Ali M and Islam M Q 2014 *Procedia Engineering* **90** 308-313
[11] Hetawal S, Gophane M, Ajay B K and Mukkamala Y 2014 *Procedia Engineering* **97** 1198-1207

Calculations of the Racah parameter *B* for Mn^{4+} and Mn^{2+} ions doped in $CaAl_2O_4$

Le Ngoc Liem[1], **Ngo Tran**[2]

[1]Office of Academic Investigation, Duy Tan University, Danang, Vietnam

[2]Department of Physics and Oxide Research Center, Hankuk University of Foreign Studies, Yongin 449-791, Korea

Email: liemle09@gmail.com

Abstract. The luminescent materials $CaAl_2O_4$:Mn^{4+} and $CaAl_2O_4$:Mn^{2+} were synthesized by solid state reaction method. The red luminescence of $CaAl_2O_4$:Mn^{4+} shows an asymmetric broad band peaking at 650 nm, respectively to the characteristic luminescence of Mn^{4+} ions. The emission spectrum of $CaAl_2O_4$:Mn^{2+} shows a broad green emission band centering at 545 nm respectively to the characteristic luminescence of Mn^{2+} ions. By doping different outer shells in the same host lattice, the ionic interaction between ions and the host lattice are also different. The differences are reflected in Racah parameter B, and the nephelauxetic effect β. The β coefficient of $CaAl_2O_4$:Mn^{4+} is small, indicating that its orbitals expand greatly, the interaction force between Mn^{4+} and the host lattice is greater than that of Mn^{2+}.

1. Introduction

Nowadays, luminescent materials are being studied and applied widely in engineering and in daily life. Therefore, more and more group of materials are being synthesized, especially transition metal doped aluminate materials. The luminescent wavelength of doped ions depends on the strength of crystal field. In weak crystal field, the emission spectrum of Mn^{2+} shows a broad band in the green, while in strong crystal field, the emission spectrum is from the orange to red. In weak crystal field, Mn^{4+} emits infrared radiation while in strong crystal field, it emits light in the red spectral range [1, 2]. The Racah parameter B or the nephelauxetic effect β must be determined to better understand the interaction between the doped ions and the crystal field. However, almost no works has been done to compare crystal field force of a host lattice when doped with different valences ions. Therefore, to understand clearly these crystal field forces, we compares the crystal field of Mn^{2+} doped and Mn^{4+} doped in $CaAl_2O_4$ (CAO) host lattice.

2. Experiments

The CAO materials (CAO:Mn^{2+} or CAO:Mn^{4+}) were synthesized by solid state reaction. Commecial powders of Al_2O_3, B_2O_3, $CaCO_3$, and MnO_2 with purity of 99.9% were use as precursors. These precursors are weighted by molar ratio and mixed with 4 *weight* percent B_2O_3 (used as fluxing agents). The mixtures are well grinded for two hours by using an agate pestle and mortar. Then, the mixtures were anneled at 1250 °C for 2h. The formation of CAO:Mn^{2+} is described by following chemical reaction:

$$(1-x)\ CaCO_3 + Al_2O_3 + x\ MnO_2 \rightarrow Ca_{(1-x)}\ Al_2O_4{:}xMn^{2+} + CO$$

The CAO:Mn^{4+} sample is annealed in the air environment. Particularly, the CAO:Mn^{2+} was annealed in an environment in which CO gas is used as the reductant to generate Mn^{2+} from Mn^{4+}. In the synthesis of CAO:Mn^{2+}, coal pellets are put in the bottom of the crucible to continuously generate CO.

The X-ray difraction (XRD) spectra at room temperature were measured by using the diffractometer Bruker D8-Advance at Faculty of Chemistry, Hanoi University of Science. Photoluminescence (PL) and Photoluminescence excitation (PLE) spectra were measured by Flourescent spectrometer FL3-22 Horiba at Duy Tan University, Vietnam. Samples were measured at room temperature and a 450W Xenon lamp was used to excite the sample.

3. Results and Discussion

3.1. X-ray diffraction

XRD pattern of CAO:Mn^{2+} (2 mole%) and CAO:Mn^{4+} (2 mole%) are shown in Figure 1. Characteristic peaks of $CaAl_2O_4$ with 2θ angle of 22°, 30°, 35.5°, 37.5° shows that $CaAl_2O_4$ phase are the dominating phase, the $Ca_3Al_2O_6$ phase virtually does not exist. With low Mn doping concentration, XRD pattern of CAO:Mn^{2+} (2 mole%) and CAO:Mn^{4+} (2 mole%) do not change compared with based-material $CaAl_2O_4$ (not shown). $CaAl_2O_4$ has orthorhombic crystal structure and its lattice parameters are $a = 8.740$ Å, $b = 8.100$ Å, $c = 15.130$ Å và $\alpha = \gamma = \beta = 90^o$.

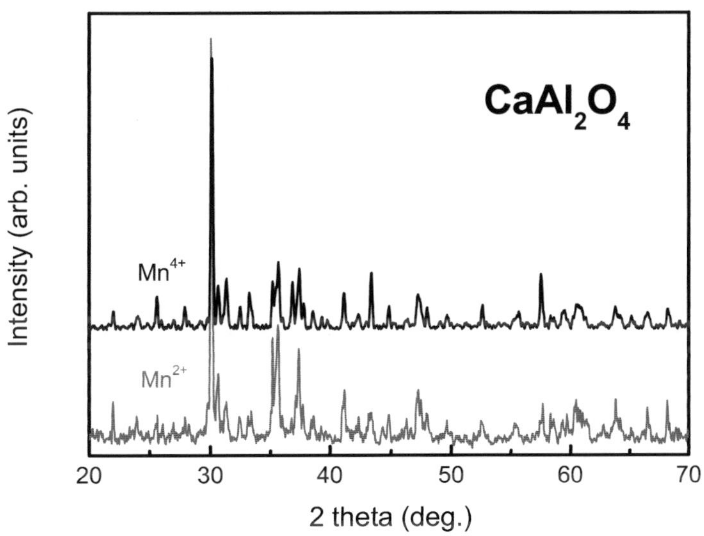

Figure 1. X-ray diffraction of CAO:Mn^{2+} (2 mole%) and CAO:Mn^{4+} (2 mole%).

3.2. Photoluminescence and Photoluminescence excitation spectrum of CAO:Mn^{2+}

The PL and PLE spectrums of CAO:Mn^{2+} (2 mole%) were measured at the wavelengths $\lambda_{ex} = 425$ nm and $\lambda_{em} = 545$ nm, respectively, shown in Figure 2. PLE spectrum shows narrow peaks at wavelengths of 456, 426, 385 and 360 nm. The emission spectrum showed a broad band peaked at 545 nm. This peak is attributed to the spin-forbidden $^4T_{1g}(^4G)$ to $^6A_{1g}$ [2]. In addition there is a low-intensity peaks at ~650 nm, correspond to luminescence characteristic of a small amount of Mn^{4+}. Because during the synthesis of material, not all of Mn^{4+} ions had been reduced to Mn^{2+}.

PLE peaks of Mn^{2+} correspond to the transition from ground state $^6A_1$ to excited states of $^4T_{2g}(^4G)$, $^4E(^4G)$'$^4A_{1g}(^4G)$, $^4T_{2g}(D)$, $^4E(D)$ [3]... There is a narrow and high intensity peak at 425 nm because according to Tanabe-Sugano (T-S) diagram, the ground level $^6A_1$ and the excited levels $^4E(^4G)$' $^4A_1(^4G)$

are illustrated in two virtually parallel lines (Figure 3). The transition from $^6A_1$ to $^4E(D)$ is shown at the wavelength of 360 nm in Figure 2.

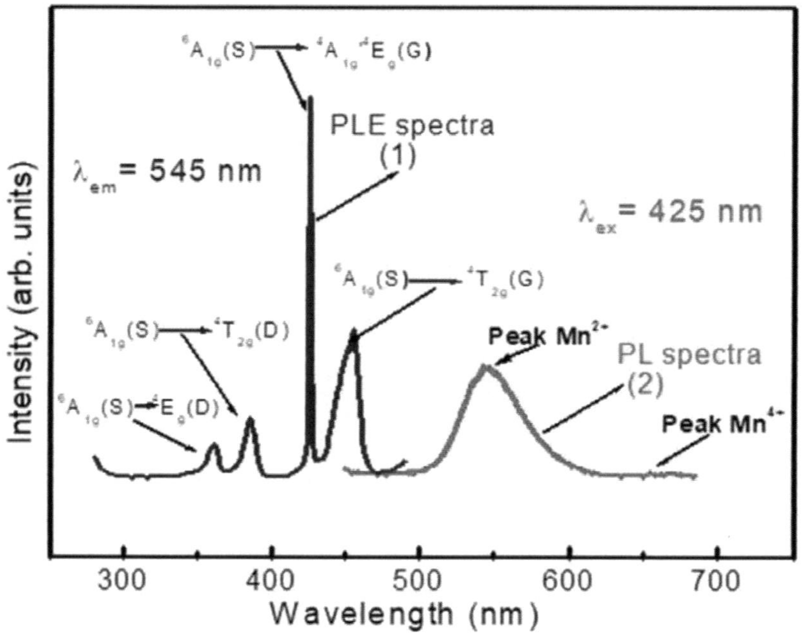

Figure 2. PLE and PL spectra of CAO:Mn^{2+}

The Racah parameter can be determined by the following formula [4]:

$$\upsilon_1 = E\,[^4E(^4G)'\,^4A_1(^4G)] - E\,[(^6A_1)] \quad (1)$$

$$= 10B+5C$$

$$\upsilon_2 = E\,[^4E(D)] - E\,[(^6A_1)] \quad (2)$$

$$= 17B+5C$$

By subtracting (1) by (2), we have

$$\upsilon_2 - \upsilon_1 = 7B. \quad (3)$$

We have υ_2 = 27777 cm^{-1}, and υ_1 = 23474 cm^{-1}. Substituting υ_1 and υ_2 into (3), we found that B = 615 cm^{-1} and C = 3464 cm^{-1}. The nephelauxetic effect β is calculated by the formula $\beta = B/B_0$, where B_0 = 785cm^{-1} [4, 5] is the Racah parameters for free Mn^{2+} ion. By substituting the B and B_0 (as mentioned above), we find that $\beta = 0.78$.

From PLE spectrum, we obtained transition energy from $^6A_1$ to $^4T_{2g}(^4G)$ is υ_3 = 22123 cm^{-1}. T-S diagram (Figure 3) can be used to determine splitting parameters Dq. We draw a line, parallel to the vertical axis (E) and it intersects excited state levels of $^4T_{2g}(^4G)$, $^4E(^4G)'\,^4A_1(^4G)$, $^4T_{2g}(D)$, $^4E(D)$ at ordinates: 22123, 23474, 25974, and 27777 cm^{-1}, respectively and intersects the horizontal axis (Dq) at 650cm^{-1}. Therefore Dq = 650cm^{-1} or the crystal field of Mn^{2+} doped CAO is Δ = 6500 cm^{-1} (Δ = 10Dq).

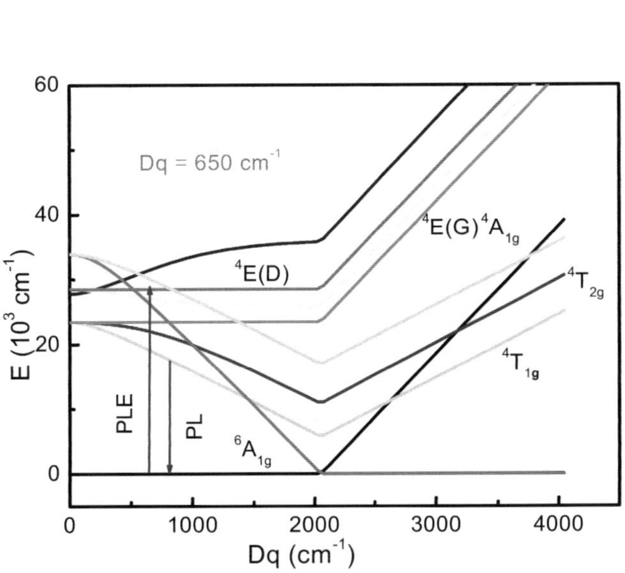

Figure 3. Determine *Dq* from T-S diagram [3]

3.3. Photoluminescence and Photoluminescence excitation of CAO:Mn^{4+}

PL and PLE spectra of CAO:Mn^{4+} are shown in Figure 4. The PLE spectra has two broad band with peaks at 450 and 328 nm, correspond to the transition from the ground state $^4A_2$ to excited states $^4T_1$ and $^4T_2$ of Mn^{4+} ions. The PL spectrum reveals peaks at 643, 650 and 668 nm which are due to the ^2E to $^4A_2$ states transition of Mn^{4+} ions. In addition, there is a low intensity peak at 545 nm, correspond to the luminescent characteristic of a small amount of Mn^{2+}. Because during the synthesis process, some Mn^{4+} ions are reduced to Mn^{2+}.

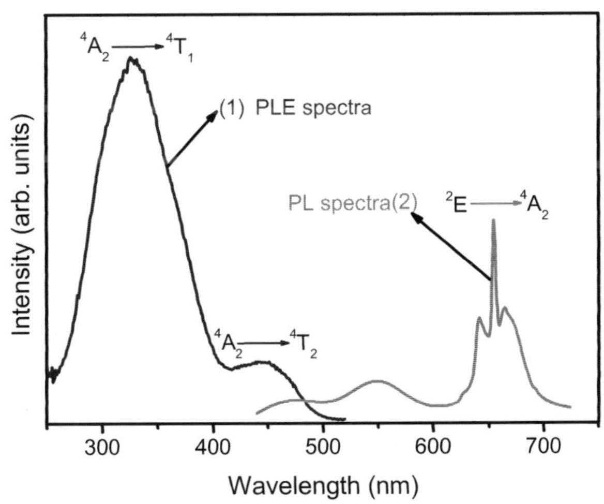

Figure 4. PL and PLE spectra of CAO:Mn^{4+} with λ_{ex} = 450 nm and λ_{em} = 650 nm

Determine Racah parameter B from diagrams and formula.

The wavelength λ = 450 nm and λ = 328 nm correspond to wavenumbers of υ_1 = 22222 cm^{-1} and υ_2 = 30487 cm^{-1}, respectively. The ratio between υ_2 and υ_1 is υ_2/υ_1 = 1.37 [5, 6]. From the Figure 5, we determined that Δ/B = 27.4. From the Figure 6 (T-S diagram with only necessary levels indicated), we draw a line parallel to E/B axis at position Δ/B = 27.4. It intersects $^4T_{2g}$ level curve at A. From A, we draw the a line parallel to Δ/B axis and its intersects E/B axis at E/B = 27,4. From E = 22222 cm^{-1}, we found that B = 811cm^{-1}, Δ = 22222 cm^{-1} and Dq = 222 cm^{-1} [7] (the υ and E signs could be used interchangably).

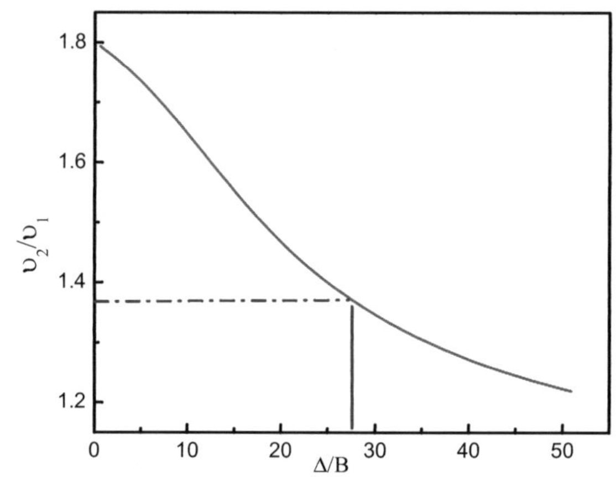

Figure 5. The relationship between v_2/v_1 and Δ/B [8].

From the formula $B = \dfrac{2v_1^2 + 2v_2^2 - 3v_1v_2}{15v_2 - 27v_1}$ [8, 9] we found that B = 808 cm^{-1}. The nephelauxetic effect coefficient $\beta = B/B_0$, where B_0 = 1160cm^{-1}, [9] is the Racah parameters for free Mn^{4+} ions. The calculated coefficient is β = 0.7. The small β indicates that the d-orbitals of Mn^{4+} ions expands greatly.

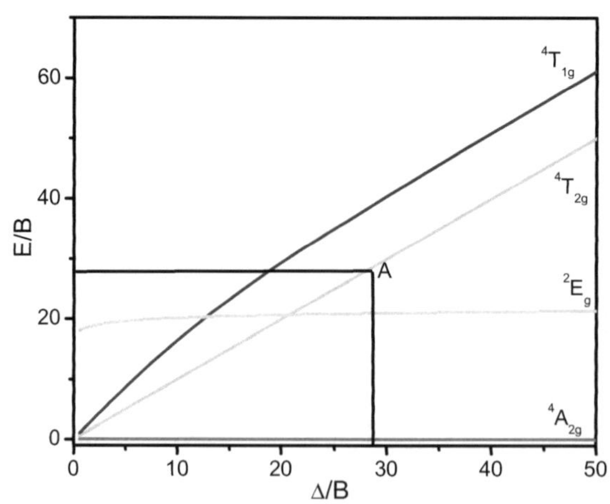

Figure 6. Determination of B for d^3 configuration from T-S diagram.

4. Conclusion

Based on the experimental photoluminescence excitation spectrum are provided to calculate the Racah parameter B and the nephelauxetic effect β. Various methods have been applied to determine the B parameter, the nephelauxetic effect coefficient β, as well as crystal field Dq or Δ. For the same CAO host lattice, the expansion coefficient of *d*-orbilals when it is doped with Mn^{2+} is smaller than when it is doped with Mn^{4+}, indicating that the outer electron shell of Mn^{4+} expands greatly. The small values of coefficient β (= 0.7) mean that should strong interaction between the ligand field with the *d*-orbitals of Mn^{4+} ions. For electrons in d^3 configuration (Mn^{4+}), both calculation methods yield similar B parameters. The crystal field force of CAO:Mn^{4+}, and CAO:Mn^{2+} are $\Delta = 22222$ cm^{-1} and $\Delta = 6500$cm^{-1}, respectively. Therefore, the crystal field of CAO doped with Mn^{4+} is 3.4 times greater than that of CAO doped with Mn^{2+}. The researchers have deliberately control the content of Mn^{2+} or Mn^{4+} by adding reductant to the reagents when they are sintered.

References

[1] Blasse G, and Geiabmaier B C 1994 *Luminescent Materials* (Springer-Verlag, Berlin)
[2] Andre Bleise, Holger Winkler and Thomas Jüstel 1999 *On the Luminescence of* Mn^{4+} *and* Mn^{2+} (University of Applied Sciences Münster, Stegerwaldstrabe 39, D-48565)
[3] Yukito Tanabe and Satoru Sugano 1954 On the absorption spectra of complex Ions *J. Phys. Soc. Jpn.* **9** 753-766
[4] Adam Bartecki and Krzysztof Kurzak 1981 *Calculations of the Racah Parameter B and Crystal Field Parameters* D_s *and* D_t *for* d^2, d^3, d^7, d^8 *-Metal Complexes* vol 24 no 5-6 (Bulletin de l'académie polonaise des sciences serie le des sciences chimiques)
[5] König E *The Nephelauxetic Effect — Calculation and Accuracy of the Interelectronic Repulsion Parameters*, (Zeitschrift Für Naturforschung, pp.175-212)
[6] Wei Q 2010 Investigations of the Optical and EPR Spectra for Cr^{3+} Ions in Diammonium Hexaaqua Magnesium Sulphate Single Crystal, *Acta Phys. Pol.* Vol. 118
[7] Casalboni M, Ciafardone V, Giuli G, Izzi B, Paris E and Prosposito P 1996 An optical study of silicate glass containing ions Cr^{3+} and Cr^{6+}, *J. Phys. Cond. Matter* 8 9059–9069
[8] Richard L 1963 Electronic spectra of transition metal complexes, *J. Chem. Edu.* 40 (3), pp 135.
[9] Simon A, Vander Sluis J H and Biedenhan L C 1954 *Tables of the racah coefficients* (Yale University New Haven Connecticut).

IOP Conf. Series: Materials Science and Engineering **343** (2018) 012027 doi:10.1088/1757-899X/343/1/012027

Effect of Phosphine-Free Selenium Precursor Reactivity on The Optical and Vibrational properties of Colloidal CdSe Nanocrystals

L A Thi[1,6], N T T Lieu[2], N M Hoa[3], N Tran[4], N T Binh[5], V X Quang[1], N X Nghia[1,5*]

[1]Institute of Research and Development, Duy Tan University, 3 Quang Trung, Da Nang, Vietnam
[2]Posts & Telecommunications Institute of Technology, km10 Nguyen Trai, Hanoi, Vietnam
[3]Hue College of Medicine and Pharmacy, 6 Ngo Quyen, Hue, Vietnam
[4]Department of Physics and Oxide Research Center, Hankuk University of Foreign Studies, Yongin, Gyeonggi 449-791, South Korea
[5]Graduate University of Science and Technology, Vietnam Academy of Science and Technology, 18 Hoang Quoc Viet, Hanoi
[6]Institute of Physics, Vietnam Academy of Science and Technology, 10 Dao Tan, Hanoi, Vietnam

nghianx.ims@gmail.com

Abstract. Phosphine-free selenium precursor solutions have been prepared by heating at temperatures ranging from 160 °C to 240 °C and studied by means of infrared absorption spectroscopy. The colloidal CdSe nanocrystals (NCs) synthesized from all those solutions by the wet chemical method. The influence of heating temperature on the chemical reactivity of selenium precursor and its role on the optical and vibrational properties of CdSe NCs are discussed in details. Their morphology, particle size, structural, optical and vibrational properties were investigated using transmission electron microscopy, X-ray diffraction, UV-Vis, fluorescence and Raman spectroscopy, respectively.

1. Introduction

Colloidal nanocrystals (NCs) of groups II-VI have attracted wide interest in many fields of physical, chemical, biochemical,... [1]. Many of these applications rely on the fluorescence emission of the quantum dots (QDs), and by proper surface engineering, quantum yields approaching unity are now being achieved with type II-VI NCs based on CdSe. By regulating the particle size, the Photoluminescence (PL) emission peaks of CdSe NCs could be tuned spanning most of the visible spectrum [2,3]. CdSe NCs have been providing an ideal stage to explore the nucleation and growth mechanisms of NCs satisfying more complicated application, various structured NCs such as alloys [4], core/shell [5], doping [6] or heterostructure [7].

The first synthesis of high-quality CdSe NCs involved high-temperature synthesis using dimethyl cadmium and phosphine based selenium precursors [8,9]. However, for their application as active materials in photovoltaics or sensors, the presence of more ligand, surface-bound organic species may be a disadvantage to device performance. Moreover, trialkylphosphines are toxic, expensive, and oxidize easily. It is interesting to investigate the development of stable selenium precursors in the absence of strongly bound phosphine [1,5].

Jacek Jasieniak and co-workers reported the first phosphine-free synthesis of CdSe, based on the dissolution of selenium (Se) in 1-octadecene (1-ODE) after prolonged periods of heating at high temperatures [11]. Subsequently, other research groups have employed similar approaches to synthesize phosphine-free NCs. Solvents such as olive oil and oleylamine (OLA) have also been implemented [10]. Using OLA is another candidate with a high boiling point which forms a Se-OLA precursor complex, but it contains an amine group which effects to characterizations and structure of the NCs [3]. So the OLA and ODE were chosen as solvents with either the double bond and amine or the double bond only functionality for complex formation. In addition to, a long time heating protocols aimed at facilitating the dissolution of Se, additives including $NaBH_4$ and alkylthiols have been introduced. The disadvantage of the latter approaches is the introduction of additional elements (Na, B, and Se) which could appear as impurities in the final product [12]. It also affects to the characterizations of NCs. Recent studies have demonstrated that the synthesis of QDs can be carried out by employing insolubilized Se or SeO_2 powder with good yields [12]. To now, the use of a stable phosphine-free solution prepared by the dissolution of elemental Se in ODE (Se-ODE) has been explored as a viable selenium precursor for CdSe, ZnSe NCs synthesis [6,13]. All research groups have been developed the Se–ODE solution approach in various forms and investigated the high quality of binary NCs growth as Se precursor. However, the mechanism of formation and nature of the Se precursors in these solvents remain to be fully elucidated.

Herein, we report a detailed study of the phosphine-free selenium precursor solutions system. We establish that the efficacy of Se-ODE is highly dependent on preparation temperature. Further, we have investigated the role of chemical reactivity of them for the characterizations of CdSe NCs. The obtained results will helpful to available choice the suitable Se precursor for fine control the homogeneously and gradient alloyed semiconductor NCs.

2. Experimental

2.1. Phosphine-free selenium precursor solutions

For a typical preparation of Se-ODE solution (0.1 M), 3mmol of Se powder (99.99%) was transferred to a three-necked round-bottom flask, and 20 mL of ODE (99.99%) was added. This mixture have heated at temperatures ranging from 160 °C to 240 °C and stirred for 30 minutes (min) under nitrogen.

2.2. Synthesis of CdSe QDs

A mixture of oxide cadmium (CdO, 99.99%) (0.6 mmol), 10 mL of ODE and 1.2 mmol of oleic acid (OA-99.99%) was loaded into a three-neck flask and heated up to 250 °C, set under nitrogen gas until the mixture turned clear. Further, the Se precursor was quickly injected into the cadmium solution which was being stirred at 280 °C. The color of this mixture turned from yellow to red depending on the size of the CdSe nanocrystals formed. They were allowed to grow for different times to 60 min. The QDs were purified by dissolving in isopropanol, toluene and centrifuged to remove all residual products.

2.3. Characterization

Transmission electron microscopy (TEM) images were collected using a JoelJEM1010 operating at 80 kV. Absorption spectra were recorded using a Varian-Cary-5000 Ultraviolet-visible (UV–Vis) spectrophotometer. PL spectra were measured by a JobinYvonFlurolog FL3-22 fluorescence spectrophotometer with the wavelength of the exciting radiation of 400 nm. Fourier transforms infrared (FTIR) spectra were taken with a Nicolet IS 10 FTIR Spectrometer. Raman spectra (Ra) were recorded by a XploRA spectrometer (HoribaJobin-Yvon) with a laser wavelength of 532 nm. X-ray powder diffraction (XRD) was conducted on a Siemens D5005 X-ray diffractometer using Cu-Kα1 radiation with a wavelength of 0.15406 nm. For Ra and XRD measurements, the samples were deposited on glass slides. All of these measurements were performed at room temperature

3. Results and Discussion

3.1. Characterization of Se-ODE Precursor

The dissolution time of Se in ODE was strongly dependent on the temperature. It was found to decrease when the temperature increases. Such as, it took 30 hours (h) at 160 °C, 6 h at 180 °C, 3 h at 220 °C, 1 h at 220 °C and 45 min at 240 °C. Those times were taken to be the point at which observing Se dissolved completely in an ODE. The color of the Se-ODE solution changes after heats at the different temperature. The strongest coloration observed changing yellowish to orange-yellow color when the temperature heating increases from 160 °C to 200 °C. However, at a higher temperature (from 220 °C to 240 °C), it became increasingly colorless. The changing of color due to the formation of solvated Se in ODE when heat elemental Se in an ODE. In order to investigate the ODE–Se complex and predict the reaction mechanism of them, we analyzed the FTIR spectra (shown in Fig.1(a)).

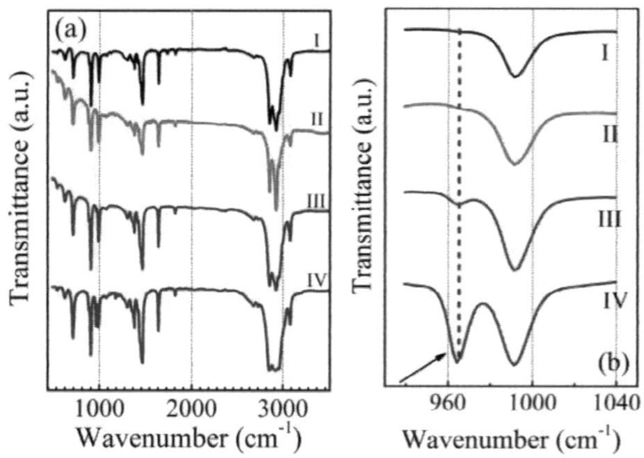

Figure 1. (a) Normalized FTIR spectra of ODE non-heating (I), ODE heating 180 °C (II), Se-ODE 180 °C (III), Se-ODE 240 °C (IV) for 30 min; (b) The main features of the various spectral responses are identical to those highlighted in Fig.1 (a)

We have been collected of ODE non-heating, ODE heating at 180 °C, Se-ODE heating at 180 °C and 240 °C for 30 min to measure. Fig.1(a) shows that for ODE non-heating the feature IR peaks of the vinyl group out of plane C-H bend located at v = 908.2 (cm^{-1}), v = 991.3 (cm^{-1}), the –C=C– stretch is located at 1642 cm^{-1} and C–H stretch located at 3077.6 (cm^{-1}) [14].

After heating ODE at 180 °C, the Se-ODE solution at 180 °C and 240 °C, the emergence of a band at v = 964.4 cm^{-1} appear and intensity is increased correlated higher temperature while it has disappeared in ODE's FTIR spectra non-heating and heating at 180 °C (see arrow Fig.1(b)). Thus, the kinetics of colorimetric changes of Se-ODE mixtures heated at a different temperature to confirm a reaction between Se and ODE that forms a poly-sulfur-alkyl complex.

3.2. The influence of Selenium Precursors on the characterization of CdSe Nanocrystals

3.2.1. Morphologies and structures

For this investigation, the CdSe NCs synthesized using Se-ODE precursors heated at three temperatures of 160 °C, 180 °C, and 240 °C. A series of CdSe NCs samples prepared at the growth temperature of 280 °C. Their name is denoted A160, A180, and A240 NCs.

The series TEM images of them with growth time of 15 min of NCs are shown in Fig.2a, b, c. The insets show the size distribution histograms of Fig. 2a, b, c, respectively. The shape and size of the CdSe QDs depend on the varied Se-ODE precursor. A160 and A180 NCs are spherical shapes with an average diameter about 11 ± 0.6 nm and about 5 ± 0.5 nm, respectively. Whereas, A240 is shown the cubic and triangles shapes with an average diameter about 10 ± 0.5 nm. A160 presents less chemical reactivity of Se precursor led to the low stabilizer, higher precursor concentration promotes the formation of larger nuclei and lower quality QDs (Fig.2a). While a higher chemical reactivity of Se precursor was found in A180 sample because the growth of QDs is faster and homogeneous. The A180 NCs particle size could be close to monodisperse (narrow size distributions) (see Fig. 2b). However, with A240 sample, we predicted possible too strong or less chemical reactivity of Se precursor so the growth of NCs is inhomogeneous (see Fig. 3c). Here, the less chemical reactivity Se precursor is possible some Se loss can occur.

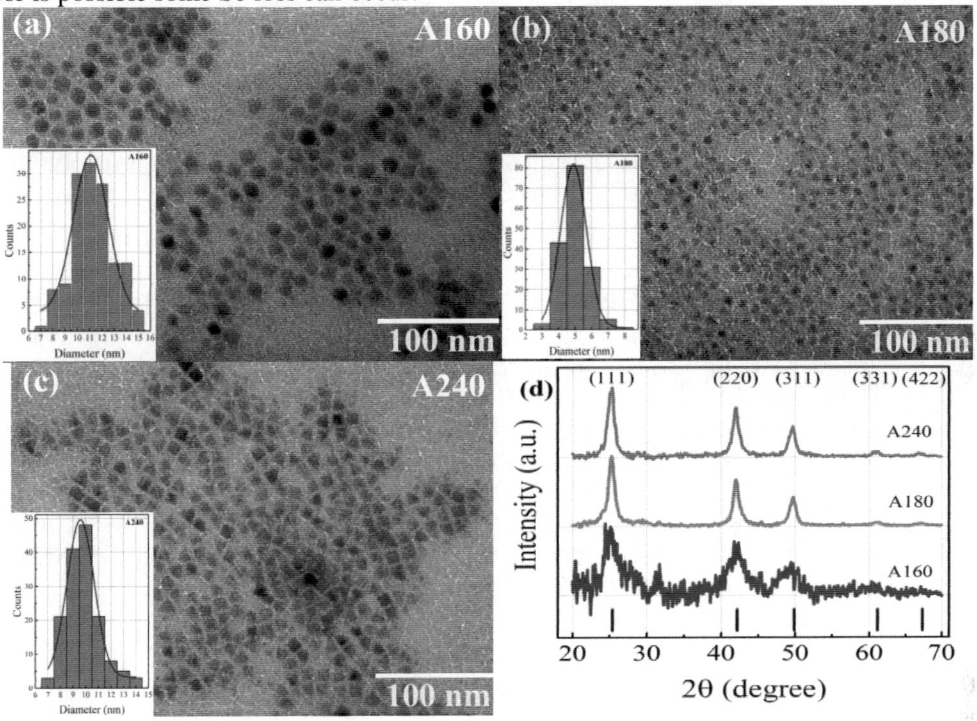

Figure 2. TEM images of (a) A160, (b) A180 (c) A240 NCs, the inset shows size distribution histogram of NCs (d) The XRD patterns of them.

To further characterization of the structures of CdSe NCs, their crystallographic properties were determined using powder XRD. The XRD patterns obtained from CdSe NCs with growth time of 30 min prepared using varied Se-ODE precursors are shown in Fig. 2d. According to the XRD pattern, three obvious diffraction peaks located near $2\theta \approx 25.41^{\circ}$, 42.1° and 49.1° corresponded to the (111), (220), and (311) planes of zinc blende phase CdSe. The broadening of the full width at half maximum (FWHM) in XRD patterns of A160 is observed (see Fig.3d.). It due to the interior defects produced in the lower quality QDs.

3.2.2. Optical properties

The time evolution absorption (Abs) and PL spectra of NCs are shown in Fig. 3a, b, c. It shows that in all reactions a significant red shift of the band-edge transition peak was observed after injection of the Se precursor.

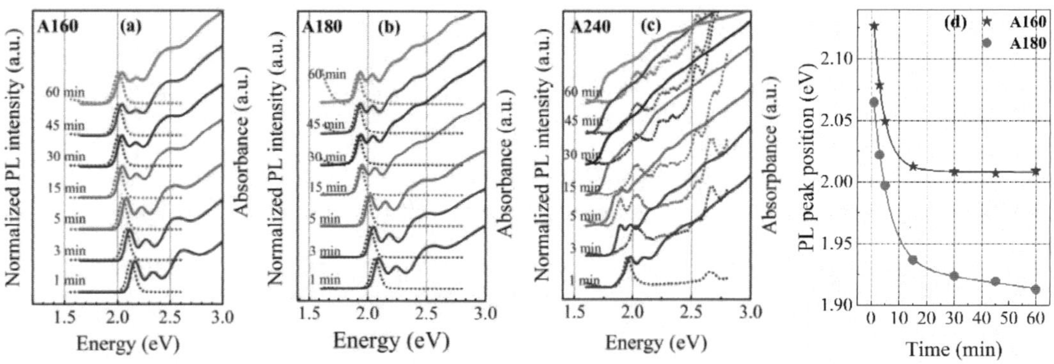

Figure 3. The temporal evolution of absorption spectra and PL spectra of CdSe nanocrystals using different Se precursors (a) A160, (b) A180, and (c) A240 NCs and (d) The PL peak positionof A160, A180 growth at 280 °C

The Abs peaks of NCs shift from 2.16 eV to 2.04 eV for A160 and from 2.11 eV to 1.94 eV for A180 (Fig. 3d). However, the optical properties of A240 NCs are presented badly quality after injected the Se-ODE precursors see in Fig. 3c. This is the reason that the Se-ODE prepared at high temperature is probably too strong chemical activity and ability Se loss had occurred [15]. Ability Se loss had occurred led to low of the reaction yield. We can see that these different kinds of Se precursors have been used to collect information on the reactivity of CdSe NCs and get optimized conditions to synthesize different sizes of high-quality CdSe QDs.

In general, the reaction speed was relatively slow with A160 and faster with A180. It has been presented by the evolution of the concentration of NCs and their PL's FWHM which shown in Fig. 4. These results indicate that the high quality of NCs is prepared with Se precursor heated at a higher temperature.

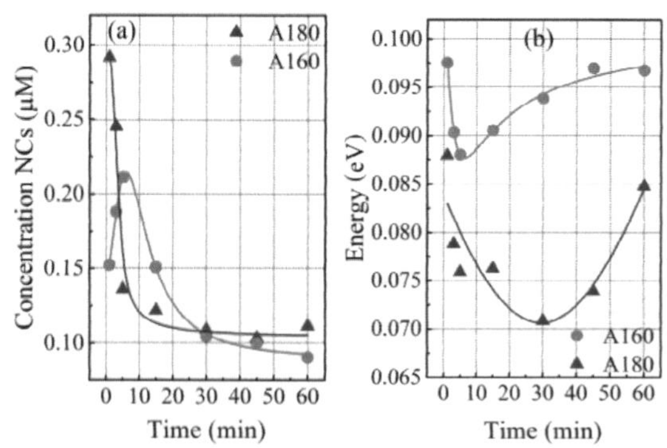

Figure 4. (a) Concentration NCs and (b) FWHM of the PL as functions of time for the A160 (circle) and A180 (triangle) NCs grew at 280 °C

This is due to a possible mechanism for the monomer formation in reactions. The number of NCs formed and their final size depends on both the rate of monomer generation and on the takeover of nucleation by growth. With the same reaction conditions for A160 and A180 NCs, the rate of nucleation and growth for A160 NCs is slow because of the monomer concentration is rich, so leads to the formation of larger particles size and NCs concentration is lower. For A180 NCs, these rates and the amount of precipitate formed of NCs are stronger, leads to the higher of NCs concentration (Fig.4a). Hence, the chemical reactivity of Se Precursor heating at 180 °C is higher than 160 °C. It might also affect the size distribution of NCs and led to change PL's FWHM (Fig.4b). It turned broader when the reaction lasted longer. For A160 NCs was changed strongly between 26 and 34 nm in the whole reaction while 23 nm and 27 nm for A180 NCs. According to the relation between the FWHM and the real nanocrystal size, the small FWHM of the NCs reveals that the sizes of them could be close to monodisperse (narrow size distributions). This means the chemicals reactivity of Se precursors plays an important role in forming the CdSe NCs. Adding different active Se-ODE into reaction solution changes the diffusion rates of the monomer and therefore controls nanocrystal growth so that the various size distributions can be achieved.

3.2.3. Vibrational properties

The room-temperature Raman spectra of CdSe NCs are shown in Fig. 5a. These two components can be assigned to surface optical (SO) and longitudinal optical (LO) phonons in agreement with previous observations for NCs [16].

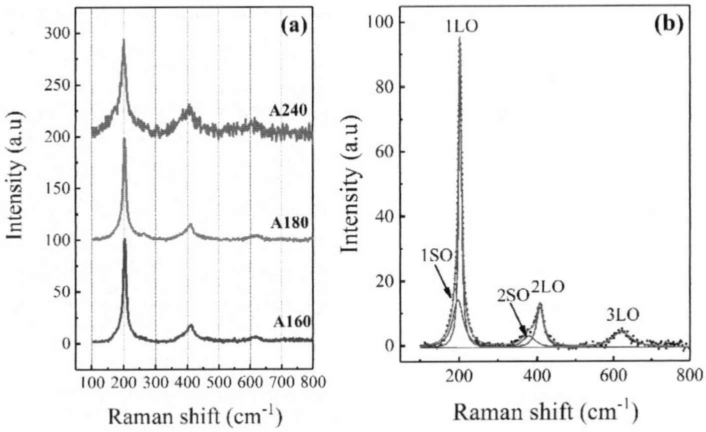

Figure 5. (a) Raman spectra and (b) Fitting of the main Raman band for A160, A180, and A240NCs

To evaluate the LO and the SO phonons, the elemental Raman bands are fitted by a sum of several Lorentzian peaks as shown in Fig. 5b. The peak parameters of A160, A180 and A240 NCs are shown in Table 1. The results show that the 1LO frequency increases from A160 to A180 and then decreases with A240. Whereas, the 1SO frequency decrease from A160 to A240. Herein, we also observed the phonon overtones up to the third order of LO phonons (see Fig. 5b). On the other hand, the FWHM of 1LO is decreased from A160 to A240. It is reasonable that can be related to the internal strain and phonon confinement in NCs.

Table 1. The Raman peak parameters of A160, A180 and A240 NCs

CdSe	1SO (cm^{-1})	1LO (cm^{-1})	2SO (cm^{-1})	2LO (cm^{-1})	3LO (cm^{-1})	FWHM 1LO
A160	197.39	203.0	379.25	407.39	621.10	17.6
A180	195.72	203.58	381.95	409.55	616.19	10.1
A240	172.99	200.37	379.59	406.53	613.7	9.3

4. Conclusions

In summary, the different heating of selenium powder dissolved in 1-octadecene has been made. The chemical reactivity of the Se-ODE is found to be highly dependent on heating temperature. The dissolution of the elemental selenium within 1-ODE proceeds through the fragmentation of the selenium chains to selenium bridges between alkene solubilizing molecules.

We evaluated the results of the influence of chemical reactivity of Se precursor on the characterization of colloidal CdSe NCs. It's high-quality NCs for A180 with fairly narrow size distributions.

We observe multiple Raman spectra by characteristical phonons up to the third order. The first order optical phonon peak that has allowed to be well-deconvoluted into two components assigned to SO and LO phonon modes. The band shifts and significantly broadens with respect varial Se precursor due to were found to result from a subtle balance between phonon confinement effects and lattice contraction due to internal strain. Moreover, this finding highlights that this easily prepared, cheap, and stable Se-ODE precursor is a highly viable selenium source for the synthesis of semiconductor nanocrystals.

References

[1] Dolai S, Dutta P, Muhoberac B B, Irving C D and Sardar R 2015 Mechanistic study of the formation of bright white light-emitting ultrasmall CdSe nanocrystals: Role of phosphine free selenium precursors *Chem. Mater.***27** 1057–70

[2] Yu W W, Qu L, Guo W and Peng X 2003 Experimental determination of the extinction coefficient of CdTe, CdSe, and CdS nanocrystals *Chem. Mater.***15** 2854–60

[3] Zimdars J and Bredol M 2015 Reduction of Selenium for the Phosphine-Free Synthesis of Metal Selenide Nanoparticles † *New J. Chem.***40** 1137–42

[4] Aldakov D, Lefrançois A and Reiss P 2013 Ternary and quaternary metal chalcogenide nanocrystals: synthesis, properties, and applications *J. Mater. Chem. C***1** 3756

[5] Shen H, Wang H, Tang Z, Niu J Z, Lou S, Du Z and Li L S 2009 High-quality synthesis of monodisperse zinc-blende CdSe and CdSe/ZnS nanocrystals with a phosphine-free method *CrystEngComm***11** 1733

[6] Shen H, Wang H, Li X, Niu J Z, Wang H, Chen X and Li L S 2009 Phosphine-free synthesis of high-quality ZnSe, ZnSe/ZnS, and Cu-, Mn-doped ZnSe nanocrystals *Dalt. Trans.* 10534

[7] Manthiram K, Beberwyck B J, Talapin D V. and Alivisatos A P 2013 Seeded Synthesis of CdSe/CdS Rod and Tetrapod Nanocrystals *J. Vis. Exp.* e50731

[8] Wang F, Richards V N, Shields S P and Buhro W E 2014 Kinetics and mechanisms of aggregative nanocrystal growth *Chem. Mater.***26** 5–21

[9] Flamee S, Cirillo M, Abe S, De Nolf K, Gomes R, Aubert T and Hens Z 2013 Fast, high yield, and high solid loading synthesis of metal selenide nanocrystals *Chem. Mater.***25** 2476–83

[10] Sapra S, Rogach A L and Feldmann J 2006 Phosphine-free synthesis of monodisperse CdSe nanocrystals in olive oil *J. Mater. Chem.***16** 3391

[11] Jasieniak J, Bullen C, Van Embden J and Mulvaney P 2005 Phosphine-free synthesis of CdSe nanocrystals *J. Phys. Chem. B***109** 20665–8

[12] Pu C, Zhou J, Lai R, Niu Y, Nan W and Peng X 2013 Highly reactive, flexible yet green Se precursor for metal selenide nanocrystals: Se-octadecene suspension (Se-SUS) *Nano Res.***6** 652–70

[13] Wang L, Sun X, Liu W and Yu X 2010 Phosphine-free synthesis of high-quality zinc-blende CdSe nanocrystals in air *Colloids Surfaces A Physicochem. Eng. Asp.***353** 232–7

[14] Banski M, Afzaal M, Malik M A, Podhorodecki A, Misiewicz J and O'Brien P 2015 Special Role for Zinc Stearate and Octadecene in the Synthesis of Luminescent ZnSe Nanocrystals *Chem. Mater.***27** 3797–800

[15] Bullen C, Van Embden J, Jasieniak J, Cosgriff J E, Mulder R J, Rizzardo E, Gu M and Raston C L 2010 High activity phosphine-free selenium precursor solution for semiconductor nanocrystal growth *Chem. Mater.***22** 4135–43

[16] Pejova B 2013 Phonon confinement and related effects in three-dimensional assemblies of cubic cadmium selenide quantum dots synthesized by conventional chemical and sonochemical routes *J. Phys. Chem. C***117** 19689–700

IOP Conf. Series: Materials Science and Engineering **343** (2018) 012028 doi:10.1088/1757-899X/343/1/012028

Effects of alkanolamide addition on crosslink density, mechanical and morphological properties of chloroprene rubber compounds

I Surya[1,2] and N Hayeemasae[3]

[1]Department of Chemical Engineering, Universitas Sumatera Utara, Medan, 20155, Indonesia
[2]Sustainable Energy and Biomaterial Center of Excellence, Faculty of Engineering, Universitas Sumatera Utara, Medan, 20155, Indonesia
[3]Department of Rubber Technology and Polymer Science, Faculty of Science and Technology, Prince of Songkla University, Pattani Campus, 94000, Pattani, Thailand

indradanas@yahoo.com

Abstract. The effects of alkanolamide (ALK) addition on crosslink density, mechanical and morphological properties of unfilled polychloroprene rubber (CR) compounds were investigated. The ALK was prepared from Refined Bleached Deodorized Palm Stearin (RBDPS) and diethanolamine and -together with magnesium and zinc oxides- incorporated into the unfilled CR compounds. The ALK loadings were 0.5, 1.0, 1.5 and 2.0 phr. It was found that ALK enhanced crosslink density, tensile modulus, tensile strength and hardness especially up to a 1.5 phr loading. Scanning Electron Microscopy (SEM) proved that the 1.5 phr of ALK exhibited the greatest matrix tearing line and surface roughness, due to the highest degree of crosslink density and mechanical properties.

1. Introduction

In curing process of polychloroprene rubber or chloroprene rubber (CR), the specialty rubber is traditionally cured by both magnesium oxide (MgO) and zinc oxide (ZnO), at satisfactory loadings of 4 and 5 phr, respectively. Besides MgO and ZnO, ethylene thiourea (ETU) is widely used as the curing accelerator for CR [1].

Nowadays, some newer curing agents such as thiophosphoryl disulfides, dimethyl l-cystine, and cetyltrimethylammonium maleate, have been reported [2-3]. Although there are many published papers on different curing agents for CR, the most practical curing agents are still ETU, MgO and ZnO, due to the superior mechanical properties of the CR vulcanisates. ETU is a toxic material and is suspected to be carcinogenic [4-5] and hence, the appearance of an alternative curing accelerator for CR, which is capable of providing CR vulcanisates equivalent or superior to those provided by the ETU and both metal oxides, has been in demand.

Some previous works revealed that alkanolamide (ALK) functioned as an accelerator in unfilled NR compounds [6-7], silica-filled NR compounds [8-9], carbon black-filled NR, SBR and ENR compounds [10-11]. Therefore, this study reports the utilising of ALK as an additive ingredient for unfilled CR compounds. The effects of ALK loading on crosslink density, mechanical and morphological properties of unfilled CR compounds were investigated.

2. Experimental

2.1. Materials

Polychloroprene rubber (CR) [Skyprene B-30] was purchased from TOSOH Co. (Japan). Other compounding ingredients such as MgO, ZnO, stearic acid and sulphur were all obtained from Bayer Co. (M). Sdn. Bhd., Petaling Jaya, Selangor, Malaysia. All materials were used as supplied. The ALK was prepared using RBDPS and diethanolamine. The procedure of the ALK preparation was given in our previous report [8, 11]. The chemical formula of ALK is $CH_3(CH_2)_{14}CON(CH_2CH_2OH)_2$.

2.2. Compounding

A sulphur-accelerated curing system was used for compounding. The recipe for the preparation of the CR compounds is given in Table 1. The compounding procedure was done in accordance with the American Society for Testing and Material (ASTM) – Designation D 3184 – 80. Compounding was done on a two-roll mill. Table 1 also shows the designation and composition of the CR-based recipes used in this study.

Table 1. The composition and designation of the CR compounds

Ingredients	Content (phr)*	Designation				
		CR-A/0.0 (Control)	CR-A/0.5	CR-A/1.0	CR-A/1.5	CR-A/2.0
CR	100.0	100.0	100.0	100.0	100.0	100.0
MgO	4.0	4.0	4.0	4.0	4.0	4.0
ZnO	5.0	5.0	5.0	5.0	5.0	5.0
Stearic acid	1.0	1.0	1.0	1.0	1.0	1.0
Sulphur	2.0	2.0	2.0	2.0	2.0	2.0
ALK	0.0 – 2.0	0.0	0.5	1.0	1.5	2.0

*parts per hundred parts of rubber

2.3. Measurement of crosslink density

Swelling tests on the CR vulcanisates were performed in toluene in accordance with ISO 1817. The cured test pieces (30 mm × 5 mm × 2 mm) were weighed using an electrical balance and swollen in toluene until equilibrium, which took 72 hours at room temperature. The samples were taken out from the liquid, the toluene was removed from the sample surfaces and the weight was determined. The samples were then dried in the oven at 60 °C until constant weights were obtained. The swelling results were used to calculate the molecular weight between two crosslinks (M_c) by applying the Flory-Rehner Equation [12-13].

$$M_c = \frac{-\rho_p V_s V_r^{1/3}}{\ln(1-V_r) + V_r + \chi V_r^2} \quad (1)$$

$$V_r = \frac{1}{1+Q_m} \quad (2)$$

Where ρ is the rubber density (ρ of CR = 0.96 g/cm^3), V_s is the molar volume of the toluene (V_s = 106.4 cm^3/mol), V_r is the volume fraction of the polymer in the swollen specimen, Q_m is the weight increase of the blends in toluene and χ is the interaction parameter of the rubber network-solvent (χ of CR = 0.4284). The degree of the crosslink density is given by;

$$V_C = \frac{1}{2M_C} \quad (3)$$

2.4. Tensile and hardness properties

Dumbbell-shaped samples were cut from the moulded sheets. Tensile tests were performed at a cross-head speed of 500 mm/min using an Instron 3366 universal tensile machine according to ISO 37. The tensile strength and stress at 100% elongation (M100), 300% elongation (M300), and elongation at break were investigated. The hardness measurements of the samples were performed according to ISO 7691-I using a Shore A type manual durometer.

2.5. Scanning Electron Microscopy (SEM)

The tensile fractured surfaces of the CR vulcanisates were examined using a Zeiss Supra-35VP scanning electron microscope (SEM) to obtain information regarding the possible presence of micro-defects. The fractured pieces were coated with a layer of gold to eliminate electrostatic charge build-up during the examination.

3. Results

3.1. The crosslink density

Table 2 shows the effect of ALK loading on crosslink density of CR compounds. As shown, the crosslink density of the CR-A/0.5 compound was higher than that of the control compound. The addition of 0.5 phr of ALK increased the crosslink density of CR compound significantly and hence, the ALK was considered as an additional vulcanising agent for the CR system.
Increases the ALK loading up to a 1.5 phr caused a further increase in crosslink density. Beyond the loading started to decrease the crosslink density.

Table 2. Crosslink density dan mechanical properties of the unfilled-CR vulcanisates with/without ALK

	Unfilled CR vulcanisates				
Mechanical Properties	CR-A/0.0 (Control)	CR-A/0.5	CR-A/1.0	CR-A/1.5	CR-A/2.0
M100, MPa	0.72	0.78	0.82	0.83	0.78
M300, MPa	1.30	1.31	1.37	1.38	1.29
TS, MPa	13.7	15.5	17.2	18.9	17.8
EB, %	1,612.3	1,590.3	1,536.8	1,487,0	1,516.3
Hardness, Shore A	55	56	57	58	56
Crosslink density, (x10^{-5} mol/cm^3)	10.37	16.95	19.92	23.79	13.96

3.2. The mechanical properties

Table 2 shows the effects of ALK loading on the mechanical properties of the CR compounds. As shown, the addition of up to 1.5 phr of ALK into the CR compound increased the M100 and M300, hardness and TS. However, further loading increases decreased those properties. The modulus (or stiffness/hardness) and tensile properties of a rubber vulcanisate are dependent mainly on the degree of crosslink density [11, 14-15]. The improvements of M100, M300, hardness and TS up to 1.5 phr were due to the function of ALK as an additional curing agent, which increased the degree of crosslink density of the CR compounds. This explanation is in line with the crosslink density trend shown in Table 2 and the morphological properties as shown later on in Fig. 1.

The deterioration of properties beyond 1.5 phr was due to the excessive loading of ALK, which decreased the total crosslink density. Presumably, the excessive amount of ALK formed a boundary layer, which dissolved and coated a part of the curatives (S, MgO and ZnO); and hence, decreased the formation of crosslinking of CR by those curatives. This explanation is again in line with the result of crosslink density shown in Table 2.

Table 2 shows the effect of ALK loading on the EB of the CR vulcanisates. As shown, ALK decreased the EB by up to 1.5 phr, and then increased it slightly as the loading further increased. EB depends mainly on the degree of crosslink density [14]. The reduction of EB by up to 1.5 phr was simply attributed to the higher degree of crosslink density; which immobilised the CR segments more pronounced. However, beyond the 1.5 phr of ALK loading, EB was found to increase. An explanation for this is again given by the excessive effect of ALK, which decreased the crosslink density and causes a more free movement of the CR segments.

3.3. The morphological properties

Figure 1 shows SEM micrographs (magnification 50 X) of the fractured surfaces of the CR vulcanisates with and without ALK. It was clearly observed that the micrographs of CR vulcanisates with ALK (CR-A/0.5, CR-A/1.0, CR-A/1.5, and CR-A/2.0) – as shown in Figs. 1b, c, d, and e - exhibited a greater matrix tearing lines and surface roughness than that of the CR vulcanisate without ALK (as shown in Fig. 1a). The rough surface of a rubber vulcanisate indicates that the matrix tearing relates to the value of tensile strength (ref). The rougher the surface, the higher is the tensile strength [16-17]. As shown in Fig. 1, the rougher surfaces also indicated a stronger interaction within the CR segments and that the micrographs of the tensile fractured surfaces were in good agreement with the results of TS as shown in Table 2.

An enhancement in rupture energy (due to a stronger interaction within the CR segments) was responsible for the roughness and matrix tearing line of the fractured surface. The 1d-micrograph of unfilled CR vulcanisate with 1.5 phr of ALK (vulcanisate CR-A/1.5) exhibited the greatest matrix tearing and surface roughness. This indicates the highest degree of crosslink density, which altered the interactions between CR segments into the strongest ones; and thus, caused an increase in tensile strength.

The micrographs of the tensile fractured surfaces were in good agreement with the results obtained by previous workers. They reported that an increase in rupture energy was responsible for the roughness and matrix tearing line of the fractured surface [17-18].

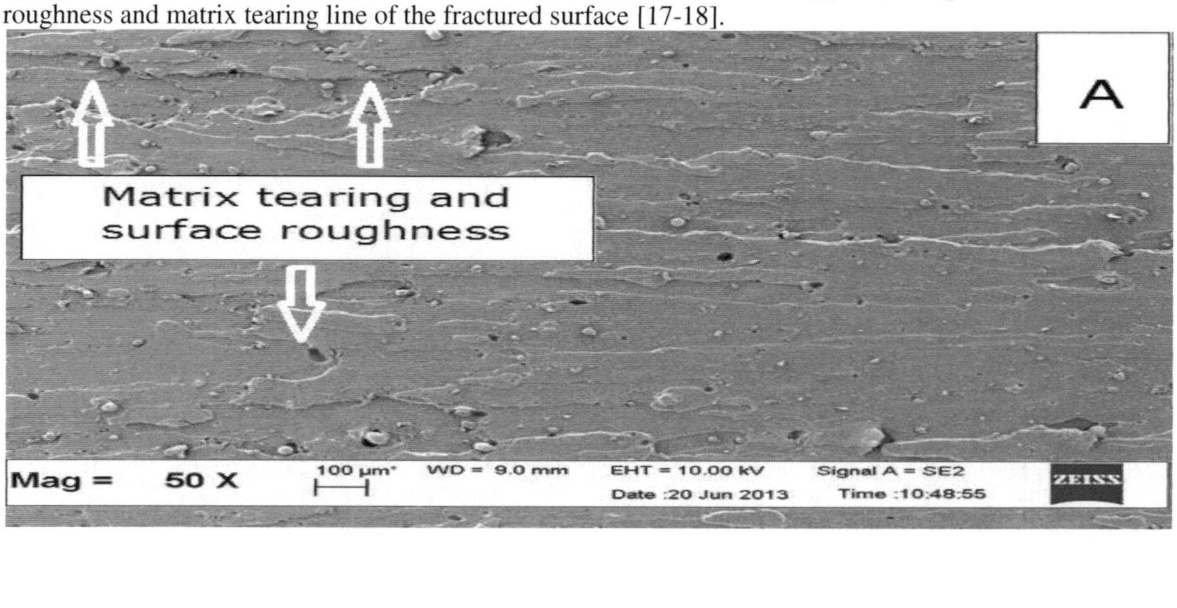

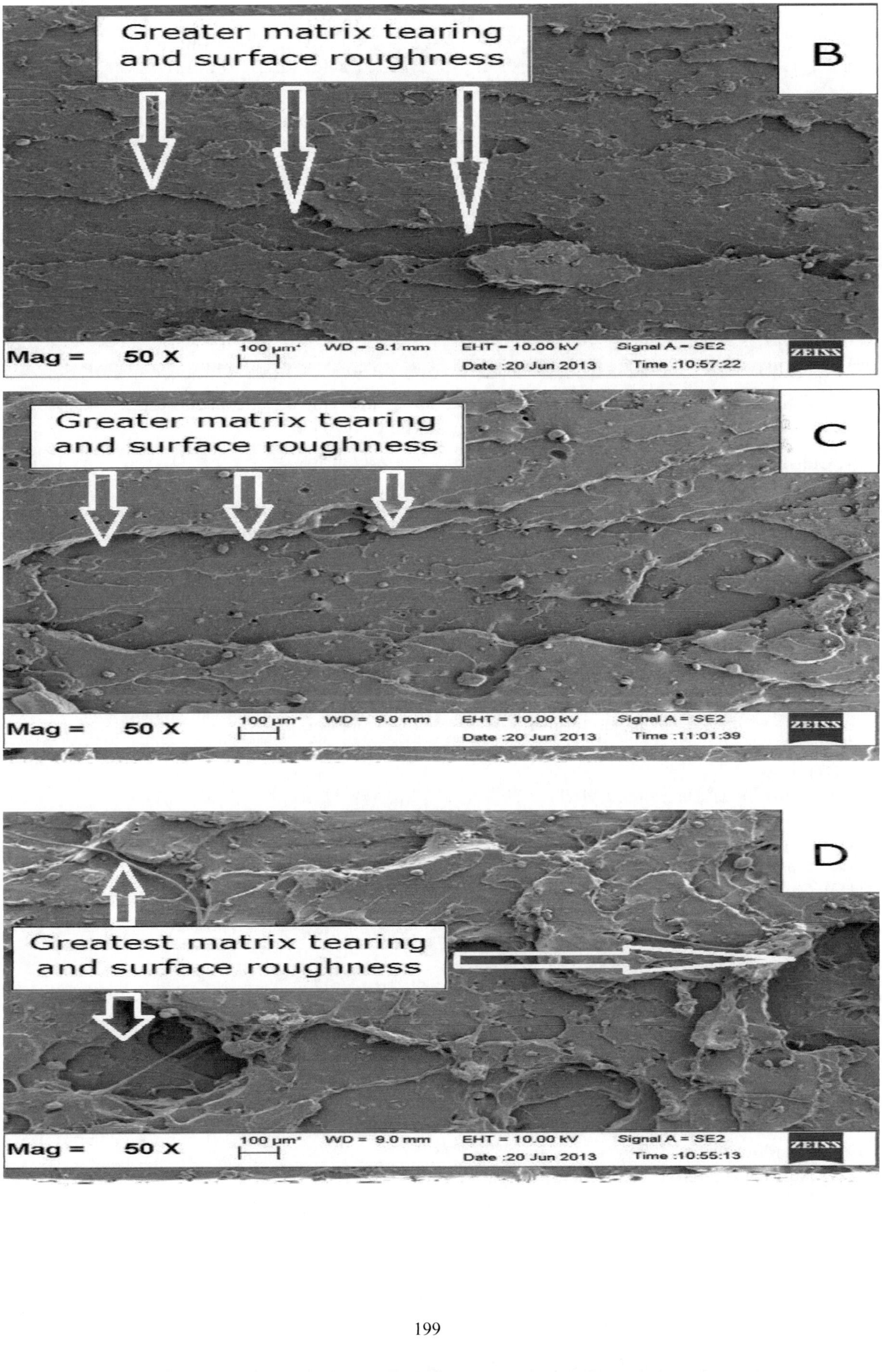
Greater matrix tearing and surface roughness
B
Mag = 50 X
100 µm
WD = 9.1 mm
EHT = 10.00 kV
Signal A = SE2
Date :20 Jun 2013
Time :10:57:22
ZEISS
Greater matrix tearing and surface roughness
C
Mag = 50 X
100 µm
WD = 9.0 mm
EHT = 10.00 kV
Signal A = SE2
Date :20 Jun 2013
Time :11:01:39
ZEISS
D
Greatest matrix tearing and surface roughness
Mag = 50 X
100 µm
WD = 9.0 mm
EHT = 10.00 kV
Signal A = SE2
Date :20 Jun 2013
Time :10:55:13
ZEISS

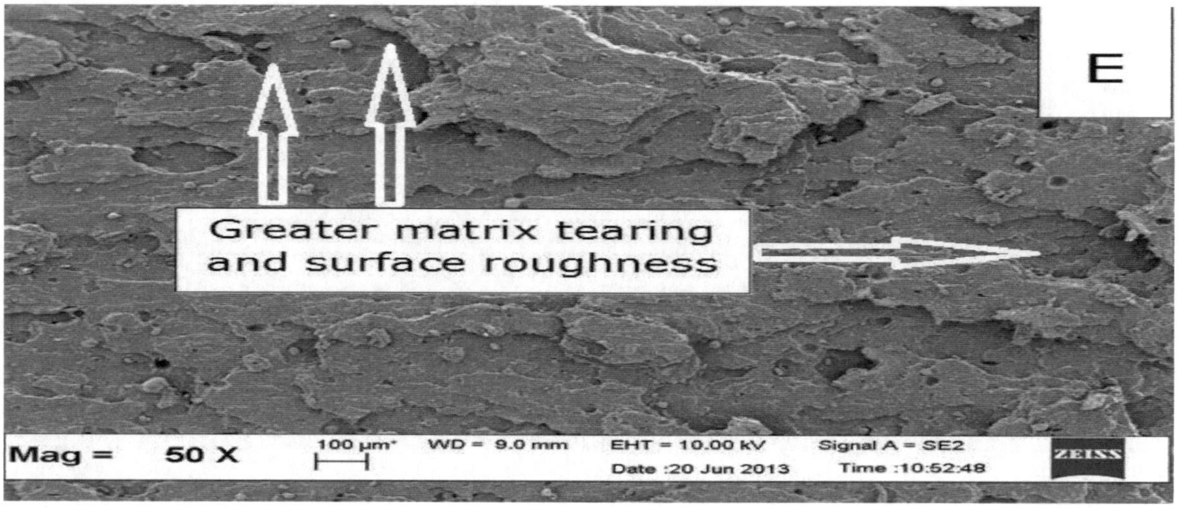

Figure 1. The SEM micrographs of CR vulcanisate failed fracture at a magnification of 50 X; (A) CR-A/0.0 (B) CR-A/0.5 (C) CR-A/1.0 (D) CR-A/1.5 (E) CR-A/2.0

4. Conclusion

From this study, the following conclusions were drawn:

1. Alkanolamide increased the crosslink density and mechanical properties of magnesium-zinc oxides vulcanised unfilled chloroprene rubber compounds, especially up to a 1.5 phr loading.
2. The enhancements of the crosslink density and mechanical properties were due to the function of alkanolamide as a curative additive in unfilled chloroprene rubber compounds.
3. Morphological studies of the tensile fractured surfaces of the unfilled chloroprene vulcanisates indicated that alkanolamide interacted well with chloroprene rubber compounds.

Acknowledgements

The authors would like to thank Universiti Sains Malaysia for providing the research facilities for carrying out the experiment and for making this research work possible.

References

[1] Akiba M and Hashim A 1997 *Prog Polym Sci* **22**(3) 475-521.
[2] Das A *et al* 2004 *J App. Polym Sci* 2004 **91**(3) 1913-19.
[3] Ismail H *et al* 2003 *Polym Test* **22**(2) 179-83.
[4] Smith DM 1984 *Bri J Ind Med* **41**(3) 362-66.
[5] Chhabra R *et al* 1992 *Fundam Appl Toxicol* **18**(3) 405-17.
[6] Surya I *et al* 2017 *IOP Conf Ser Mater Sci Eng* **223** 012006.
[7] Surya I *et al* 2017 *IOP Conf Ser Mater Sci Eng* **223** 012012.
[8] Surya I *et al* 2013 *Polym Test* **32** 1313-21.
[9] Surya I *et al* 2014 *Polym Test* **40** 24-32.
[10] Surya I *et al* 2015 *Polym Test* **42** 208-14.
[11] Surya I and Ismail H 2016 *Plast Rubber and Compos* **45(7)** 287-93.
[12] Flory PJ and Rehner JrJ 1943 *J Chem Phys* **11** 521-26.
[13] Flory PJ 1950 *J Chem Phys* **18**(1) 108-11.
[14] Ismail H and Chia H 1998 *Polym Test* **17**(3) 199-210.
[15] Surya I *et al* 2016 *Polym Test* **50** 276-82.
[16] Ismail H *et al* 2011 *Polym Test* **30**(7) 784-90.
[17] Ismail H *et al* 2011 *Polym Plast Technol Eng* **50**(14) 1421-28.
[18] Nabil H *et al* 2013 *Polym Test* **32**(2) 385-93.

Identification of Renewable Energy Potential in Ciberang River, Cisarua Village, Bogor, West Java

Ika Sari Damayanthi Sebayang[1], Acep Hidayat[1], Nur Indah[2]

[1] Department of Civil Engineering, Faculty of Engineering, Universitas Mercu Buana, 11650 Jakarta, Indonesia
[2] Department of Mechanical Engineering, Faculty of Engineering, Universitas Mercu Buana, 11650 Jakarta, Indonesia

ikasari.damayanthi@mercubuana.ac.id

Abstract. This paper presented the analysis of potential energy in Ciberang River, Cisarua Village, Bogor West Java. The objective of this work is to ascertain the availability of water due to rainfall (discharge simulation). The simulation required data in the form of rainfall intensity, climate, evapotranspiration and water discharge. The rainfall station is determined by Thiessen Method and located in Cisalak-Baru Station. Rainfall data from 1997-2012 was used. The area of Ciberang River basin is about 46.19 km^2 and not influenced by the change of land area used per year. The height of waterfall allowed was 160 meters due to its topography. The result of water availability was analysed using NRECA method and calibration was done using water data recording (Automatic Water Level Recorder) in downstream of Ciberang-Sabagi station. Calibrated result analysis was then plotted to show the Flow Duration Curve (FDC). Potential capacity of power was obtained from the amount of discharge with the reliability level of 50-60%. At 60% reliability level, calculation of discharge equal to 5.8 m^3/s and then was used for design parameter. The generated power capacity is 45,813,400.21 kWh/year with the assumptions of net head 159 meters, generator's efficiency of 0.95 and turbine's efficiency of 0.85. The result shows that Ciberang River has a potential to be developed as a hydro power plant.

Keywords: Ciberang, Rainfall-Discharge Simulation, Flow Duration Curve, Discharge, Hydropower Potential

1. Introduction

The Government of Indonesia has launched a 35,000 MW electricity development program by employing river flow kinetic energy. The Ministry of Energy and Mineral Resources (ESDM) reported that the potential of small to large-scale energy in Indonesia reached about 75,670 MW, and only 4,200 MW or 5,6% was used. On the other hand, the potential of mini to micro-scale energy (hydropower) has been estimated for about 500 MW, and approximately 210 MW (or equivalent to 42%) has been built [7]. The potential of mini and micro scale hydropower was apparently much larger than the estimated number of 500 MW if the hydro power energy of small and large-scale will being used. The report shown that the development of many PLTMH (*Pembangkit Listrik Tenaga Mikro Hidro*) in Indonesia have likely to be sizable capacity due to several reasons such as the amount of electrical conformity, investment cost, and local resources capability.

To fulfil the above needs, the study and analysis of the river stream's reliability is required. One of the potential source for hydro power station is Ciberang River located at Cisarua Village, Bogor, West Java. The river provides sufficient discharge every year and make it a promising source of river flow kinetic energy. The area is fairly high and not influenced by the change of land area used per year. It is one of the national park area in place. Having located at Halimun Salak Mountain National Park, which has strong current and suitable for rafting, Ciberang River is a potential candidate for the energy sources [5,7,8,9]. Two important characteristics possessed by of Ciberang Rivers are the head and flow current. The height of its location make Ciberang River has enough potential and kinetic energy. Thus, the discharge and river flow are ensured to be available throughout the years. The main objective of this work is to determine the discharge availability that can be generated by Ciberang River as input of rainfall-discharge modelling.

2. Descriptions of The Study

In this study, the upstream river flow of Ciberang River at Halimun Salak National Park area was used. The research area administratively is located in Kampung Lebak Sanab, Cisarua Village, Cigudeg District, Bogor Regency. Coordinates of the weir plant at X: 661412.21 and Y: 9267156.2. The estimated area of the catchment is about 46,19 km^2 and its topography mainly mountainous with a steep slope. There is a river gauging station with Automatic Water Level Recorder located downstream to the proposed weir position on the same river.

3. Methodology

Assessment of hydropower potential energy requires the followings: 1). topography map, 2), location of proposed weir and power house, 3). rainfall data, and 4). evapotranspiration. Location of proposed weir and power house will determine the falling height (head) [4,5,2,3] while evapotranspiration will be obtained from climate data at climate station.

3.1 Topography Map

Figure 1 shows the topography map obtained from Bakosurtanal with the scale of 1:25.000 (unit in mm). The catchment area was then calculated from the figure and approximately equal to 46,19 km^2. Falling height was determined from measurement on site and the value is 160 m.

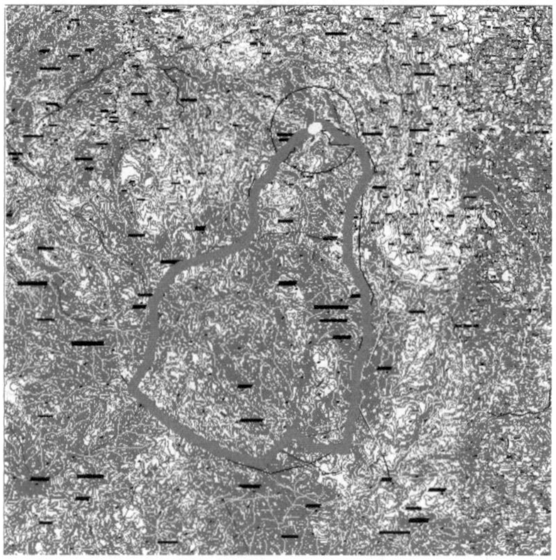

Figure 1. Estimated Catchment Area

3.2 Rainfall Data

Discharge analysis on rainfall data was conducted on Cisalak Baru Station, where the location is considered to be the closest to the area being studied. The rainfall statistics was obtained from the recorded data since 1997-2012, it was statistics showing the information of daily rain.

3.3 Evapotranspiration Data
To obtain the evapotranspiration data, Penman Modification Method [1] was used by considering the following parameters:

a) Air temperature
b) Sunshine radiation
c) Humidity
d) Wind velocity

The equation used to calculate the magnitude of the evapotranspiration is written below.

$$E = C.(W.R_n + (1 - W).F(u).(e_a - e_d)) \quad (1)$$

Where:
E = daily potential evapotranspiration (mm/day)
C = regulator factor
W = weighting factor
Rn = Radiation net
F(u)= function of wind velocity (m/s)
ea = saturated vapour pressure
ed = actual vapour pressure

3.4 NRECA Water Balance
One of the models used in obtaining water availability is the NRECA model [1,2,3]. The water equilibrium model in NRECA is based on a common water equilibrium processes. It was determined by the rainfall over the soil surface and evaporation of crops that cover the area, some will flow on the surface and the rest will seep into the soil. If the water infiltration saturates the soil surface, it will creep into percolation and then come out into the river as base flow. This model can simulate the monthly water equilibrium in a catchment area in order to calculate the total run-off of values: monthly rainfall, evapotranspiration, soil moisture and groundwater availability. The difference of the NRECA model with other water equilibrium models is only the number of parameters taken. The input data required for this NRECA model include:

- The monthly rain
- Evapotranspiration
- Monthly average temperature
- Intensity of sunlight
- Relative humidity
- Wind speed
- Initial conditions of moisture content
- Ground count of ground water
- Index of soil moisture storage capacity in the catchment area
- Percentage of run off that flows on the subsurface path
- Percentage of water entering into groundwater flow.

3.5 Flow Duration Curve
To calculate the level of debit reliability, the flow duration curve was used [1,2,3]. The flow duration curve is a graph showing the river flow for a certain period of time in a year [2,3,4,5]. Its value was determined from the arc at flow duration graph plotted form discharge data recorded from the last 10 years.

The mainstay debit is the minimum flow of the stream with the possibility of a certain percentage of debit being fulfilled, e.g. 90%, 80% or any other value, so it can be used for a particular need. The level of discharge reliability is selected based on the probability of occurrence following the Weibull formula below [1,9,10].

$$P = {}^{i}/_{(n+1)}\, x\, 100\% \quad (2)$$

Where:
i = Order number of discharge,
n = amount of data,
P = Probability of the set of values during the observation period (%).

4. Result and Discussion

The reliable discharge data was estimated by developing the model of Ciberang River based on the available data.

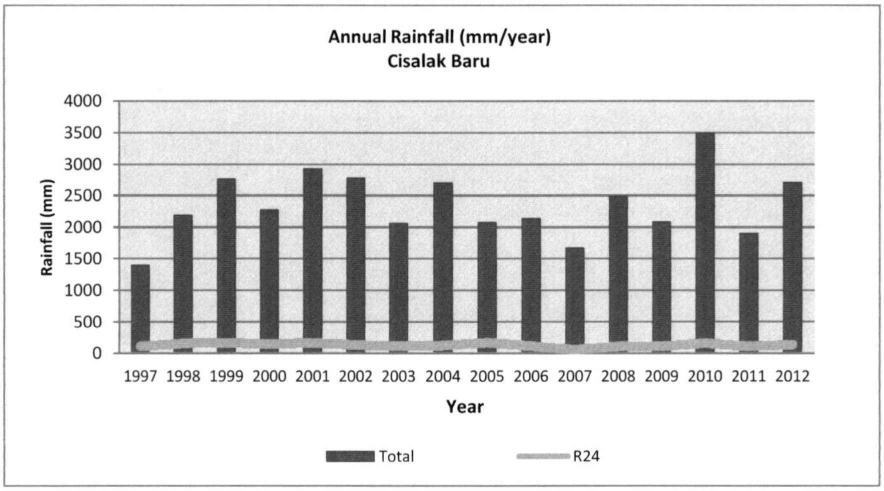

Figure 2 Cisalak Baru annual rainfall

To calculate potential evapotranspiration, the data used was derived from climatological data at Taktakan Climatological Station, the station closest to the location of the study. Using the Penman modification method (eq.1), the calculation result is presented in Table 1.

Table 1. Result of Potential Evapotranspiration

Month		Eto (mm)
		normal
Jan	1	46.7
	2	62.1
Feb	1	39.4
	2	56.0
Mar	1	69.1
	2	73.4
Apr	1	53.7
	2	70.5
Mei	1	49.8
	2	64.3
Jun	1	66.5
	2	54.4
Jul	1	66.6

	2	71.5
August	1	65.8
	2	81.0
Sep	1	75.1
	2	86.2
Oct	1	64.7
	2	90.3
Nov	1	68.7
	2	62.8
Des	1	70.1
	2	56.7

The data and analysed results that have obtained from the previous step, the reliable discharge then can be modelled. The model was calibrated using logging debit data at AWLR Ciberang-Sabagi Station with a wide range of watersheds of 301.2 km^2. The recorded discharge data used for the calibration model was taken from year 2010 to 2012 and compared with the discharge data from Cibeureum's rainfall station from the same year.

Table 2. Parameter Calibration

Catchment Area	**46.19**	**km^2**	**R^2**	**0.26144**	
Average Annual Rainfall	2382.83	mm			
Coefficient C	0.50		Storage	2400.00	mm
Nominal	1291.42		IGWS	220.00	mm
PSUB	0.50		Rain Factor	3.57	
GWF	0.60				

From the model that has been calibrated, the initial model of Ciberang River is depicted as follows:

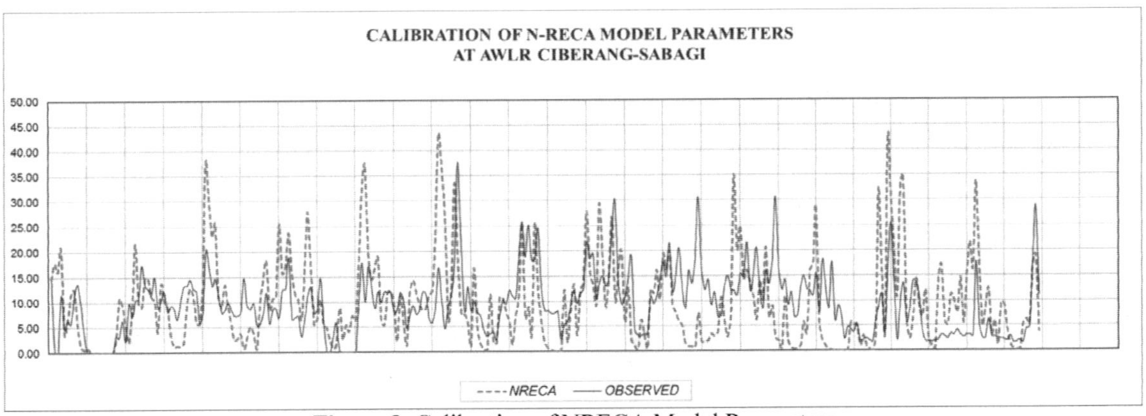

Figure 3. Calibration of NRECA Model Parameters

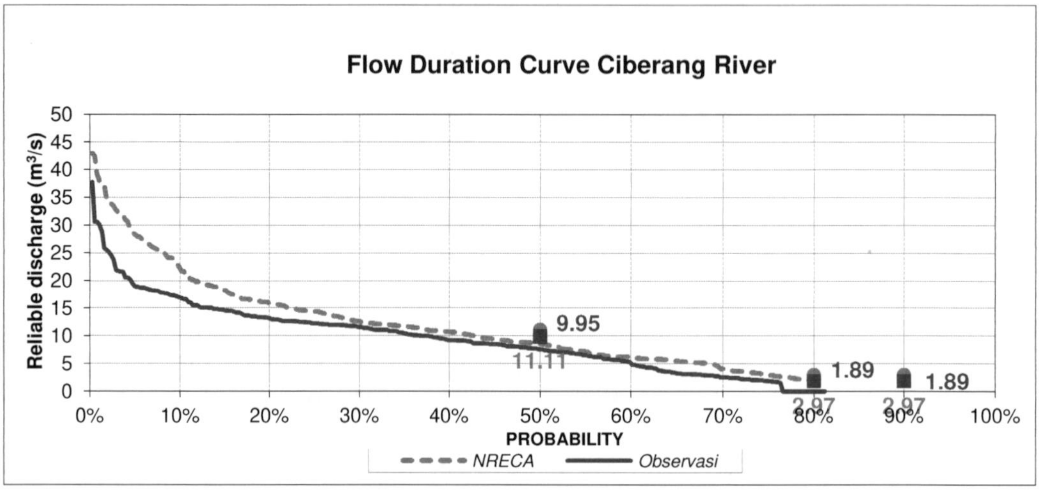

Figure 4. Flow Duration Curve Ciberang River

To know the capacity of power that can be utilized from Ciberang River, it can be taken the amount of discharge with the reliability level of 50% - 60%. In this study, the reliability taken for power capacity calculation is a 60% reliability level equal to 5,8 m^3/s (Table 3.)

Table 3. Result of Reliable Discharge Calculation

Percentage	Debit	NRECA	Observed	Difference
5%	49.49%	19.2	28.25	18.9
10%	30.84%	38.4	22.12	16.9
20%	20.40%	76.8	15.87	13.18
30%	8.87%	115.2	12.6	11.58
40%	15.87%	153.6	10.72	9.25
50%	11.68%	192	11.11	9.95
60%	13.90%	230.4	5.8	5.09
70%	61.23%	268.8	4.07	2.52
80%	57.63%	307.2	2.97	1.89
90%	57.63%	345.6	2.97	1.89
95%	57.63%	364.8	2.97	1.89

The result of reliable discharge analysis from NRECA method with probability of 60% was equal to 5,8 m^3/s. Its value was used as input for discharge design to determine the energy that can be produced for one year. Capacity of power plant was generated with the following three assumptions: the height of 159 meters (net), efficiency of generator 0.95 and efficiency of turbine 0,85 (based on the FDC) were presented in Table 4.

Table 4. Result of Calculation of Energy Capacity Per Year

Qmax	5.80	m^3/s
Head	159.00	m
Eff. gen	0.95	
Eff. turbine	0.85	

Prob.	FDC	Qgen	Power (kW)	Energy (kWh)
5%	28.95	5.80	7,305.28	3,155,879.43
10%	20.67	5.80	7,305.28	3,155,879.43
15%	18.24	5.80	7,305.28	3,155,879.43
20%	15.81	5.80	7,305.28	3,155,879.43

25%	13.99	5.80	7,305.28	3,155,879.43
30%	12.18	5.80	7,305.28	3,155,879.43
35%	11.23	5.80	7,305.28	3,155,879.43
40%	10.27	5.80	7,305.28	3,155,879.43
45%	9.57	5.80	7,305.28	3,155,879.43
50%	8.87	5.80	7,305.28	3,155,879.43
55%	7.34	5.80	7,305.28	3,155,879.43
60%	5.80	5.80	7,305.28	3,155,879.43
65%	4.63	4.63	5,830.82	2,518,914.12
70%	3.46	3.46	4,355.43	1,881,545.80
75%	2.61	2.61	3,290.80	1,421,626.07
80%	1.77	1.77	2,226.17	961,706.35
85%	1.18	1.18	1,488.08	642,849.47
90%	0.60	0.60	749.98	323,992.59
95%	0.35	0.35	444.94	192,212.61
100%	0.00	0.00	0.00	0.00
			TOTAL	**45,813,400.21**
			capacity	**63,994,221.83**
			plant factor	**71.59%**

From the above table, the total capacity equal to 45.813.400,21 kwH/year. This result concludes that the development of Ciberang River has potential for hydro power plant.

5. Conclusion

This work has concluded that Ciberang River has potential as a renewable energy source with 60% discharge of 5,8 m^3/sec. Capacity of energy that can be produced with the assumption of height of 159 meters, efficiency of generator 0,95 and efficiency of turbine 0,85 is equal to 45.813.400,21 kwH/year. Some parameters need to be considered for future work: a better analysis result and recorded data of water discharge at specific location of Ciberang River for one year. Further surveys and analysis need to be done to determine a more precise falling height.

6. References

[1] Chow V. T., 1964. *Handbook Of Applied Hydrology*, Mc Graw-Hill. New York.

[2] Emeribe, C.N. et al, 2016, *Hydrological Assessments Of Some Rivers In Edo State, Nigeria For Small-Scale Hydropower Development,* Nigerian Journal of Technology (NIJOTECH) Vol. 35, No. 3, pp. 656 – 668.

[3] H. Rintis et al, 2016, *Rainfall-Discharge Simulation in Bah Bolon Catchment Area by Mock Method, NRECA Method, and GR2M Method*, Applied Mechanics and Materials, Vol. 845, pp. 24-29.

[4] Koteia R., et al, 2016., *Estimation of Flow-Duration and Low-Flow Frequency Parameters for the Sumanpa Stream at Mampong Ashanti in Ghana for the 1985-2009 Period,* American Scientific Research Journal for Engineering, Technology, and Sciences (ASRJETS), Volume 15, No 1, pp 62-75.

[5] M. Mohibullah, A. Radzi, and M. Hakim, "*Basic design aspects of microhydro power plant and its potential development in malaysia*," in Power and Energy Conference, 220–223.

[6] N. Ijumba and C. Wekesah, 1996, *Application potential of solar and minihydro energy sources in rural electrification*, AFRICON, 1996., IEEE AFRICON 4th, vol. 2, pp. 720–723, Vol. 2.

[7] PLN, Rencana Umum Penyediaan Tenaga Listrik 2016-2025.

[8] S. Adhau, R. Moharil, and P. Adhau, 2012. *Design of mini hydro power plant on small weir: Need based technology for rural electrification*, in Power Electronics (IICPE), IEEE 5th India International Conference on, pp. 1–4.

[9] S. Singh and M. Upadhyay, 2014. *Study of different issues and challenges of small hydro power plants operation*, Advances in Energy Conversion Technologies (ICAECT), International

Conference on, pp. 88–91.
[10] R. Salman, M. Riaz, Yawar H K, Iftikhar J K and Javed Ali Khan, 2015. *Design and Installation of Mini Hydropower Plant in Rural Areas of Khyber Pakhtunkhwa Pakistan* International Journal Of Multidisciplinary Sciences And Engineering, Vol. 6, No. 8.

IOP Conf. Series: Materials Science and Engineering **343** (2018) 012030 doi:10.1088/1757-899X/343/1/012030

Study Improving Performance of Centrifugal Compressor In Paiton Coal Fired Power Plant Unit 1 And 2

Yuriadi Kusuma, Dadang S Permana
Department of Mechanical Engeneering, Faculty of Engineering, Universitas Mercu Buana, 11650 Jakarta, Indonesia

kusuma.yuriadi@gmail.com

Abstract. The compressed air system becomes part of a very important utility system in a Plant, including the Steam Power Plant. In PLN'S coal fired power plant, Paiton units 1 and 2, there are four Centrifugal air compressor types, which produce compressed air as much as 5.652 cfm and with electric power capacity of 1200 kW. Electricity consumption to operate centrifugal compressor is 7.104.117 kWh per year. This study aims to measure the performance of Centrifugal Compressors operating in Paiton's coal fired power plant units 1 and 2. Performance Compressor is expressed by Specific Power Consumption (SPC) in kW/100 cfm. For this purpose, we measure the compressed air flow rate generated by each compressor and the power consumed by each compressor. The result is as follows Air Compressor SAC 2B : 15.1 kW/100 cfm, Air Compressor SAC 1B : 15.31 kW/100 cfm,Air Compressor SAC 1A : 16.3 kW/100 cfm and air Compressor SAC 2C : 18.19 kW/100 cfm. From the measurement result, air compressor SAC 2B has the best performance that is 15.1 kW / 100 cfm. In this study we analyze efforts to improve the performance of other compressors to at least match the performance of the SAC 2B air compressor. By increasing the Specific Power Consumption from others Compressor, it will get energy saving up to 284,165 kWh per year.

1. Introduction

This future market requires flexible systems that can react to rapid variations in energy demand and supply. Moreover, because renewable sources like solar energy have their peak power production during the day, it is expected that coal-fired power plants will be in full operation during the night and part load operation during the day. Until a economical efficient way of storing huge amounts of energy is developed, coal-fired power plants will keep playing a central role during the transition to renewable energy [1].

The coal-fired power plant in the road project is designed for rapid changes in power supply. This ramping up and down of the power plant has a direct effect on the amount of CO2 captured in the capture unit, which could range between 40-100 % of the nominal mass-flow[2].Paiton PLN power plant is located in Paiton, Probolinggo, east java, Indonesia. The Paiton power plant is a coal-fired power plant (PLTU) with an installed capacity of 800 MW consisting of two units (2 x 400). Both units were commissioned in 1993 and operated by the Java Bali Plant (PJB). Each year the Paiton power plant generates electricity at an average of 5606.18 GW transmitted through an extra 500kV high-voltage network connecting the Java power system. Paiton PJB units 1 and 2 have two Compressor houses located each in units 1 and 2. Lay out of the compressed air system in each Compressor house is schematically described as follows;

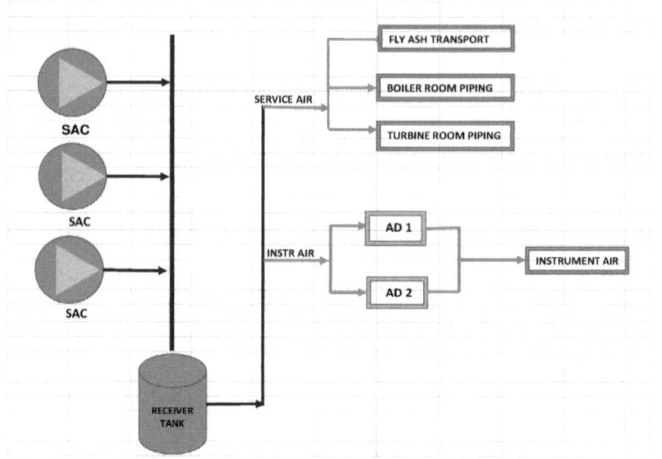

Figure 1. Lay Out Compressor House

The P and ID Layout of the Compressor house are shown in the following figure.

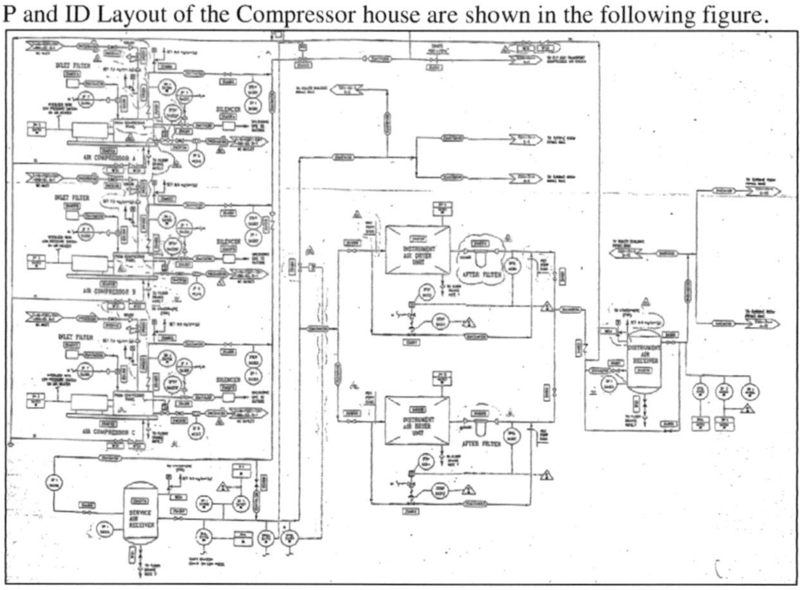

Figure 2. P & ID of Compressed Air System

Three Ingersoll-Rand Centrifugal Compactor Centacts operate alternately supplying compressed air. In practice, only one compressor that operates alternately, because one compressor is considered sufficient in Plant needs. The system is equipped with two Receiver tanks placed outside the Compressor House. Air from Receiver tanks flows into two pipelines each for the need of "water service" and "water instrument". Air for Instrument Air needs to be passed to the air treatment unit consisting of Refrigerated Dryer and Filter. In each Compressor House, there are two Refrigerated Dryers, although daily operations only one Dryer.

Compressor house equipped with a good ventilation system, consisting of ducting to remove hot air, as well as Fan exhaust. However, the exhaust heat from the electric motor Fan driving the compressor is not delivered to the Ducting. The temperature of the exhaust air from the motor cooling fan is measured between 50 - 51 ^{o}C. Fresh air flows into the compressor through a large door that is always wide open and through a gap in the wall. The outside air can go inside because of the negative pressure generated by the Exhaust Fan placed on the roof of the compressor room.

2. Problems

In the Paiton power plant units 1 and 2, six air compressors are installed, each composed of 4 2826 cfm compressors with the same brand and type, and 2 pieces of air compressor with a capacity of 1557 cfm of the same type and brand. However, the energy consumption of compressors of the same type is different, indicating that the performance of each compressor is different even though the type, brand and size are the same [2]. Therefore, it is necessary to measure each air compressor and then calculated the energy saving potential that can be obtained if the performance compressor can be upgraded to the best level owned by one of the machines in Paiton units 1 and 2.

3. Basic Theory

A compressor is a fluid handling mechanical device capable of efficiently transferring energy to the fluid medium so that it can be delivered in large quantities at elevated pressure conditions. Compressors have numerous applications ranging from aircraft and process industries to household appliances such as refrigerators and air conditioners [4].

The modern practise to determine the impeller meridional blade angle distributions is to calculate the compressible inviscid flow through the impeller passage, to determine the blade loading along the blade, and then to modify the meridional blade angle distribution to produce more favourable blade loading[5-6].

The flow in a centrifugal compressor is very complicated, with strong three-dimensional features. The subject of designing impellers is widely covered in the literature[7-12]

The achievement of the air compressor in general can be expressed by Specific Power consumption (SPC) ie how much power is required to produce 100 cfm of compressed air, thus SPC is expressed in kW / 100cfm. It is therefore necessary to measure the consumption of electrical energy and the rate of air flow generated by each air compressor as well as the cooling water temperature and environmental air temperature.

Performance of Centrifugal Compressor depends on several parameters such as;

a. Water Inlet Temperature
b. Inlet Air Pressure
c. Relative Humidity of Air
d. Temperature of cooling water
e. Cooling water flow rate

Usually the manufacturer of air compressor engine (Ingersoll-Rand) provides a performance curve of each centrifugal compressor, which can be used as a reference benchmark between actual operating conditions and current conditions at commissioning test (when the curve is made). Since the performance curve is not available, the performance compressors in Paiton units 1 and 2 will be compared between actual and rating and are also compared between them respectively.

4. Measurement Methods

To calculate the performance of each compressor needs to be measured electrical energy consumption of compressor machine using power meter and measurement of air rate produced by each compressor using pitot tube flow meter.

Since both of these measuring devices are not installed in the compressor machine, a portable measuring device is used, and specially for measuring the air flow rate of the compressor, it is necessary to prepare the insertion point in the pipeline first. Making insertion point is done by drilling on pipe and welding to install valve.

Figure 3. Installation of Flow Meter

Further, the compression of air pressure measurements are made using pressure gauge mounted on the pipe by piercing the pipe.

IOP Conf. Series: Materials Science and Engineering **343** (2018) 012030 doi:10.1088/1757-899X/343/1/012030

5. Measurement Results

Based on the result of measurement on all compressors, the performance are shown in Figures 4, 5, 6 and 7 in graphs and resumes shown in table 1, 2, 3 and 4.

Compressor SAC 1A

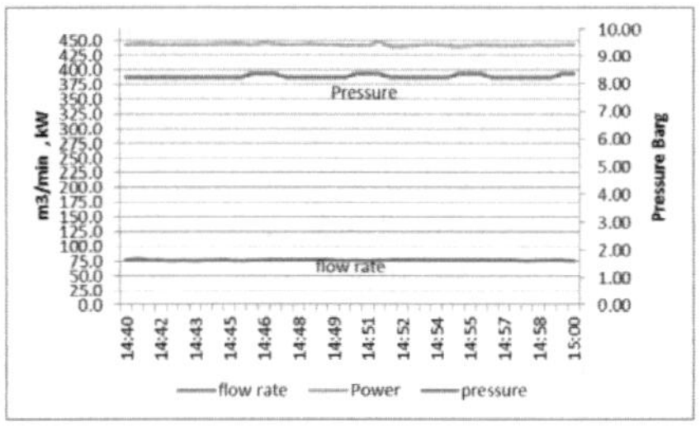

Figure 4. Compressor Performance of SAC 1A

PERFORMANCE KOMPRESSOR SAC 1A			
		DESIGN	ACTUAL
Air Capacity	cfm	2826	2702
Barometric Pressure	Psia	13.94	14.64
Relative Humidity	%	78	82
Inlet Air Temperature	°F	82.4	89.6
Discharge Presssure	Psig	120.4	119.5
Cooling Water Inlet Temperature	°F	104	
Specific Power Consumption	kW/100cfm		16.3

Table 1 Resume of Compressor Performance SAC 1A

Compressor SAC 1B

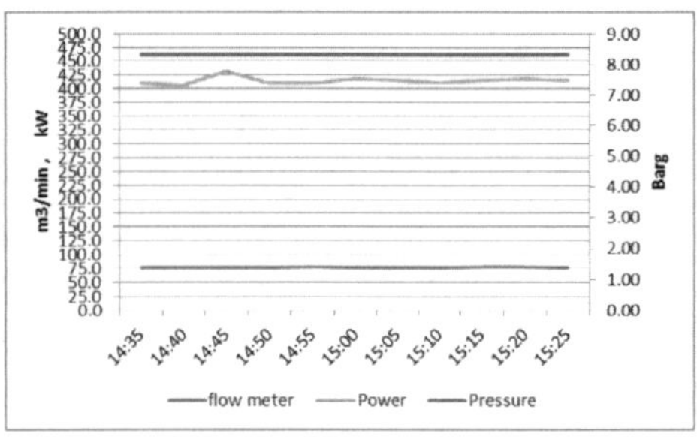

Figure 5. Compressor Performance SAC 1B

PERFORMANCE KOMPRESSOR SAC 1B			
		DESIGN	ACTUAL
Air Capacity	cfm	2826	2706
Barometric Pressure	Psia	13.94	14.64
Relative Humidity	%	78	82
Inlet Air Temperature	°F	82.4	89.6
Discharge Presssure	Psig	120.4	120
Cooling Water Inlet Temperature	°F	104	90
Specific Power Consumption	kW/100cfm		15.31

Table 2. Resume of Compressor Performance SAC 1B

Compressor SAC 2B

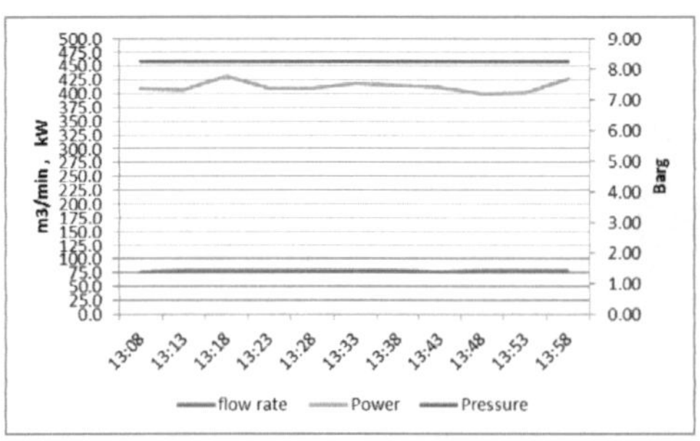

PERFORMANCE KOMPRESSOR SAC 2B			
		DESIGN	ACTUAL
Air Capacity	cfm	2826	2730
Barometric Pressure	Psia	13.94	14.64
Relative Humidity	%	78	82
Inlet Air Temperature	°F	82.4	89.6
Discharge Presssure	Psig	120.4	119
Cooling Water Inlet Temperature	°F		
Specific Power Consumption	kW/100cfm		15.1

Table 3. Resumes of Compressor Performance SAC 2B

Figure 6. Performance of Compressor SAC 2B

Compressor SAC 2C

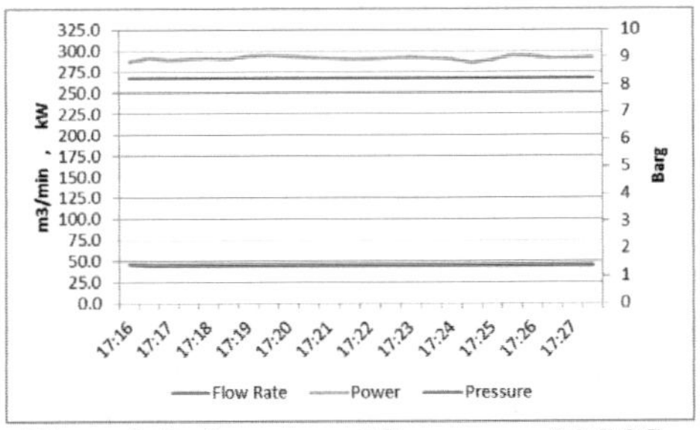

Figure 7. Performance of Compressor SAC 2C

PERFORMANCE KOMPRESSOR SAC 2C			
		DESIGN	ACTUAL
Air Capacity	cfm	1557	1599
Barometric Pressure	Psia	13.94	14.64
Relative Humidity	%	78	82
Inlet Air Temperature	°F	82.4	89.6
Discharge Presssure	Psig	120.4	120
Cooling Water Inlet Temperature	°F		
Specific Power Consumption	kW/100cfm		18.19

Table 4. Resume of Compressor Performance SAC 2C

Based on the survey results shown in the graph and table above can be concluded that the performance compressor varied, if sorted by ranking as follows

1. Compressor SAC 2B: 15.1 kW / 100 cfm
2. Compressor SAC 1B: 15.31 kW / 100 cfm
3. Compressor SAC 1A: 16.3 kW / 100 cfm
4. Compressor SAC 2C: 18.19 kW / 100 cfm

Measurement on Air Compressor Cooler
The results of measurements at the air-cooling temperature of the air compressor are as follows;

SAC 1B
T in = 90 °F = 32.2 °C
Tout = 100 °F = 37.8 °C
P water in = 76 Psig
P water out = 70 Psig

SAC 2B
T in = 84 °F = 28.8 °C
Tout = 104 °F = 40 °C
P water in = 76 Psig
P water out = 72 Psig

Based on the measurement result can be concluded the things as follows;
SAC 1 B
Delta T on the cooling water side = 10 ° F or 5.6 ° C
Delta P on the cooling water side = 6 kg / cm2

SAC 2B
Delta T on the cooling water side = 20 ° F or 11.2 ° C
Delta P on the cooling water side = 4 kg / cm2

Terms of manufacturer Ingersoll Rand, for water Cooler; temperature rise 25 ° F, thus generally the two Coolers are not yet in the desired condition. Seen Cooler on SAC 2B is better than Cooler in SAC 1B, because its delta T is higher which means to throw more heat with lower pressure drop. There is a possibility that the water pipeline in SCA 1B is dirty, It is advisable to check the condition of cooling water pipes in SAC 1B.

Analysis
It appears that for compressor type CV2, Specific power consumption number is owned by Compressor SAC 2B with the number 15.1 kW / 100 cfm. So the SAC 2B compressor can be used as a reference for improvement. Some things that can be done to improve performance compressor is;

1. Maintain compressor room temperature is low, by repairing ventilation system, install ducting to remove hot air from electric motors, and install insulation on hot air ducting.

2. Clean the filter water inlet regularly, and check with more frequencies around the compressor room due to the dust potential from fly ash Coal, especially the inlet side of the air in the outside facing the coal conveyor.

3. Perform the routine cleaning of the cooling water pipe. This is evidenced from the results of pressure measurements on Cooling water system, the pressure drop of cooling water on SAC 2B compressor only 4 Psig. Compared to the pressure drop on the Compressor SAC 1B of 6 Psig. The pressure drop is associated with the resistance to the piping caused by one of them by the dirt on the pipe surface. The effectiveness of the Air Cooler can also be seen from the delta Temperature that is the difference temperature outlet and temperature inlet of the cooling Water.

4. Ensure the Preventive Maintenance program runs well in accordance with the SOP of the vendor. It is also important of doing predictive maintenance activities such as the level of vibration checking, oil analysis and the temperature checking of the electric motor as a compressor drive.

5. Keep the coolant flow rate in the standard specified by the vendor can be seen in table 5, that is;

Table 5 Cooling Water Rate

STAGES	GPM
First Stage	36
Second Stage	41
After Cooler	43
Oil Cooler	50
Total Cooler	170

Based on the simulation we can compare the annual operating cost between SAC 2B compressor with Compressor SAC 1A, then the difference of energy cost is Rp. 292.408.704, as shown in the following figure 8;

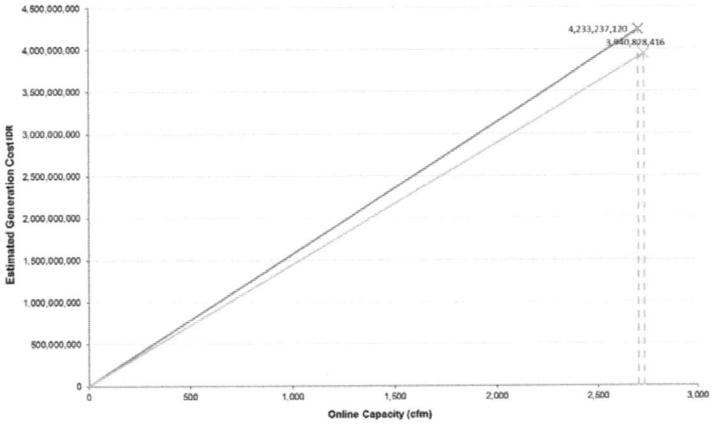

Figure 8. Simulation of Operating Cost Compressor

Conclusion

The measurement results show that the air compressor has different performance levels. The SAC 2B compressor has the highest performance of 15.1 kW / 100 cfm, while the SAC 2C Compressor has the worst performance of 18.19 kW / 100 cfm. If several improvement steps are taken to raise the compressor's performance to the level of SAC 2B compressor, energy saving will be obtained. For example, if the SAC 1A compressor is upgraded from 16.3 to 15.1 kW / 100 cfm, saving energy of Rp. 292.408.704.

Reference :

[1] H. Chalmers, M. Leach, M. Lucquiaud, and J. Gibbins, "Valuing flexible operation of power plants with CO2 capture," in Energy Procedia, vol. 1, pp. 4289–4296, 2009.

[2] M.H.L. Ogink, "Analyses and dynamic modelling of the compressor section in a PCC-process for coal-fired power plants with offshore storage", Master of Science Thesis For the degree of Master of Science in Mechanical Engineering at Delft University of Technology, June 2015.

[3] Hazel Ariantara, The Solution for Optimal Power Flow (OPF) Method Using Differential Evolution Algorithm, ISSN 2550-0554 JITEE, Vol. 1, No. 1, March 2017.

[4] Meherwan P. Boyce Dr. "Principles Of Operation And Performance Estimation Of Centrifugal Compressors" Proceedings Of The Twenty-Second Turbomachinery Symposium, 1993.

[5] M. Zangeneh, M. Schleer, F. Pløger et al., "Investigation of an inversely designed centrifugal compressor stage—part I: design and numerical verification," Journal of Turbomachinery, vol. 126, no. 1, pp. 73–81, 2004.

[6] Pekka Röyttä, Aki Grönman, Ahti Jaatinen, Teemu Turunen-Saaresti, and Jari Backman, "Effects of Different Blade Angle Distributions on Centrifugal Compressor Performance" International Journal of Rotating Machinery Volume 2009 (2009), Article ID 537802, 9 pages http://dx.doi.org/10.1155/2009/537802

[7] O. E. Balje, Turbomachines: A Guide to Design, Selection and Theory, John Wiley & Sons, New York, NY, USA, 1981.

[8] P. M. Came and C. J. Robinson, "Centrifugal compressor design," Journal of Mechanical Engineering Science, vol. 213, no. 2, pp. 139–155, 1998. View at Google Scholar

[9] S. L. Dixon, Fluid Mechanics, Thermodynamics of Turbomachinery, Pergamon Press, Oxford, UK, 3rd edition, 1989.

[10] D. Japikse, Centrifugal Compressor Design and Performance, Concepts ETI, Wilder, Vt, USA, 1996.

[11] H. I. H. Saravanamuttoo, G. F. C. Rogers, and H. Cohen, Gas Turbine Theory, Prentice-Hall, Upper Saddle River, NJ, USA, 5th edition, 2001.

IOP Conf. Series: Materials Science and Engineering **343** (2018) 012031 doi:10.1088/1757-899X/343/1/012031

Compressed Air Quality, A Case Study In Paiton Coal Fired Power Plant Unit 1 And 2

Nur Indah, Yuriadi Kusuma, Mardani

Department of Mechanical Engeneering, Faculty of Engineering, Universitas Mercu Buana, 11650 Jakarta, Indonesia

nur.indah@mercubuana.ac.id

Abstract: The compressed air system becomes part of a very important utility system in a Plant, including the Steam Power Plant. In PLN'S coal fired power plant, Paiton units 1 and 2, there are four Centrifugal air compressor types, which produce compressed air as much as 5.652 cfm and with electric power capacity of 1200 kW. Electricity consumption to operate centrifugal compressor is 7.104.117 kWh per year. Compressed air generation is not only sufficient in quantity (flow rate) but also meets the required air quality standards. compressed air at Steam Power Plant is used for; service air, Instrument air, and for fly Ash. This study aims to measure some important parameters related to air quality, followed by potential disturbance analysis, equipment breakdown or reduction of energy consumption from existing compressed air conditions. These measurements include counting the number of dust particles, moisture content, relative humidity, and also compressed air pressure. From the measurements, the compressed air pressure generated by the compressor is about 8.4 barg and decreased to 7.7 barg at the furthest point, so the pressure drop is 0.63 barg, this number satisfies the needs in the end user. The measurement of the number of particles contained in compressed air, for particle of 0.3 micron reaches 170,752 particles, while for the particle size 0.5 micron reaches 45,245 particles. Measurements of particles conducted at several points of measurement. For some point measurements the number of dust particle exceeds the standard set by ISO 8573.1-2010 and also NACE Code, so it needs to be improved on the air treatment process. To see the amount of moisture content in compressed air, it is done by measuring pressure dew point temperature (PDP). Measurements were made at several points with results ranging from -28.4 to 30.9 ^{o}C.
The recommendation of improving compressed air quality in steam power plant, Paiton unit 1 and 2 has the potential to extend the life of instrumentation equipment, improve the reliability of equipment, and reduce the amount of energy consumption up to 502,579 kWh per year.

1. Introduction

Paiton PJB units 1 and 2 have two Compressor houses located each in units 1 and 2. Lay out of the compressed air system in each Compressor house is schematically described as follows;

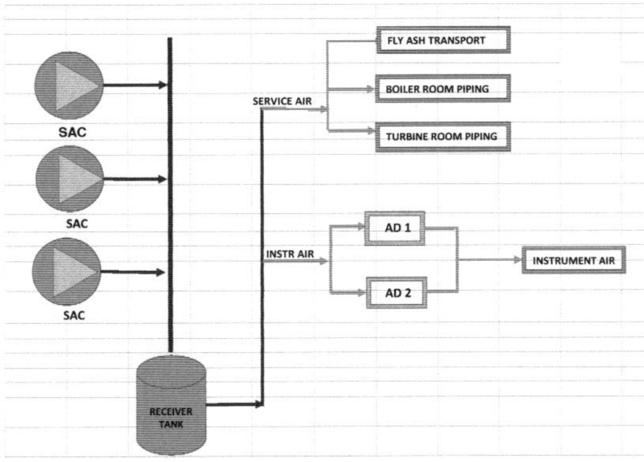

Figure 1. Lay Out Compressor House

Three Centers Ingersoll-Rand Centrifugal Compactor Centacts in turn operate pressurized air. In practice, only one compressor operates alternately, because one compressor is considered sufficient for the Plant. The system is equipped with two Receiver tanks placed outside the Compressor House. Air from Receiver tanks flows into two pipelines each for the need of "water service" and "water instrument". Air for Instrument Air needs to be passed to the air treatment unit consisting of Refrigerated Dryer and Filter. In each Compressor House, there are two Refrigerated Dryers, although daily operations only one Dryer.
Compressor house equipped with a good ventilation system, consisting of ducting to remove hot air, as well as Fan exhaust. However, the exhaust heat from the electric motor Fan driving the compressor is not delivered to the Ducting. The temperature of the exhaust air from the motor cooling fan is measured between 50 and 51 ^{0}C.
Fresh air flows into the compressor through a large door that is always wide open and through a gap in the wall. The outside air can go inside because of the negative pressure generated by the Exhaust Fan placed on the roof of the compressor room.
The capacity of an air-conditioning unit must be able to remove the maximum cooling load, which is the full load condition of an air-conditioning unit. The centrifugal chiller is mostly used for large buildings and usually works all year round[1,2].

2. Problems
In the Paiton Power Plant the pressurized air quality is considered to be one of the causes of the low life time of the instrument equipment and solenoid valve using pressurized air.

3. Basic Theory
An air compressor is a device that converts power (using an electric motor, diesel or gasoline engine etc...) in to potential energy stored in pressurized air. By one of several methods, an air compressor forces more and more air into a storage tank, increasing the pressure. When tank pressure reaches its upper limit the air compressor huts off. The compressed air in the tank is held until used [3].

Each day people are exposed to millions of bioaerosols, including whole microorganisms, which can have both beneficial and detrimental effects[4]. Environmental air contains a lot of water vapor, dust particles, microorganisms, and others. In air compression should be taken into account the oil content that comes with the air flow. But on Centrifugal Compressor, oil content is not a problem, because this compressor pertained to free oil compressor. Thus the biggest issue of contaminants on the Centrifugal compressor is the moisture content and dust particles. The quality of compressed air is regulated by the standards, among which the most widely referred is ISO 8573.1

Recent advances in high-throughput sequencing have generated a rush to characterize the microbiome of various environments, including indoor and outdoor air[5-8]. The built environment is of particular interest because humans spend over 90 % of their time indoors[9]

The function of air compression system is to produce dry air, oil free, and dust particles. The remaining oil and dust particles can be removed by a complex filter media. However, water vapor should be reduced by the medium Dryers (refrigeration dryers, membrane dryers, adsorption dryers and others) that ideally work independently of the load. Air can bind more moisture if the temperature is higher and the volume is larger. In contrast, air has poor capacity when air is compressed. A compressor compresses atmospheric air into a small part of the volume. At one point the compression process will reach the limit of the air's ability to bind water vapor.

The air will become saturated and some of the air will turn into water droplets or condensate. With the added effort of reducing the temperature will even increase the amount of water droplets that come out in the exit air. The amount of air coming out of the compressed air under pressure conditions can become even greater. For example, a 30 kW compressor removes about 20 liters of water into the compressed air system under a Humidity condition of 60% and an ambient temperature of 20 ° C. On larger compressors, the numbers will be much higher.

Generally compressed air piping lines are made of steel or steel without zinc lining. Therefore the corrosion rate will increase quite high from the relative humidity 50%. Water vapor will cause corrosion in the case of pipes made of non-zinc coated steel. Rust will release gradually and carry over to the point of use. This will cause problems such as compressed nozzles, defective control elements and disturbed production. Impact Poor air quality can be seen in Figure 2.

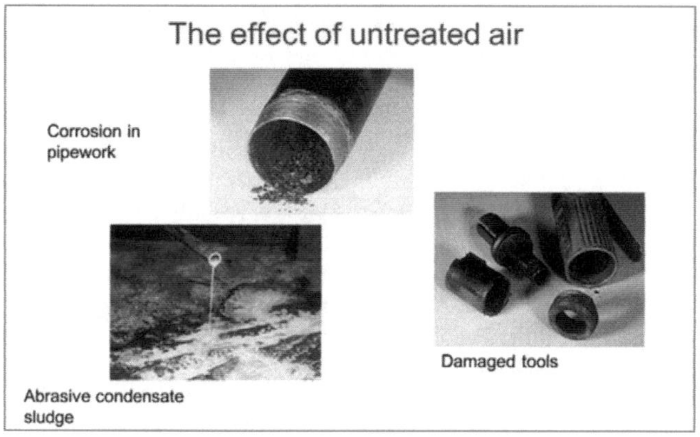

Figure 2. Impact of Poor Air Quality

In practice, different types of Dryes are used to control problems due to the high moisture content. The compressor used the pressure dew point parameter to indicate the dryness level of the compressed air. Pressure Dew Point is the temperature where the water vapor contained by air will condense into water liquids (saturated with 100% relative humidity). The lower the pressure dew point, the lower the amount of water vapor in the compressed air. Refrigeration dryers have a dew point value of about + 2°Ctd.

The existing air dryer is not suitable. For example, the traditional air dryer has many defects due to the pure machine structure, such as the inaccurate control of pressure, the exceeded waiting time for the regeneration control of molecular sieve, eventually, leading to the untimely failure of the air dryer[10-13].

There are several types of Air Dryers, the most famous are refrigeration dryers and adsorption dryers. Refrigeration Dryers cool compressed air up to 2 to 5 °C. Excess amount of water vapor will be condensed and discarded. In this case pressure dew point also ranges from 2 to 5 °C, while adsorption dryers have greater capability with pressure dew point can reach between -40 to -70 °C.

The air quality needed depends on the application of compressed air usage itself. As example, table 1 and table 2 below can be used as reference, the kind of compression air quality needed.

Table 1. Relationship between Air Quality and Application of its Usage

APLICATION	COMPRESSED AIR QUALITY CLASSES ACCRODING TO DIN ISO 8573-1			
	PARTICLE		RESIDUAL FLOW	
	CLASS	μm	CLASS	DEW POINT
Respiration Air	1	0.1	1 – 3	-70/-20 °C
Spray Gun	1	0.1	2	-40 °C
Medical Technology	1	0.1	3 -4	-20/+3°C
Measurement & control Technology	1	0.1	4	+3 °C
Transportation food & beverages	2	1	3	+3 °C
General Factory Air	3	5	4	+3 °C

Table 2. ISO 8573.1-2010

Solid Particles/Dust				Humidity and Liqud Water			Oil		
Particles per m^3, by size				Pressure Dewpoint			Total Concentration (Liquid, Aeroso and Vapor		
Class	0.1 to 0.5 Micron	0.1 to 0.5 Micron	0.1 to 0.5 Micron	Class	Degress C	Degress C	Class	Mm/m^3	Ppm w/w
0	As Specified			0	As Specified		0	As Specified	
1	≤ 20.000	≤ 400	≤ 10	1	≤ -70	≤ -94	1	≤ 0.01	≤ 0.008
2	≤ 400,000	≤ 6,000	≤ 100	2	≤ -40	≤ -40	2	≤ 0. 1	≤ 0.08
3	Not specified	Not specified	≤ 1,000	3	≤ -20	≤ -4	3	≤ 1.0	≤ 0.8
4	Not specified	Not specified	≤ 10,000	4	≤ -3	≤ -38	4	≤ 5.0	≤ 4
5	Not specified	Not specified	≤ 100,000	5	≤ -7	≤ -45	x	> 5.0	> 4
	Particle Concentration ($mg/m^{3)}$			6	≤ -10	≤ -50			
6	0 to 5			Liquid Water Concentration ($g/m^{3)}$					
7	5 to 10			7	≤ 0.5				
x	>10			8	0.5 to 5				
				9	5 to 10				
				X	>10				

4. Measurement Methods

To prove the hypothesis at the beginning, that the pressurized air quality becomes the main factor of the low life time of instrumentation equipment and solenoid valve, it is necessary to measure the quality of compressed air. There are three air quality parameters that must be obtained from this measurement namely; pressure dew point temperature (Pdp), the number of particles in the air and the relative humidity, and the amount of lubricating oil in pressurized air. However, in this case because Paiton uses Centrifugal Compressor free oil, the measurement of oil content in the air is not done.

Equipment used in this measurement is the Portable Dew Point model 500 CS and Particle Counter CS Instrument equipped with data logger to record the measurement results.

5. Measurement Results

Measurement of dew point pressure is done at some point of compressed air usage. The results are shown in the following table.

From the measurement results, the dew point numbers ranged from -28.4 to 30.9 oC. With Realtif humidity ranges from 0.85 to 99.95%. The 100% figure means the air is saturated and the moisture will condense.

Table 3. Dew Point Measurement Results

NO	LOCATION	DEW POINT	PRESSURE	TEMPERATURE	RELATIVE HUMIDITY
1	After Dryer and Reciever Tank Instrument	9.3 °Ctd	7.83 bar	27.7 °C	31.75 %RH
2	Area Turbin UNIT 2	27.6 °Ctd	8.00 bar	33.4 °C	71.94 %RH
3	Line WTP After Dryer	24.5 °Ctd	4.87 bar	33.7 °C	58.60 %RH
4	Area Turbin UNIT 1	8.5 °Ctd	7.89 bar	34.5 °C	20.32 %RH
5	Supply Shoot Boiler UNIT 2	29.2 °Ctd	7.94 bar	33.1 °C	80.24 %RH
6	Before Dryer SAC2	29.0 °Ctd	7.97 bar	29.0 °C	99.95 %RH
7	Supply Shoot Boiler UNIT 1	13.2 °Ctd	7.82 bar	35.5 °C	26.29 %RH
8	Area Fly Ash	6.7 °Ctd	7.42 bar	32.0 °C	20.68 %RH
9	Supply Shoot Boiler UNIT 2	29.0 °Ctd	7.95 bar	33.2 °C	79.01 %RH
10	Chems Analisis UNIT 2	-28.4°Ctd	8.00 bar	33.8 °C	0.85 %RH
11	Line WTP	27.7 °Ctd	7.50 bar	27.7 °C	99.95 %RH
12	After Dryer and Reciever Tank Inst. SAC 2	7.3 °Ctd	7.97 bar	29.6 °C	24.68 %RH
13	After Dryer and Reciever Tank Inst. SAC 1	9.2 °Ctd	7.83 bar	27.7 °C	31.44 %RH
14	02 Analisis UNIT 1	-1.2 °Ctd	4.25 bar	34.9 °C	9.88 %RH
15	02 Analisis UNIT 2	8.1 °Ctd	7.61 bar	32.5 °C	22.18 %RH
16	Before 02 Analisis UNIT 2	24.0 °Ctd	7.13 bar	33.1 °C	58 %RH
17	Line WTP	27.6 °Ctd	7.50 bar	27.6 °C	99.93 %RH
18	Before Dryer SAC 1	30.9 °Ctd	8.02 bar	30.9 °C	99.94 %RH
19	Area Fly Ash UNIT !	24.8 °Ctd	7.80 bar	32.4 °C	64.16 %RH

Measurement of the number of particles in the air using a Portable particle counter, carried out at several important points includes:

- Measurement after Centrifugal Compressor
- Measurements in PAPs
- Measurements in O2 analysis lines
- Measurements on fly ash lines

The results are displayed on the following charts:

a. Measurements on Centrifugal Compressor SAC 1:

The number of particles of 0.3 microns reaches the the highest level at 18,000, while for particles of 0.5 microns reaches 12,500

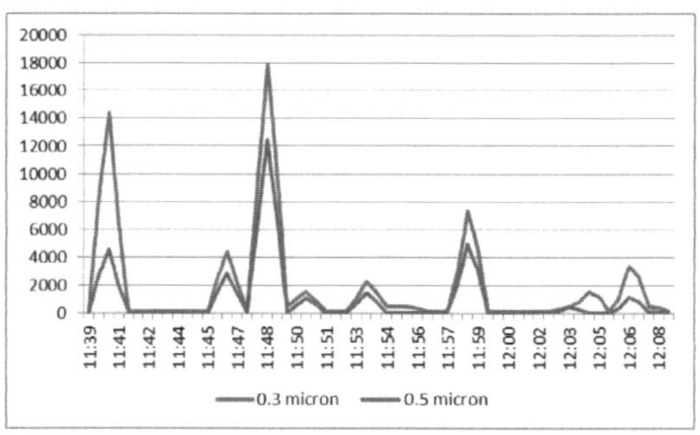

Figure 3. Results of Measurements of Air Particles in SAC1

b. Measurements on the side after Centrifugal Compressor SAC 2

The number of particles of 0.3 microns reached the highest level at 97.458 while for particles of 0.5 microns reached 37,712.

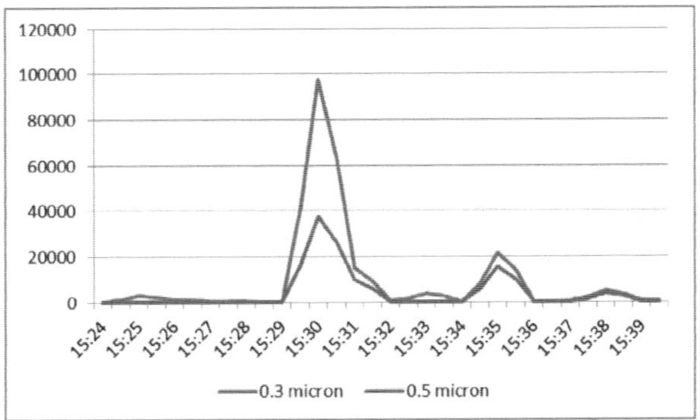

Figure 4. Results of air particle measurements in SAC 2

c. Measurements in the WTP Path

The number of particles measuring 0.3 microns reaches the highest at 81,552 while for 0.5 micron particles it reaches 28,384.

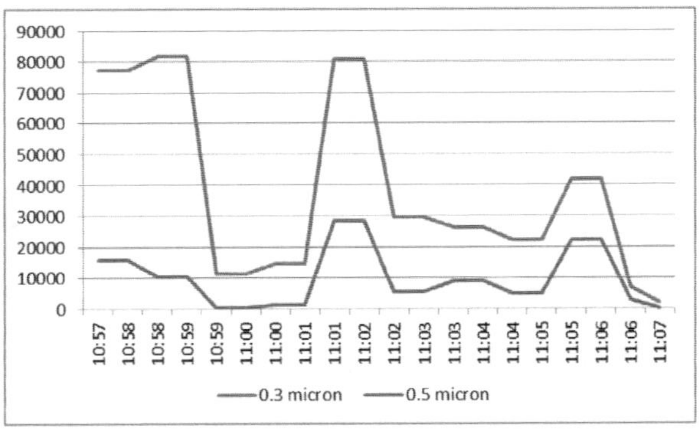

Figure 5. Results of measurements of air particles in WTP Line

d. Measurement on the side of O2 Analysis Unit 1

The number of particles measuring 0.3 microns reaches the highest at 16,066 while for the 0.5 micron particles it reaches 11,663.

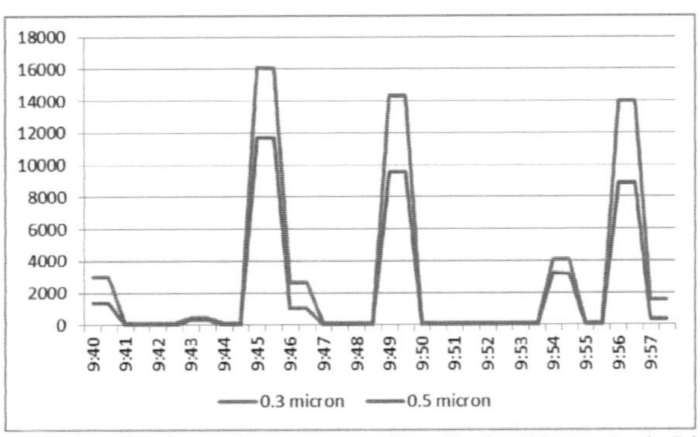

Figure 6. Air Particle Measurement Results in O2 Analysis Line

e. Measurements in the fly Ash Line of Unit 1

The number of particles of 0.3 microns reaches the highest level at 170,752 while for 0.5 micron particles reaches 45,245.

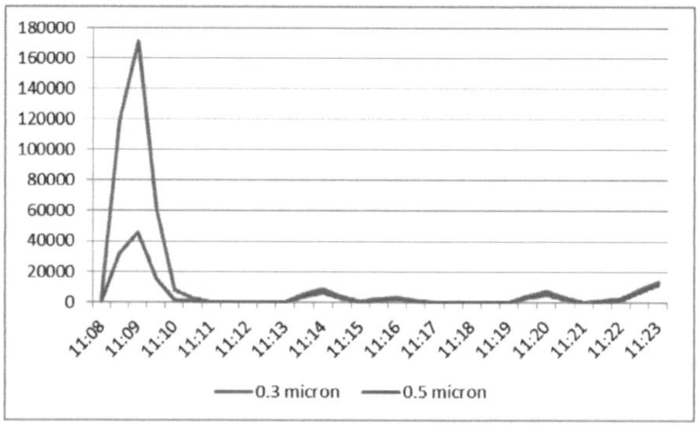

Figure 7. Results of fly ash line of unit 1

6. Analysis

Water Content Analysis

The content of water vapor in pressurized air is directly proportional to the Pressure Dew Point Temperature, as shown in the following table 4:

Table 4. Relationship between Dew Point with Water Content in Air

Dewpoint	g/m³	Dewpoint	g/m³
+10	588.20	+	7.24
+9	417.93	+	6.35
+8	290.01	+	5.57
+7	196.21	+	4.86
+6	129.02	-10	2.15
+5	82.25	-20	0.8
+4	50.67	-30	0.3
+3	30.07	-40	0.11
+2	17.14	-50	0.03
+1	9.35	-60	0.01
+	8.34	-70	0.003

The result of measurement of pressure dewpoint compared to the standard shown in table 5 as follows:

Table 5. Dew Point measurement results compared to standard.

LOCATION	AKTUAL		STANDARD ISO 8573-1		
	DEW POINT °Ctd	Water Vapour g/m3	DEW POINT °Ctd	Water Vapour g/m3	KELAS
After Dryer and Reciever Tank Instrument	9.3	417.93	3	30.7	4
Area Turbin UNIT 2	27.6	> 588.20	3	30.7	4
Line WTP After Dryer	24.5	> 588.20	3	30.7	4
Area Turbin UNIT 1	8.5	290.01	3	30.7	4
Supply Shoot Boiler UNIT 2	29.2	> 588.20	3	30.7	4
Before Dryer SAC2	29	> 588.20	3	30.7	4
Supply Shoot Boiler UNIT 1	13.2	> 588.20	3	30.7	4
Area Fly Ash	6.7	129.02	3	30.7	4
Supply Shoot Boiler UNIT 2	29	> 588.20	3	30.7	4
Chems Analisis UNIT 2	-28.4	0.8	-20	0.8	3
Line WTP	27.7	> 588.20	3	30.7	4
After Dryer and Reciever Tank Inst. SAC 2	7.3	196.21	3	30.7	4
After Dryer and Reciever Tank Inst. SAC 1	9.2	417.93	3	30.7	4
02 Analisis UNIT 1	-1.2	7.24	-20	0.8	3
02 Analisis UNIT 2	8.1	290.01	-20	0.8	3
Before 02 Analisis UNIT 2	24	> 588.20	3	30.7	4
Line WTP	27.6	> 588.20	3	30.7	4
Before Dryer SAC 1	30.9	> 588.20	7	196	5
Area Fly Ash UNIT 1	24.8	> 588.20	3	30.7	4

Based on the measurements in Paiton Units 1 and 2. The dew points vary from -28.4 to 30.9 ° C, the relative humidity ranges from 0.85 to 99.95%. The 100% figure means the air is saturated and the moisture will condense. This condition is quite bad and is still far from the standard required for compressed air usage especially for use in the water instrument sector. Since the water instrument requires grade 3 air quality from the water content side, it means the dew point should be less than -20 ° C with a water content of no more than 0.8 g / m3.

The solution of this problem is to improve the water treatment system on the user side, some efforts can be done:

a. Install the Water Separator on the user side
b. Install Electronic Zero loss Drain, to dispose of water.
c. Change filter coalescing elements periodically
d. Evaluate the existing Dryer system.
   Based on the Dryer measurement results on the WTP, the dew point indicates a 24.5 ° C number which should be between 3 - 7 ° C, so it should be considered to replace the two Dryers, as it no longer shows the expected performance, and is physically unfeasible.
e. Replace and or repair an existing Desiccant Dryer.

The results of measurements after the Desiccant Dryer showed results varying from -28.7 ° C to over 20 ° C, it is known that some Desiccant Dryers are not functioning properly, as Desiccant Dryers should produce air with dew point pressure ranging between -40 to -73 ° C .

Air Particle Analysis

From the particle measurement results can be summarized in the following table.

Table 6. Summary Measurements of Air Particles

No	Location	Maximal Number of Particle			
		Actual		Standard ISO	
		0.3 mikron	0.5 mikron	0.3 mikron	0.5 mikron
1	SAC 1	18.000	12.500	20.000	400
2	SAC 2	97.458	37.612	20.000	400
3	WTP Line	81.552	28.384	20.000	400
4	O2 Analysis Line Unit 1	9.421	7.136	20.000	400
5	O2 Analysis Line Unit 2	16.066	11.663	20.000	400
6	Fly Ash Line Unit 1	170.752	45.245	20.000	400
7	Fly Ash Line Unit 2	92.529	6.790	20.000	400

The particle content of the SAC 2 measurements shows a considerable number of 97,458 well above the standard 1st air standard requirements of ISO standards. However, on O2 line on units 1 and 2 have met the criteria, because the particle content measuring 0.3 microns is less than 20,000. While in the WTP path, the number of particles both measuring 0.3 microns and 0.5 microns far above the standard ISO for air class 1 is much greater than 20,000 for 3 micron size.

So far, it seems necessary to evaluate the filter system both in terms of frequency replacement and technology. It is suggested that the filter rating is set based on the air conditions standard ie. inlet air temperature of not more than 21 ° C. The Standard life time is 6000 hours or a year while the active carbon element is 1000 hours. If the default condition is exceeded then the element's age will be shorter.

Based on the analysis of water vapor and the number of particles, some improvement recommendations may be proposed as follows:

- Due to the lack of Dryer capacity in the Compressor house, as only one compressor operates, it is advisable to run both refrigerated dryers. Operation of two Dryers will ensure the decrease of Dew Point pressure value, air quality will be better because the amount of water present in the compressed air is reduced. To calculate the cost-benefit, it is necessary to calculate the losses due to damage to end-users (solenoid valve and instrument) due to high water content.
- Investigation of some Desiccant Dryers in point of use, as the result of dew point measurement at this point shows poor results.
- Further investigation of the filter system and its elements is required associated with replacement schedule and technology used. It is recommended to install Coalessing filter before end-users especially for applications on instrumentation.
- Periodic monitoring of compressed air quality is required. If possible mounted measuring tools that monitor the dew point and particle counter.

7. Conclusion

Based on the results of compressed air quality measurements at several points in the usage line, the results obtained that most air quality parameters such as pressure dew point temperature associated with the amount of water vapor in the air and the number of particles in the pressurized air has not met the desired standard by ISO 8573.1-2010.

This issue impacts the lifetime of instrumentation equipment and solenoid valve. Since both types of equipment will be exposed to high pressure water and dust in unacceptable quantities, then it will lead to accelerated damage and it will result in declining lifetime of pressurized air user equipment.

From these findings, it can be done several precautions such as repairing the filter system, dryer system and installing some new equipment such as desiccant dryers, coalescing filters and water separator.

Reference :

[1] Hazel Ariantara, The Solution for Optimal Power Flow (OPF) Method Using Differential Evolution Algorithm, ISSN 2550-0554 JITEE, Vol. 1, No. 1, March 2017

[2] Kuo-Shu Hung, Jenn-Chyi Chung, Chung-Che Liu1 and Jeng-Min Huang, A study of off-design performance improvement for a centrifugal refrigerant compressor, Advances in Mechanical Engineering 2017, Vol. 9(3) 1–10, 2017. DOI: 10.1177/1687814017696224 journals.sagepub.com/home/ade

[3] Vishnu R, Prof. Deepika Vasanthakumar. Performance Analysis And Modification Of Air Compressor System, International Research Journal of Engineering and Technology (IRJET), Volume: 02 Issue: 05 | Aug-2015, e-ISSN: 2395-0056 p-ISSN: 2395-0072. Department of Electrical and Electronics Engineering, NCERC, Kerala, India

[4] Aaron J. Prussin. Sources of airborne microorganisms in the built environment Microbiome, 2015; 3;78 Publish online 2015. doi: 10.1186/s40168-015-0144-z

[5] Kelley ST, Gilbert JA. Studying the microbiology of the indoor environment. Genome Biol. 2013;14:202. doi: 10.1186/gb-2013-14-2-202. [PMC free article] [PubMed] [Cross Ref]

[6] Smith DJ, Jaffe DA, Birmele MN, Griffin DW, Schuerger AC, Hee J, et al. Free tropospheric transport of microorganisms from Asia to North America. Microb Ecol. 2012;64:973–985. doi: 10.1007/s00248-012-0088-9. [PubMed] [Cross Ref]

[7] Smith DJ, Timonen HJ, Jaffe DA, Griffin DW, Birmele MN, Perry KD, et al. Intercontinental dispersal of bacteria and archaea by transpacific winds. Appl Environ Microb. 2013;79:1134–1139. doi: 10.1128/AEM.03029-12. [PMC free article] [PubMed] [Cross Ref]

[8] DeLeon-Rodriguez N, Lathem TL, Rodriguez-R LM, Barazesh JM, Anderson BE, Beyersdorf AJ, et al. Microbiome of the upper troposphere: species composition and prevalence, effects of tropical storms, and atmospheric implications. Proc Natl Acad Sci U S A. 2013;110:2575–2580. doi: 10.1073/pnas.1212089110. [PMC free article] [PubMed] [Cross Ref]

[9] Klepeis NE, Nelson WC, Ott WR, Robinson JP, Tsang AM, Switzer P, et al. The National Human Activity Pattern Survey (NHAPS): a resource for assessing exposure to environmental pollutants. J Expo Anal Env Epid. 2001;11:231–252. doi: 10.1038/sj.jea.7500165. [PubMed] [Cross Ref]

[10] Xuan Zhang, a, Liu Yang, b,Xian Yan Wang1, c,Xiao Yan Tan, d,Wei Wang. Machine & electrical double control air dryer for vehicle air braking system. IOP Conf. Series: Materials Science and Engineering 231 (2017) 012156 doi:10.1088/1757-899X/231/1/012156. SORL Auto Parts, Inc., Wenzhou 325200, China.

[11] C.W. Li, X. Zhang, L. Yan: Intelligent air dryer, CHINA Patent 200910177514.3. (2009)

[12] C.W. Li, X. Zhang: Multifunctional air dyrer cartridge , CHINA Patent 200910162325.9. (2009)

[13] C.W. Li, X.Y. Tan, X. Zhang: Electric air direr, CHINA Patent 201010193182.0. (2010)

IOP Conf. Series: Materials Science and Engineering **343** (2018) 012032 doi:10.1088/1757-899X/343/1/012032

Surface martensitization of Carbon steel using Arc Plasma Sintering

Haris Wahyudi[1], Arbi Dimyati[2], Darwin Sebayang[1]

[1]Mechanical Engineering Department, Faculty of Engineering, Mercu Buana University, Jakarta – Indonesia
[2]Center for Science and Technology for Advanced Materials, PSTBM-Batan – Indonesia

haris.wahyudi@mercubuana.ac.id

Abstract. In this paper new technology of surface structure modification of steel by short plasma exposure in Arc Plasma Sintering (APS) device is presented. APS is an apparatus working based on plasma generated by DC pulsed current originally used for synthesizing materials via sintering and melting. Plasma exposure in APS was applied into the specimens for 1 and 3 seconds which generate temperature approximately about 1300-1500°C. The SUP9, pearlitic carbon steel samples were used. The hardness, hardening depth and microstructure of the specimens have been investigated by Vickers micro hardness test and Scanning Electron Microscopy (SEM) supported by Energy Dispersive X-Ray Spectroscopy (EDX). The results have showed that the mechanical property was significantly improved due to the formation of single martensitic structures as identified by SEM. The hardness of treated surface evaluated by Vickers hardness test showed significant improvement nearly three time from 190 VHN before to 524 VHN after treatment. Furthermore, EDX confirmed that the formation of martensite layer occurred without altering its composition. The APS also produced uniform hardened layer up to 250 μm. The experiment has demonstrated that arc plasma process was successfully improved the mechanical properties of steel in relatively very short time.

1. Introduction

Various methods to attain the desired surface hardness of carbon steel generally differ in the way the surface is brought to elevated temperature. Surface hardening is generally used to improve tribological properties such as wear resistance and surface fatigue but maintain its ductility [1] . Different processes have own pros and cons. Using ordinary methods for instance flame hardening [[2] , [3]], rapid cooling in quenching medium is required. Temperature of flame hardening process is difficult to control [[2]], since hardening's result is controlled by the distance and exposure time. While duration time for induction hardening process is relatively long [[4] , [5]]. The current methods such as electron beam [6] and laser modification hardening [[7] , 8] enabled a very rapid heating of the surface area above the austenitic temperature.

Selective surface hardening methods such as induction hardening, electron beam, arc and plasma process were used to enhance surface layer without affecting the bulk of the workpiece. Thermal energy absorbed by surface layer was quickly distributed to the entire workpiece, generate very large temperature gradients. This phenomenon lead to self-quenching mechanism resulted in a fine-grained martensitic structure and/or retained austenite [9] with very high hardness. The hardness and the depth of hardened layer vary in in different process, for example in the induction process of steel the hardness reported as high as 650-700 HVN [4], while in plasma hardening and electron beam the hardness improve significantly, 130-150 % [6, 7].

In this study, Arc Plasma Sintering was used to improve the mechanical properties of SUP9 metal. The APS generally used to synthesize powder into bulk material. The APS process operates at very short time and expected to produce little distortion, induced less residual compressive stresses on the surface. The objectives of this work were to improve mechanical properties of the samples and study its microstructure before and after arc plasma hardening process.

2. Experimental Detail

The samples of JIS SUP9 (leaf spring steel), Table 1 [10] , were cut into rectangular shape, 15*x*15*x*5 mm (*w* x *l* x *t*). Sample was then grinded as shown in Figure 1 and ready for hardening process. The APS is powered by 12 Volt and 80 Ampere DC current to form electrical arc within Argon gas. The electrically ionized Argon gas (plasma) will exit through the nozzle and applied onto the surface of the samples statically, the nozzle nor samples were not moved. Figure 2 depicts the APS mechanism where temperature of process reached approximately 1300-1500 ^{o}C. The Argon gas flow with the velocity of 10 L/minute.

Table 1. Chemical composition of spring steels SUP9

Material	C	Si	Mn	P	S	Cr	Cu
SUP 9	0.53	0.31	0.71	0.023	0.012	0.67	0.14

Figure 1. Grinded sample of SUP9

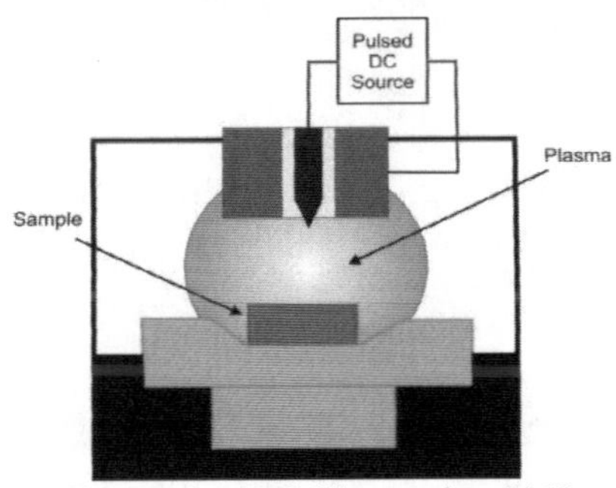

Figure 2. APS illustration [11]

The arc plasma was applied on the surface of SUP9 samples in three different exposures' time 0.5, 1 and 3 second, respectively. After surface hardening process, samples were cut in half, cleaned and polished. Two halves of samples were adhered side by side (Figure 3) and mounted into cylindrical resin and fine polished, followed by etching in 2 % nital. The samples were then observed under Scanning Electron Microscope (SEM) to expose the microstructure. Energy Dispersive X-Ray Spectroscopy (EDX) were used to determine the composition of samples, to resolve whether the composition will alter or not. The hardness of the treated layer was evaluated using Vickers micro hardness. The points of hardness measurement were approximately selected as shown in Figure 4.

IOP Conf. Series: Materials Science and Engineering **343** (2018) 012032 doi:10.1088/1757-899X/343/1/012032

Figure 3. Two samples were adhered side by side

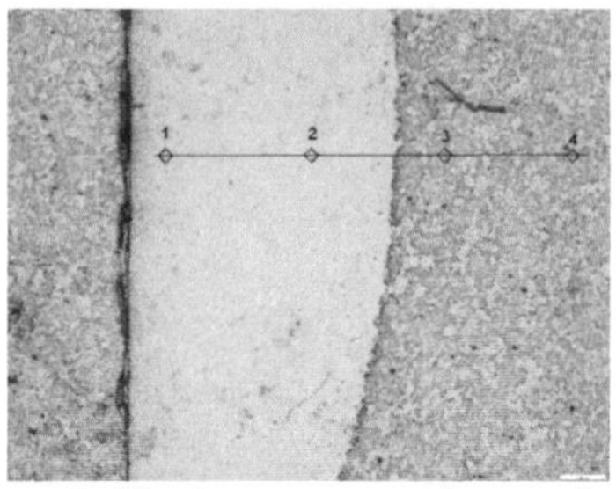

Figure 4. Hardness measurement (x100)

3. Result and Discussion

3.1. Microstructure of samples

Figure 5(a) and 5(b) below shows the interface between base metal and hardened metal for 1s and 3s exposure, respectively. The figures were obtained from secondary electron imaging (SEI) from SEM. The vivid color depicts the pearlite and ferrite structure and the darker color 5(c) and 5(d) show typical martensite structure. The martensite structure was produced by rapid cooling from austenite phase of samples due to heating by arc plasma.

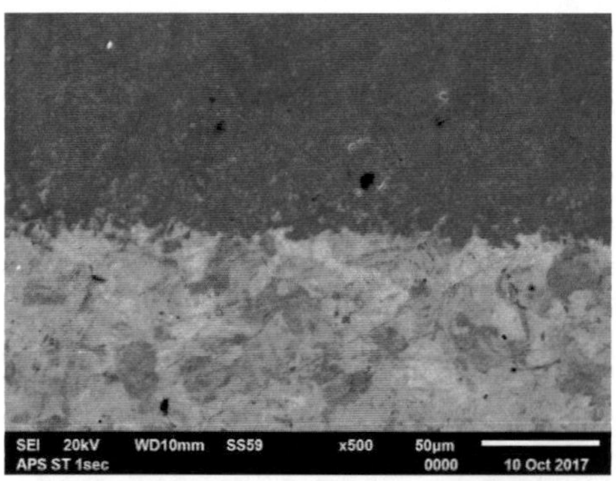

(a) Interface of 1s exposure (x500)

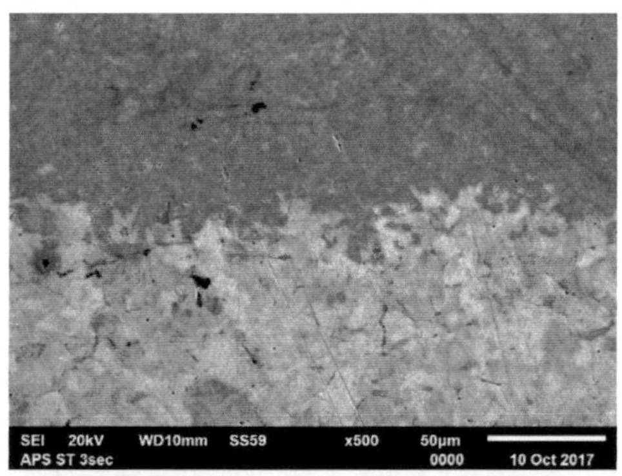

(b) Interface of 3s exposure (x500)

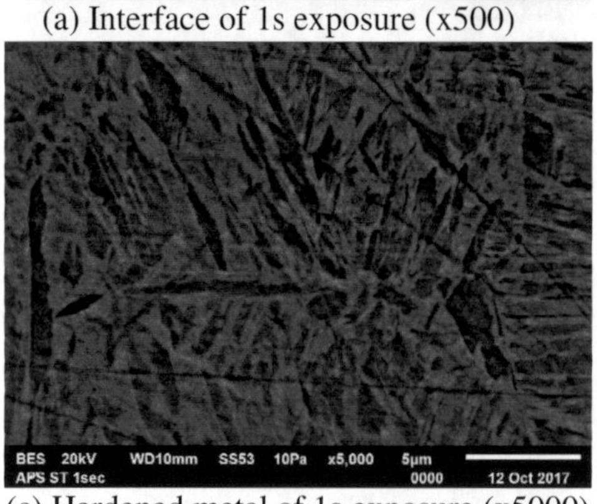

(c) Hardened metal of 1s exposure (x5000)

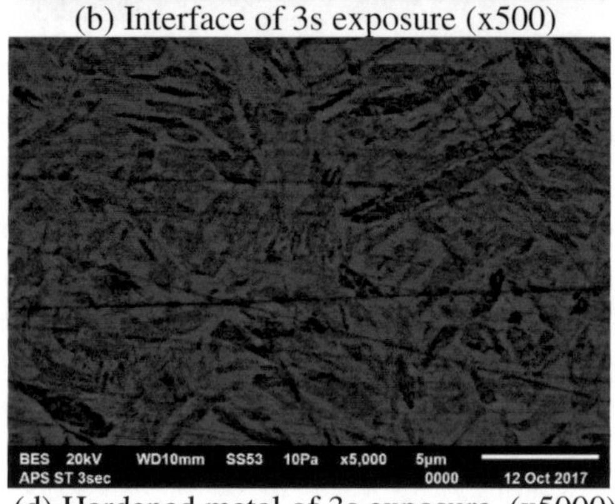

(d) Hardened metal of 3s exposure, (x5000)

Figure 5. Interface and hardened layer

3.2. Composition and Depth of Hardened Layer

Two samples were selected to examine its composition, 1s and 3 s exposure time respectively. The composition of hardened layer resulted from EDX are depicted in Figure 6. The EDX scanned along the straight path about 70.37µm in first sample and 70.16µm in second sample. Relation of composition of elements in the mass percentage and distance are shown in vertical and horizontal axis, respectively. Fe element has the highest composition in both figures followed by Cr, Si, and C. Figure 6 shows that the composition of its elements was not change along the scanning path. By combining the above result, it concludes that arc plasma process will not alter the composition of the elements but changed its phase and microstructure. The experiment showed that the surface hardness was increased solely due to phase transformation.

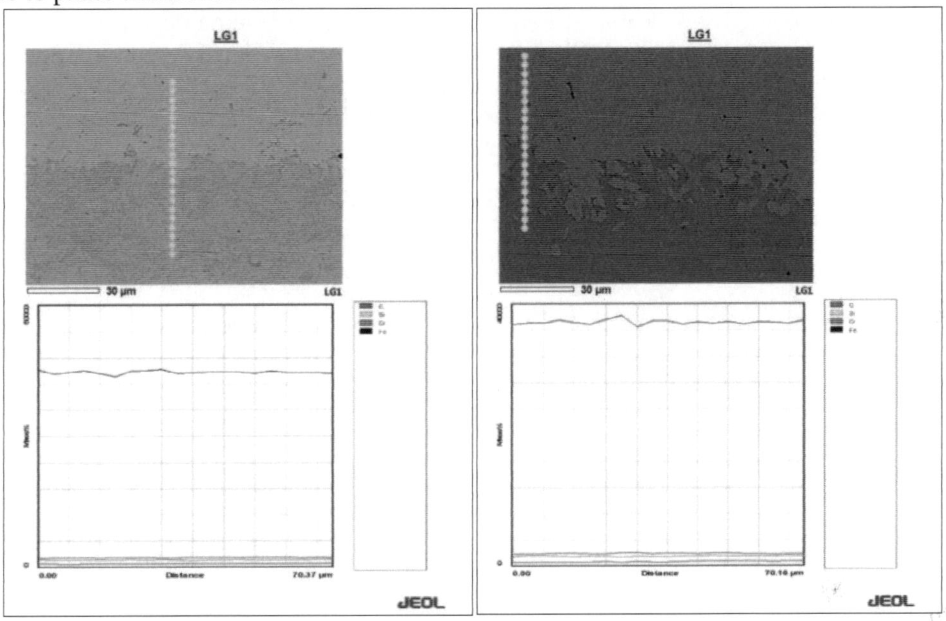

(a) 1s exposure (b) 3s exposure

Figure 6. Composition of elements in hardened layer

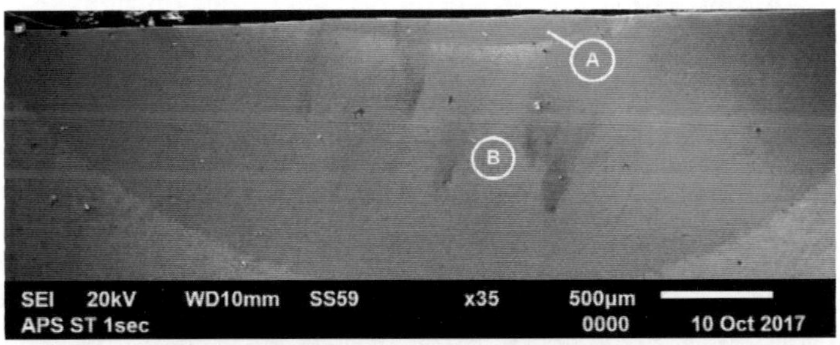

(a) Depth of 1s exposure (x35)

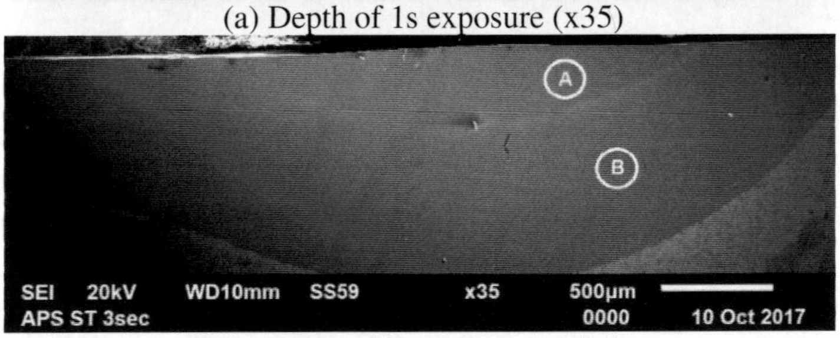

(a) Depth of 3s exposure (x35)

Figure 7. Depth of hardened layer (upper 3s, lower 1s exposure, respectively)

The depth of hardened layer in Figure 7 (x35 magnification) shows that 3s exposure resulting deeper layer (Figure 7(b)) compare to 1s exposure (Figure 7(a)). There are two layers in both cross section where the label A is the hardened layer while label B is the heat affected zone (HAZ). Depth of hardened layer of 1s exposure obtained from carefully compare with the magnification scale is about 100 μm while the 3s exposure yield the depth of 250 μm. It concludes that exposure time affected the depth of hardened layer, the longer the exposure time the deeper the hardened layer.

3.3. Hardness

The hardness of specimens was measured using Vickers hardness test at four-point locations as shown in Figure 4. Each sample was hardness tested four times for approximately at the same point location and the results were recorded as in Table 2. Point 1 and 2 are the point at hardened metal, while point 3 and 4 are base metal. The average VHN value then be plotted in Fig.8. At point 1 the average hardness is about 433-524 VHN, slightly lower than the hardness at point 2. At point 3 and 4, where the metal was not affected by heat, the hardness is about 190-199 VHN. The result shows that hardness of hardened layer increased nearly 3x higher compare to base metal.

Table 2. Hardness test measurement result

Sample 1		VHN	Avg. VHN	Sample 2		VHN	Avg. VHN
1 second		447.7199		3 second		518.6831	
	1	395.6053	433.5751		1	518.6831	513.5224
		457.4001				503.2009	
		474.2458				526.6935	
	2	481.2457	476.5791		2	526.6935	524.0234
		474.2458				518.6831	
		197.1182				185.5908	
	3	203.7756	199.6476		3	215.994	199.6476
		198.0491				203.7756	
		185.5908				185.6108	
	4	190.7819	190.8555		4	192.7025	190.7428
		196.1939				192.3295	

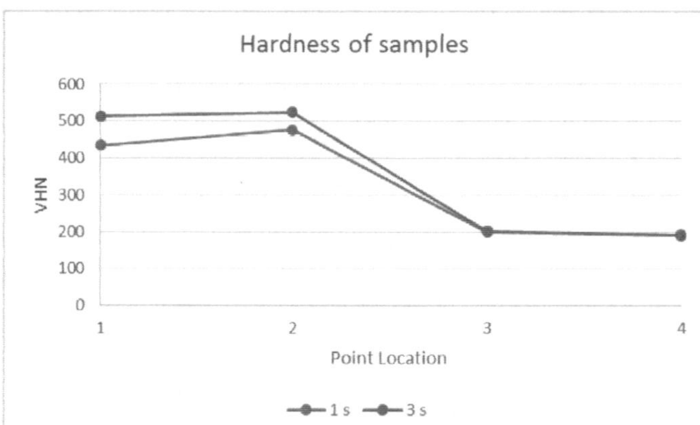

Figure 8. Hardness measurement of samples

4. Conclusion

This experiment provides some significant result in the arc plasma surface hardening process of SUP9 material and potentially can be used for others steel. The results showed that hardness of steel increase nearly three time higher, from 190 VHN in base metal to 524 VHN in hardened layer. Hardness of surface layer increased due to the formation of martensite phase as observed under SEM. EDX confirmed that the composition of elements (Fe, Cr, Si, and C) remains constant, there was no change in mass percentage before and after hardening process. Depth of hardened layer that have been obtained from Vickers hardness test increased as exposure time increased from 1s to 3s, 100 μm to 250 μm respectively. In summary, arc plasma surface hardening process has demonstrated has significantly improved mechanical properties of steel in relatively very short time.

5. References

[1] DeGarmo EP, Black, JT & Kohser, RA 1988. *Materials and processes in manufacturing*. New York: Macmillan Pub. Co.
[2] Lee MK, Kim G H, Kim K H, Kim W W 2006 *J. Materials Processing Tech.* **176** 140
[3] Huang CA, Lin CK, Chiou WA, Hsu FY 2011 *Surf. & Coating Tech.* **206** 325
[4] Slatter T, Lewis R, Jones AH 2011 *Wear* **270** 302
[5] Gao K, Qin X, Wang Z, Zhu S 2016 *Materials Sci. & Eng.* **651** 535
[6] Fu Y, Hu J, Shen X, Wang Y, Zhao W 2017 *Nuclear Instruments and Methods in Physics Research B* **410** 207
[7] Babu PD, Buvanashekaran G, Balasubramanian KR 2012 *Transactions of the Canadian Society for Mech. Eng.* **36**-3 241
[8] Syed B, Shariff SM, Padmanabham G, Lenka S, Bhattacharya B, Kundu S 2017 *Materials Sci. & Eng. A* **685** 168-177
[9] Roy S, Zhao J, Shrotriya P, Sundararajan S 2017. *Tribology Int.* **112** 94
[10] Murakami, Y. (2002). Metal Fatigue: *Effects of Small Defects and Nonmetallic Inclusions*. Elsevier Science Ltd, Oxford.
[11] Bandriyana et al 2017 IOP Conf. Ser.: *Mater. Sci. Eng.* **176** 012047

IOP Conf. Series: Materials Science and Engineering **343** (2018) 012033 doi:10.1088/1757-899X/343/1/012033

Analysis of electrical audit and energy efficiency in building Hotel BC, North Jakarta

Agung Wahyudi Biantoro

Mechanical Engineering Lecturer, Mercu Buana University, Jakarta, Indonesia

agung_wahyudi@mercubuana.ac.id

Abstract. The Hotel BC is using power source from PLN with capacity of 4300 kVA which is divided into 3 units of 2000 kVA transformer. Transformers are used to supply the load of Mall tenants, and Utility loads, such as Chiller, pumps and others. Problems found in the field are complaints from the hotel regarding the safety of electrical installations and wasteful, inefficient electrical costs. The purpose of this study is to check the electrical installation in the building and determine the *Energy Use Intensity (EUI)* and the cost of payment according to usage based on historical data of the building then compare it with the *EUI* standard of Ministry of Energy and Mineral Resources of Indonesia. The method used is survey measurement method and quantitative descriptive analysis by comparing in general condition of energy consumption of this building with standard issued by Ministry of Energy and Mineral Resources of Indonesia. The *EUI* is average 645.58 kWh/m^2/year, or 53.79 kWh/m^2/month, this is inefficient category, because its *EUI* value is > 24 kWh / m^2 / month. For Electrical audit on imaging thermal test at Panel Out Going of chiller pump, 200 ampere, the highest temperature is 97.3^0 C, at 200 ampere phase S termination, and this is included in the major category. The numbers of hot spots on the Capacitor bank panels are 10 major points and Chiller panel has 10 major. There are many major points and they are quite dangerous because they can cause fire hazard on the panel. The AC average temperature and humidity distribution did not meet the standard of SNI (Indonesia National Standard).

Keywords: *Electrical Audit, Energy Efficiency, Energy Use Intensity (EUI)*

1. Introduction

1.1. Profil Discovery Hotel Ancol

BC Hotel is a hotel located in the heart of North Jakarta's tourist attractions which provides a comfortable base for families, friends, and colleagues, with modern interpretations of contemporary stylish design and elegant neutral tones. This building is over 20 years old and located in Pademangan, North Jakarta. Building Hotel BC consists of a main building 10 Floors, on floor 3 to 8 is used for hotel guest rooms and on other floors are used for main lobby, ballroom, and other public facilities.

Safety, convenience, and efficiency are the most important factors for achieving customer satisfaction and operational productivity. Therefore, an awareness of the importance of maintenance and security of the building, especially in the danger of electric fires. A study showed 60% of fire incident due to an electrical short circuit, so it needs early detection and preventive maintenance to prevent the occurrence of an electric fire hazard. During this time, usually the owner is more focused on the core business and less concerned about the condition of the district, whereas the company must protect the assets and must be safe against fire hazard. Electrical hazards cannot be detected by the invisible and the layman and the limitations of instrumentation and labor tools. The company also wants a saving (efficiency) on the building considering the building is more than 20 years old. The aged building is often found to increase electricity bills over time, thus increasing maintenance costs.

One of the ways to check and monitor energy consumption is to audit the energy, so that energy consumption data can be seen from a particular process or machine and can be seen inefficient energy consumption. The parameters used are energy consumption intensity (IKE) in buildings defined in energy quantities per unit area in buildings served by energy (kWh / m2 / year or kWh / m2 / month). Meanwhile, to check the electrical safety of electrical audit required by using a thermographic tool.

1.2. Profil Discovery Hotel Ancol

BC Hotel is a hotel located in the heart of North Jakarta's tourist attractions which provides a comfortable base for families, friends, and colleagues, with modern interpretations of contemporary stylish design and elegant neutral tones. This building is over 20 years old and located in Pademangan, North Jakarta. Building Hotel BC consists of a main building 10 Floors, on floor 3 to 8 is used for hotel guest rooms and on other floors are used for main lobby, ballroom, and other public facilities.

1.3. Objective of this Project

1.3.1 To know the value of Energy Use Intensity (EUI) at Hotel BC

1.3.2 To find out if all Electrical installations and equipment are in good condition, safe and in compliance with applicable standards and regulations.

1.4. Problem Formulation

Hotel BC is a hotel that is more than 20 years old so it needs to be done step austerity and efficiency and replacement of electrical or mechanical equipment. Buildings that are more than 20 years old usually have complaints of increased energy consumption over time, so the bill is more wasteful than the new building. Some machine tools such as air conditioners, pumps have decreased performance because they are old and need to be checked again. Furthermore, some electrical equipment and cables have decreased quality and peeling so that must be wary and replaced to prevent the occurrence of fire. To manage the power and security needs of buildings, it is necessary to save energy and electrical security checks on the building.

2. Methodology Adopted

The methodology adopted for this research is:

2.1 Method of Collecting Data

Data collection is done by survey method, recording monthly report data, and measurement in an electrical panel. This research uses primary data type and secondary data. Tools used Thermografik, Tang Ampere, and Multimeter, and Cos phi meter.

2.2 Method of Analyzing Data

Data were collected and then analyzed by using quantitative descriptive analysis. The research variables include the amount of energy use based on energy audit and saving opportunities based on field conditions. In the initial of energy audit, will be calculated the amount of Energy Consumption Intensity (IKE) per unit area is conditioned (net area) according to usage based on campus historical data. In a detailed energy audit will be calculated IKE based on observations of the use of electrical energy in detail with various equipment that consume electrical energy and the time of its use. Furthermore, for electrical audit, an infrared thermographic tool used in various of panels. The data will then be analyzed the percentage of major and minor amounts, then which parts should be repaired or replaced, thus preventing risk of any fire hazard. To find the value of IUE then the formula used is :

$$IEU = \frac{The\ amount\ of\ KWH\ per\ m\ square}{building\ area} \quad (1)$$

The IEU standard values for each building are as follows :

Table 1. Value of energy consumption intensity standards in building, based on activities (Reference ASEAN USAID 1987)

No	Type of Building	EUI (Kwh/m2/year)
1	Office	240
2	Mall	330
3	Hotel	300
4	Hospital	380

If (EUI) > 240 kwh/m2/year for office building, it means Energy consumption not efficient

3. Result and Discussion

3.1 The Value of Energy Use Intensity (EUI) at Hotel BC

The energy use intensity (EUI) in the building is the value of the indicator to measure the level of energy efficiency in a building. Unit Intensity of energy consumption in the building is the amount of energy per unit area in buildings served by energy (kWh / m2 / year or kWh / m2 / month). Total Kilowatt per hour / month in Hotel BC, Jakarta could see in Figure 1.

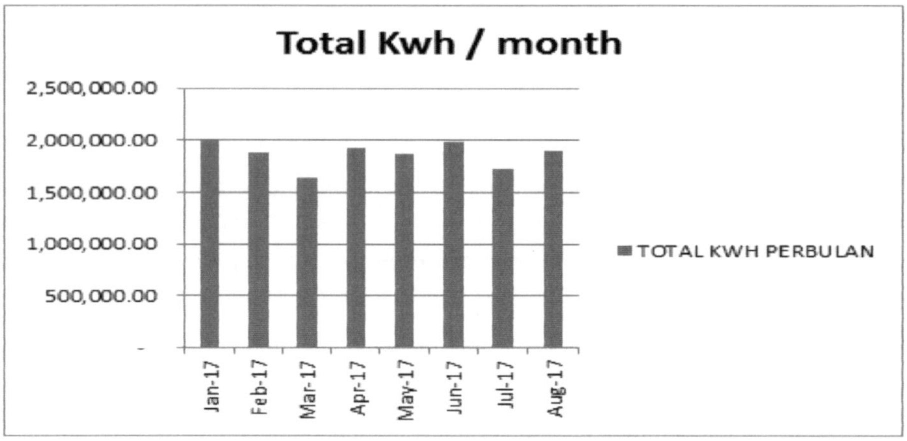

Figure 1. Total Kilowatt per hour / month in Hotel BC, Jakarta.

Then below is a graph of total cost of electricity usage BC Hotel per month.

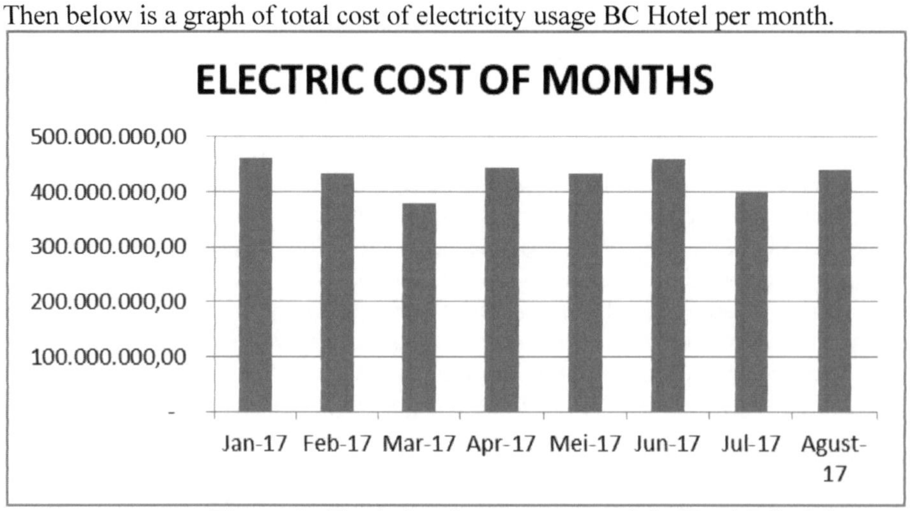

Figure 2. Total electric cost per month in Hotel BC, Jakarta.

As we know that the building area of the hotel bc is 34,616 m2, and the evarage kwh per month is 1,862,285. So the EUI value is 53.79 kWh/m^2/month or 645.58 kWh/m^2/year. As we know that the building area of the hotel bc is 34,616 m2, and the evarage kwh per month is 1,862,285. So the EUI value is 53.79 kWh/m^2/month or 645.58 kWh/m^2/year. The electrical load distribution at BC Hotel Building, the highest is air conditioning, then followed by lighting. The full percentage is as follows:

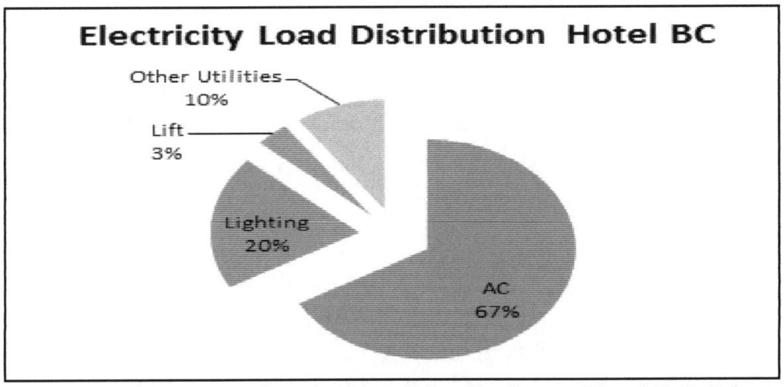

Figure 3. Estimated electrical load distribution at Hotel BC, Jakarta

From Table 1 it can be compared between the actual value of EUI in the field with reference value of ASEAN USAID standard 1987. Based on this table the value of EUI Building Hotel BC = 645.58 kwh/m2/year, included in the category inefficient / wasteful.

The potential for energy savings can be focused on the system of air and lighting in the BC Hotel Building. Use of R-410 or R-32 refrigerant on AC inverter to replace R-22 refrigerant on some Split AC units to be installed. Because the freon type R-410 or R-32 is lighter than the freon type R-22 so that the compressor work is not too heavy and electric energy is also smaller, besides that this type of freon has entered standard green because it does not damage ozone. The Simulation for Energy Saving Potential (PHE) is as follows:

Table 2. Simulation for energy saving potential

No	Description	Watt/1 PK	Energy Consumption /th	Cost (Rp)	Calculation
1	74 AC, 136,5 PK	864	118 kWatt	-	354,000 kwh / year
2	74 AC, new, using AC inverter, 136,5 PK (a)	540	73.71 kWatt	477,750,000 (a)	221,130 kwh / year
3	Potential Savings	324		179,640,240	132,870 kwh / year
4	Payback Period (a/b)				2.66 years

Description : 1 year = 3,000 hours, 1 month 25 days, 1 day 10 hours, 1 Kwh = Rp. 1,352 price AC 1 PK inverter Rp. 3,300,000

From the simulation for Energy Saving Potential (PHE) based on the above table is that by replacing the old AC into AC inverter with the latest technology then the cost incurred is Rp. 477.750.000 with the potential savings of Rp. 179,640,240 per year or with pay back perode is 2.66 years

IOP Conf. Series: Materials Science and Engineering **343** (2018) 012033 doi:10.1088/1757-899X/343/1/012033

The above analysis shows that the EUI value for BC hotels is 696 kwh / m2 / year which means greater than the standard value of 300 kwh / m2 / year, and this means very wasteful. Therefore it is necessary to do the energy saving step as follows:

1. Fix the Building Automation System (BAS) system to set the central AC system.
2. If possible, increase the AC temperature setting.
3. Perform AC maintenance (filter cleaning at least 3 months).
4. Turn off the AC and lights if the room is not in use.
5. Replace electrical appliances, air conditioners, and pumps that exceed the technical age limit (more than 20 years).
6. Control-motor control should use Variable Speed Drive(VSD).
7. The transformer capacity can be lowered to 1000 kVA to reduce transformer losses. Currently, the transformer load is only 28.3%.

3.2 The Electrical Audit at Hotel BC

Hotel BC uses a power source from PLN with a capacity of 1,730 kVA, with one unit of 2000 kVA transformer. Hotel BC à in the back up of 2 units of Generators with a capacity of each 1000 kVA parallel synchronous before supplying power to the loads. Maximum usage load à453 kW or 566 kVA, 33% of total power connection of PLN. This thermal data collection is obtained during measurement using the Thermography Infra Red tool. Category results description is Major for temperature 70°C or more, Minor for 65-69°C, Observation for 50-64°C, and Normal for 30 - 49°C. Here are the results of thermographic measurements on some electrical panels :

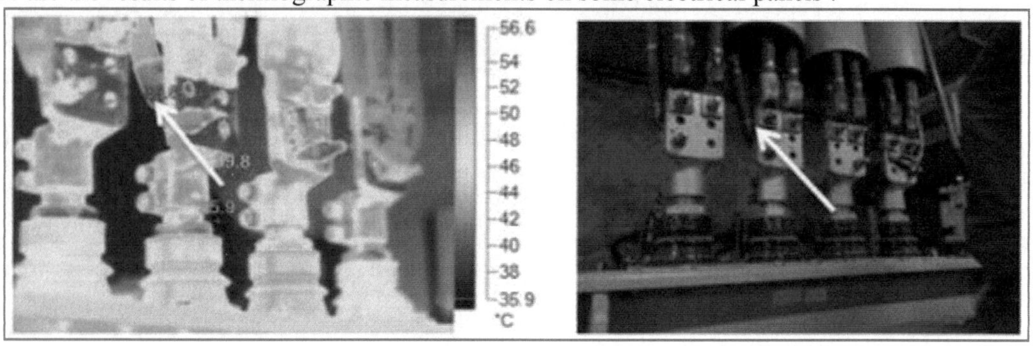

Figure 4. Thermographic test data on 2000 kVA transformer, with load 566 kVA

The result of thermal imaging at 2000 kVA transformer, Out Going Cable Spot, obtained result of the highest temperature 56.6 ° C. The highest temperature is found in phase S cable skeleton, with load at each phase is R = 929.9 A, S = 907,3 A , T = 853.9 A. The temperature in this panel is still in the normal category, no need for improvement.

Next is the results of thermographic measurements on some electrical panels at Panel Out Going of chiller pump, 200 A.

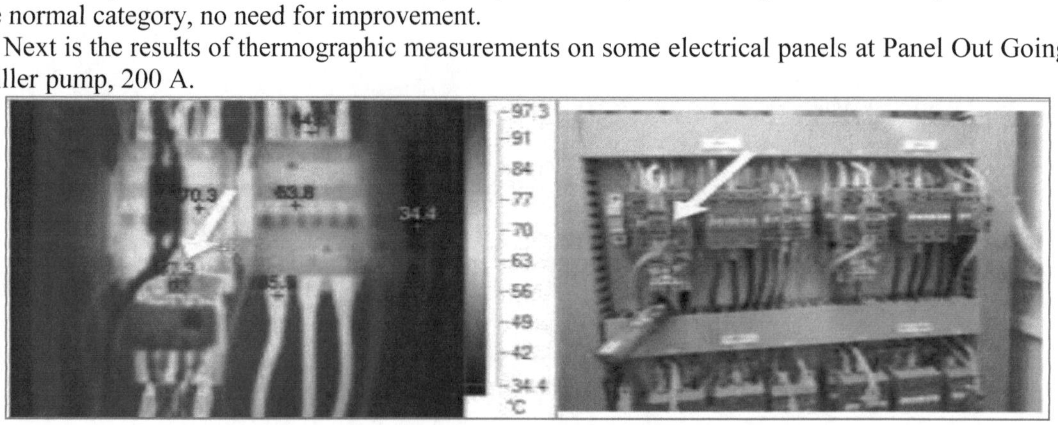

Figure 5. Thermo graphic test data at panel out going of chiller pump, 200 A

At Figure 4 there is Thermo graphic test data at Panel Out Going of chiller pump, 200 A. That Panel Out Going for chiller pumps on 200 200 C Contractor Slot, the highest temperature is 97.3 C, at 200 A phase S termination, while the load has reached 93.1 A and the temperature of the contactor body is 70.3 C, with cable using NYAF type 16mm. With the highest Temperature of 97.3 C, this is included in the major category. The panel on the cable uses NYAF type 16 mm, does not match the load. We recommend wiring NYAF 16 mm replaced with NYAF 35 mm cable. Further recapitulation of all electrical panel at Building Hotel BC is as follows:

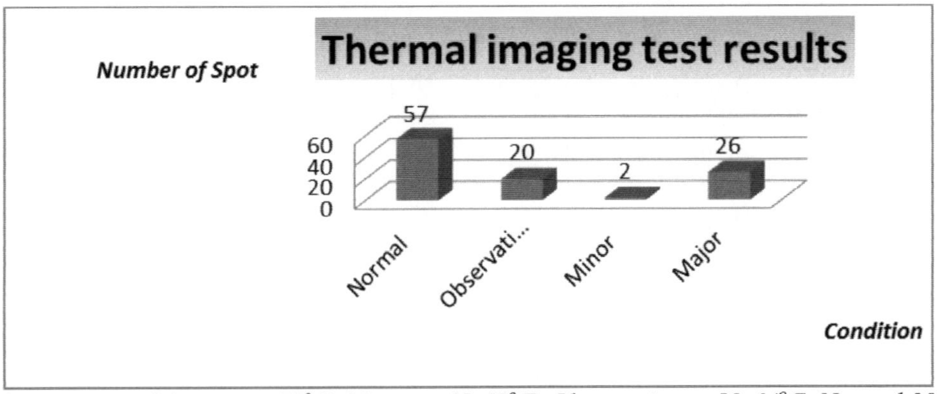

Description : Major = > 70°C, Minor = 65-69°C, Observation = 50-64°C, Normal 30 – 49°C.
Figure 6. The thermal imaging test result

Further details about panels that have major and minor hot spots are as follows:

Table 3. Number of major and minor hot spots on the building panel

No	Panel Name	Major	Minor
1	Capacitor Bank Panel	10	-
2	Chiller Panel	10	1
3	AHU Panel	1	1
4	Pump Panel 11th Floor	2	-
5	Tenant 12nd Floor	1	-
6	Panel 7th Floor	1	-
7	STP Panel	1	2
	Total	26	2

From the table above shows that total of the major categories are 26 spots, while minor are 2 spots. The number of hot spots on the capacitor bank panel there are 10 major points, and chiller panel has 10 major and 1 minor point. In addition, there are panel AHU panel, tenant panel, panel floor 7 and STP panel which also has 1 major point each, while the pump panel there are 2 major panels. The number of major points is quite dangerous because it can cause excessive heat can eventually occur short-circuit or fire hazard on the panel. It is imperative to take precautionary measures, for example by cleaning the panel, replacing the cables or tighten the termination of the contactor.

The type of air conditioner used in Hotel BC is a type of central air conditioner and partly split AC. Here is the measurement of the use of air conditioning in each room in Hotel BC.

Table 4. Humidity and air temperature, in the room of Hotel BC

No. Room	Room	Temp (^{0}C)	RH (%)	dB	
				Low	High
606	FCU-6-606	22.90	62.90	37.90	42.80
615	FCU-6-615	26.90	65.50	39.70	42.10
625	FCU 6-625	24.70	67.20	37.50	42.10
626	Living Room	24.90	65.30	38.80	45.40
215	Bedroom	24.60	59.70	44.10	44.20
628	Ruang Tamu	22.80	63.50	37.10	39.30
216	Bedroom	23.30	59.70	41.30	46.80
415	Bedroom	23.40	59.70	44.90	47.60
550	Living room	24.10	63.00	38.70	43.40
313	Bedroom	23.10	59.50	37.50	44.10
314	Bedroom	22.70	68.20	37.00	40.20
604	Living room	23.10	63.40	44.80	48.80
Avg		**23.88**	**63.47**	**39.94**	**43.90**

In Table 4 above shows that Hotel BC has an average temperature average 23.88 ° C. The distribution of temperature is approaching SNI standard (24 °C - 27° C) of total measured space. For humidity (RH) lowest value measured 59.5% and highest humidity 67.2%. It was concluded that the distribution of humidity distribution did not meet the standard of SNI (Indonesia national standard) 55% - 65%.

4. Conclusion

EUI is expressed as energy per square foot per year. It's calculated by dividing the total energy consumed by the building in one year or the month by the total gross floor area of the building. The IEU of the building 645.58 kWh/m^2/year, or 53.79 kWh/m^2/month. The analysis shows that the EUI value for BC hotels is 645.58 kwh / m2 / year which means greater than the standard value of 300 kwh / m2 / year, and this means very wasteful. Therefore it is necessary to do the energy saving step as follows : Fix the Building Automation System (BAS) system to set the central AC system. If possible, increase the AC temperature setting, Perform AC maintenance (filter cleaning at least 3 months), Turn off the AC and lights if the room is not in use, and replace electrical appliances, air conditioners, and pumps that exceed the technical age limit (more than 20 years). Distribution of temperature is close to SNI standard (24 ° C - 27 ° C) of total measured space. It is not the standard of SNI (Relative humidity (RH) lowest value measured 59.5% and highest humidity 67.2%. It was concluded that the distribution of humidity distribution did not meet the standard of SNI (55% - 65%).

References

[1.] Abdurrachim dan Aditya Budi Nugraha. 2017. *Audit Energi Pembangkit Listrik Tenaga Panas Bumi (Pltp) Menggunakan Standar Internasional*. Digilib ITB, Bandung. Sumber : https://digilib.itb.ac.id/gdl.php

[2.] Astu Pudjanarsa dan Djati Nursuhud. 2014. *Mesin Konversi Energy*. Andi Publisher. Yogyakarta.

[3.] Arismunandar, A. dan Kuahara, S. 1973. *Teknik Tegangan Tinggi.Jakarta*. PT Dainippon Gitakarya Printing.

[4.] Badan Standardisasi Nasional (BSN). 2000. *SNI 03-6390- 2000. "Konservasi energi sistem tata udara pada bangunan gedung"*. Jakarta

[5.] SNI, 2000. SNI 03-6196-2000. *Prosedur Audit Energi Pada Gedung*. Jakarta.

[6.] Capital Heat Inc. 2017. *Take Control your Energy Cost. Home Energy Audit*. Sumber : http://capitalheat.com/geothermal-heating-cooling-buffalo-rochester

[7.] Daryanto. 2016. *Teknik Pendingin : AC, Frezeer, Kulkas*. Yrama Widya Publisher. Bandung.
[8.] Gousia Sultana and Harsha.H. 2015. *Electrical Energy Audit a Case Study*. Electrical Audit a Case Study. IOSR Journal of electrical and Electronics Engineering. e-ISSN: 2278-1676,p-ISSN: 2320-3331, Volume 10, Issue 3 Ver. III (May – Jun. 2015), PP 01-06 www.iosrjournals.org
[9.] Lamba, Manoj Kumar dan Abishek Sanghi. 2015. Energy Audit on Academic Building. *International Journal of Engineering Research and General Science,* ISSN 2091-2730, vol. 3, issue 4, p. 600-604. www.ijergs.org
[10.] Swati Khare, Abhay Kumar Sharma R. K. Ranjan Shashank Khare. *Energy conservation through energy audit.*
[11.] Getu, Beza Negash and Hussain A. Attia. 2016. Electricity Audit and Reduction of Consumption: Campus Case Study. *International Journal of Applied Engineering Research,* ISSN 0973-4562, vol. 11, no. 6, p. 4423-4427. www.ripublication.com

IOP Conf. Series: Materials Science and Engineering **343** (2018) 012034 doi:10.1088/1757-899X/343/1/012034

Effect of thickness on Fatigue Crack Growth of Aluminium alloys

Haftirman[1], Haris Wahyudi[1], Julpri Andika[2], and Noor Afifah Yahadi[3]

[1]Department of Mechanical Engineering, Faculty of Engineering, Universitas Mercu Buana, 11650 Jakarta, Indonesia.
[2]Department of Electrical Engineering, Faculty of Engineering, Universitas Mercu Buana, 11650 Jakarta, Indonesia.
[3] Universiti Malaysia Perlis, Malaysia

hatirman@gmail.com

Abstract. The effect of thickness on fatigue crack growth has been investigated for aluminium alloys such as a high-strength aluminium alloy (A7075-T6), a medium strength alloy (A6063-T6), and A2024-T351. The fatigue tests were conducted using Instron 8801 fatigue testing machine on all those materials with thicknesses of 12.7, 15.8, and 20 mm. All specimens were undergoing finishing process on the surface. These tests were run under constant amplitude load for each different thickness and materials. Fatigue crack growth specimen made of A2024-T351 and A6063-T6 aluminum alloys with thickness of 12.7 mm were higher than specimen made of A7075-T6 aluminum alloy with thickness of 20 and 12.7 mm and A6063-T6 aluminum alloy with thickness of 15.88 mm. It is concluded that fatigue crack growth decreases significantly with increasing thickness, and the crack initiates very fast on surface material.

1. Introduction

In a previously paper, the fatigue crack initiation and growth of aluminium alloy A7075-T6, A6063-T6 and A2024-T351 with various stress ratios were investigated using the compact test specimens having thickness of 12.7 mm [1]. The result found that the gradients of crack growth rate increase while the stress ratio, R increase. Higher R ratio result in higher value range of minimum applied load. Using the sheet specimens having an edge notch 3 mm deep and 0.1 mm root radius at stress ratios R raging from -0.5 to 0.5, the result shows that the fatigue crack propagation rate was dependent on the stress ratio that was the higher the stress ratio, the higher the rate of fatigue crack growth for a given ΔK value [2]. The present paper is concerned with an effect of thickness on fatigue crack growth of aluminium alloy. Effect of thickness in fracture of 2219-T851, 6061-T651,7075-T6, 7075–T651, and 7079-T651aluminum alloy [3,4], the result show that value of K_{Ic} generally increase with increasing specimen thickness. Then, the amount of fatigue crack growth of 2024-T3 aluminum alloy delay increases with decreasing sheet thickness. This behavior is attributed to greater plastic strains associated with the larger plastic zone size formed under plane stress conditions [5]. With finite element analysis and experimental of critical crack-tip-opening angle (CTOA) in 2024-T351 aluminum alloy, the result shows both computationally and experimentally that the critical surface CTOA value continues to decrease for increasing specimen thickness [5,6]. The present thickness of specimens is used 20, 15.7, and 12.7 mm will investigate the fatigue crack growth of aluminum alloy.

2. Materials and Experimental Procedure

The fatigue tests were conducted using Instron 8801 fatigue testing machine on aluminium alloys such as a high-strength aluminium alloy (A7075-T6), a medium strength alloy (A6063-T6), and A2024-T351 with thickness of 12.7, 15.8, and 20 mm. All the specimen designs were referred to the ASTM E647-11 standard for compact test for fatigue crack growth rate testing. One of the specimens is shown in Figure 1. Recommended thickness and suggested minimum dimensions are provided to ensure the results or data obtained are valid and full in the ranged of predominantly elastic condition with the force applied. All specimens were undergoing finishing process on the surface. These tests were run under constant amplitude load for each different thickness and materials. The standard test method consists of determining the fatigue crack growth rate near threshold to maximum stress intensity controlled instability. The dimensions stated were provided with the tolerance. Design is done by AutoCAD software. Since the EDM wire cut machine is only support with the AutoCAD software drawing. Table 1 and Table 2 show the mechanical properties and chemical compositions of all types of aluminium alloy. Instron console software is the software that collaborates with the Instron 8800 to transfer the setting on the software successfully to the Instron 8801 Hydraulic Server Machine.

The fatigue crack growth experiment was conducted using Instron 8801 Hydraulic Server Machine. After the data acquisition system, the specimens were further analysing using the Scanning Electron Microscope (TM3000 Table top Microscope). SEM was used to analyze on specimen surface and the crack propagation area to identify the crack initiation and the fatigue behaviour on the particular locations. Paris law is used to describe the long crack behavior under constant amplitude loading with the small range of yielding. Further modification on Paris law are needed for describe the overall crack growth, the effect of stress ratio and the effect of high amplitude loading [12]. Paris-Erdogan equation can be shown as in Eq. (1) where $\frac{da}{dN}$ = fatigue crack growth rate, ΔK = Stress intensity factor range, C and m are the coefficients of material constants. $\frac{da}{dN} = C\ (\Delta K)^m$

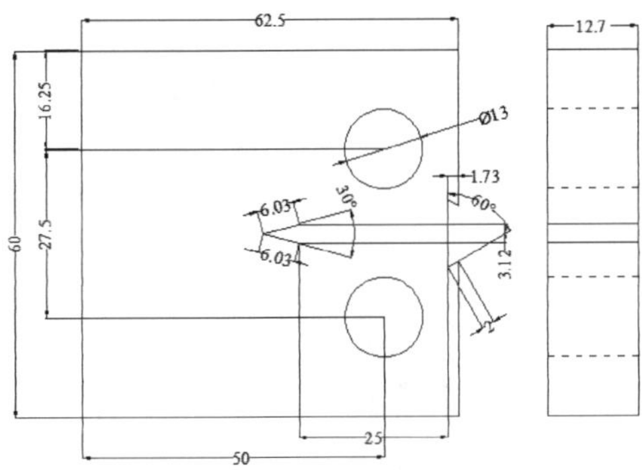

Figure 1. Standard Compact Specimen for fatigue Crack Growth Rate Testing [11], dimension is in mm and degree.

Table 1. Mechanical Properties of Aluminum Alloy

Aluminum alloy	Yield Strength (MPa)	Tensile Strength (MPa)
A2024-T351	290	444
A6063-t6	172	207
A7075-T6	535	585

Table 2. Chemical Composition of Aluminum Alloy (%wt)

Materials	Si	Fe	Cu	Mn	Mg	Cr	Zn	Ti
A2024-T351	0.11	0.15	4.58	0.66	1.54	0.02	0.04	0.032
A6063-T6	0.42	0.22	0.04	0.05	0.48	0.02	0.03	0.03
A7075-T6	0.10	0.38	1.52	0.10	2.6	0.21	5.55	0.03

3. Results and Discussion

3.1. Fatigue crack growth of A2024-T351, A6063-T6, and A7075-T6 aluminum alloy.

Fatigue crack growth experiments under constant stress amplitude have been carried out in order to investigate the effect of thickness on fatigue crack growth aluminium alloys. The fatigue crack growth curves *da/dN* vs ΔK is shown in Fig. 2 for all sizes of specimen. This figure shows as prediction of fatigue crack growth on A2024-T351, A6060-T6, and A7075-T6 aluminum alloys. Fatigue crack growth specimen made of A2024-T351 and A6063-T6 aluminum alloys with thickness of 12.7 mm were higher than specimen made of A7075-T6 aluminum alloy with thickness of 20 and 12.7 mm, and specimen made of A6063-T6 aluminum alloy with thickness of 15.88 mm. It can be seen that aluminum alloy with thicker size has lower fatigue crack growth rate in stress intensity, ΔK region (10-20 MPa $m^{1/2}$) significantly than that thinner size of specimen, which according to [8] are evidence of ΔK region the same as result for A2024-T351 aluminum alloy. Specimen with thickness of 20 mm for A7075-T6 aluminum alloy material, fatigue crack initiation and crack propagation were faster than others of aluminum alloy.

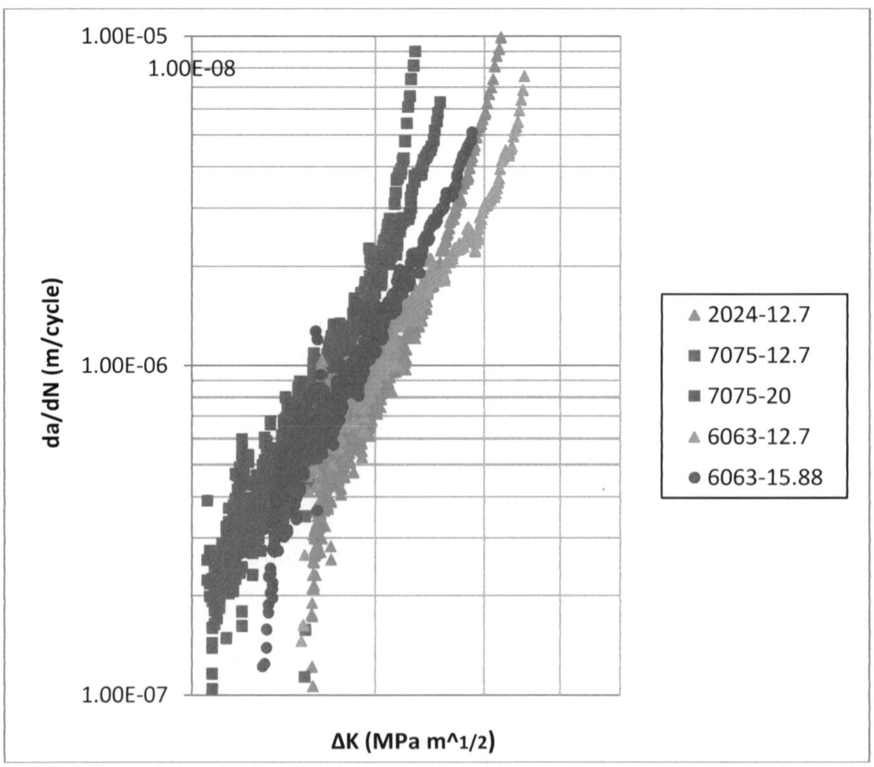

Figure 2. Fatigue crack growth rate for all types of aluminum alloy with difference specimen size

3.2. Effect of thickness on A2024-T351, A6063-T6, and A7075-T6 aluminum alloy

As mentioned earlier, fatigue crack growth of aluminium alloy specimen with thickness of 20 mm is higher than that of aluminum alloy specimen with thickness of 12.7 and 15.8 mm. However, fatigue crack growth based on crack length, the result show that specimen with thickness of 12.7 mm for A7075-T6 aluminum alloy is faster than of specimen with thickness of 20, and 15.8 mm for A7075-T6, A6063-T6, and A2024-T351 aluminum alloy. The effect of thickness on the fatigue crack growth for A7075-T6, A6063-T6, and A2024-T351 with thickness of 12.7, 15.8, and 20 mm is shown in Figure 3. Figure 3 shows that the crack length for A7075-T6 aluminum alloy materials specimen with thickness of 12.7 mm initiates at crack length of 20 mm and then crack propagate up to failure with crack length of 23.92mm. For A6063-T6 specimen with thickness of 15.88, and 12.7 mm and A2024-T351specimen with thickness of 12.7 mm, crack initiates at crack length of 20 mm and propagate up to failure at similar crack length around 27.99 mm. Fatigue crack growth of A7075-T6 specimen with thickness of 20 mm initiates slower than that of other aluminum alloy. It can be seen that retardation occurrence of this crack due to the A7075-T6 specimen was thicker than that of other specimen.

Figure 4 shows crack length versus number of cycles curve. This figure indicates that at lower number of cycles, A7075-T6 specimen with thickness of 12.7 mm crack occurs faster than that of A6063-T6, A7075-T6, A2024-T351 and A7075-T6 specimen with thickness of 12.7, 15.88, 12.7, and 20 mm, respectively. Similarity the result from Figure 3 that fatigue crack growth based on crack length, A7075-T6 specimen with thickness of 12.7 mm initiates faster than that of other specimen.

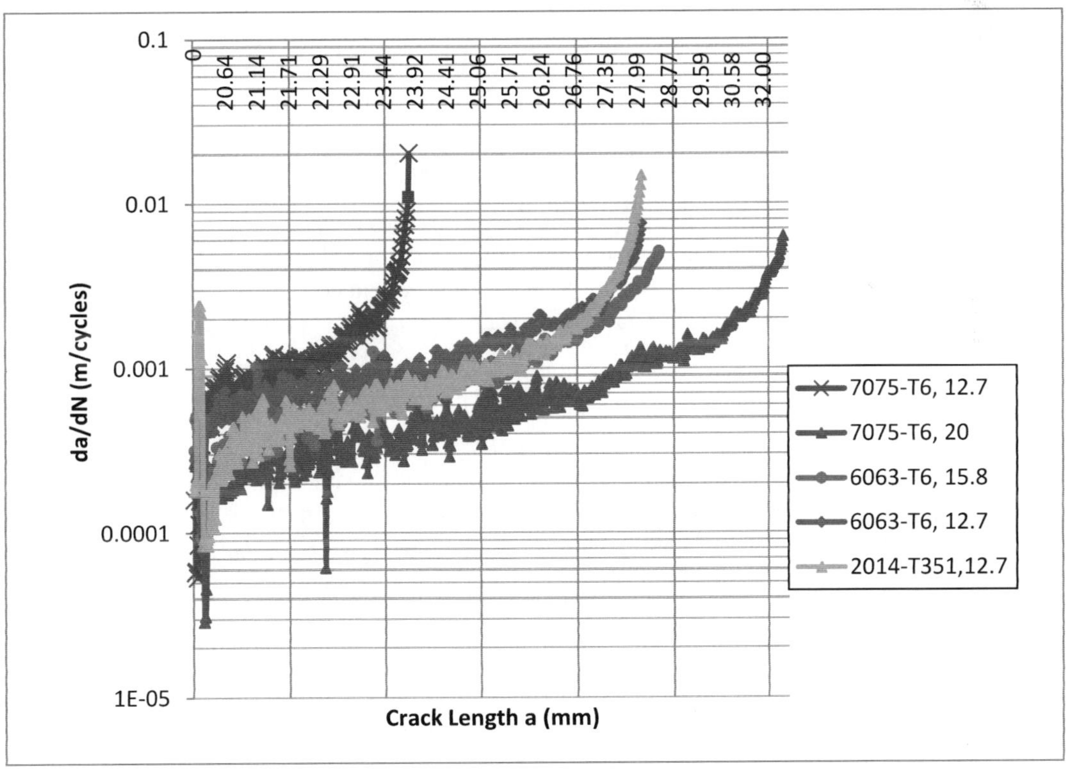

Figure 3. Effect of thickness on fatigue crack growth for all types aluminum alloy with difference specimen size

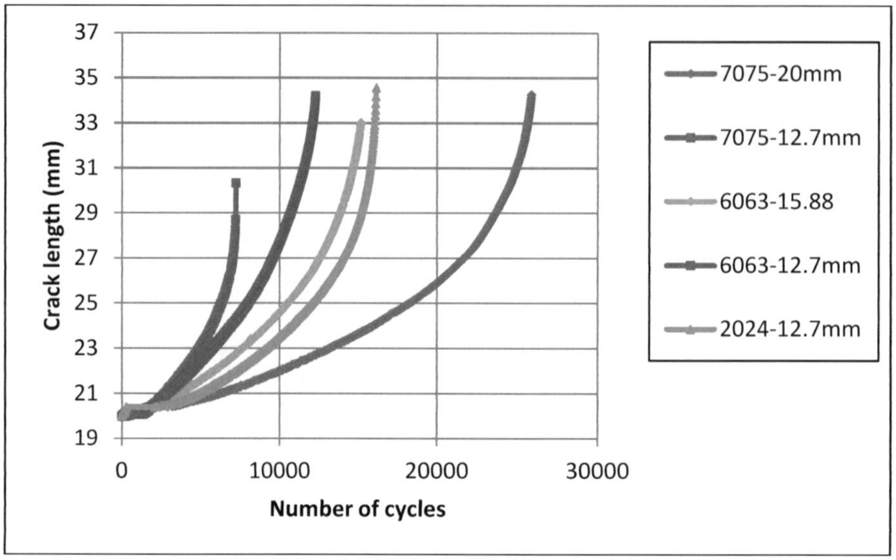

Figure 4: Crack length versus Number of cycles curve.

4. Conclusion

The effect of thickness on fatigue crack growth of aluminum alloys such as A2014-T351, A7075-T6, and A6063-T6 specimen with thickness of 12.7, 15.8, and 20 mm was investigated. Fatigue crack growths of A2024-T351 and A6063-T6 aluminum alloy specimen with thickness of 12.7 mm were higher than A7075-T aluminum alloy specimen with thickness of 20 and 12.7 mm, and A6063-T6 aluminum alloy specimen with thickness of 15.88 mm. It is concluded that fatigue crack growth of aluminum alloy decreases significantly with increasing thickness, and the crack initiates very fast on aluminum alloy material. Based on the number of cycles was found similar result that the crack size of 12.7 mm was faster than size of 20 and 15.8 mm. This conclusion should assist the designer in optimizing aluminum alloy selection for fracture resistant aluminum engineering structures.

5. References

[1] Haftirman Idrus, M. Afendi, Wong Chun Hoe,"Fatigue crack initiation and growth of aluminum alloy with aatio effects," Key Engineering Materials, vol 594-595,pp. 1105-1111, 2013.

[2] Omer G. Bilir and Metin Harun,"Effect of stress ratio on the rate of growth of fatigue cracks in 1100 Al-alloy," Engineering Fracture Mechanics, vol 37, 6, pp. 1203-1206, 1990.

[3] Robert E. Zinkham,"Anistropy and thickness effects in fracture of 7075-T6 and–T651 aluminum alloy," Engineering Fracture Mechanics, vol 1, pp. 275-289, 1968.

[4] Nelson. F. G, Schilling, P. E and Kaufman. J. G,"The effect of specimen size on th results of plane-strain fracture –toughness tests," Engineering Fracture Mechanics, vol 4, pp. 33-50, 1972.

[5] Mills. W. J, and Hertzberg. R. W,"The effect of sheet thickness on fatigue crack retardation in 2014-T3 aluminum alloy," Engineering Fracture Mechanics, vol. 7, pp. 705-711, 1975.

[6] Samer Mahmoud and Kaven Lease,"The effect of specimen thickness on the experimental characterization of critical critical crack-tip-opening angle in 2024-T351 aluminum alloy," Engineering Fracture Mechanics, vol. 70, pp. 443-456, 2003.

[7] Samer Mahmoud and Kaven Lease,"Two-dimensional and three-dimensional finite element analysis of critical crack-tip-opening angle in 2024-T351 aluminum alloy at four thicknesses," Engineering Fracture Mechanics, vol. 71, pp. 1379-1391, 2004.

[8] Kermanidis, A.T and Pantelakis, Sp. G," Prediction of crack growth following a single overload in aluminum alloy with sheet and plate microstructure", vol.78,pp. 2325-2337, 2011.

IOP Conf. Series: Materials Science and Engineering **343** (2018) 012035 doi:10.1088/1757-899X/343/1/012035

Microstructure and phase analysis of Zirconia-ODS (Oxide Dispersion Strengthen) alloy sintered by APS with milling time variation

Bandriyana B, Arbi Dimyati, Bambang Sugeng, Rohmad Salam, Sumaryo, Agus Sujatno
Center for Science and Technology for Advanced Materials, PSTBM-Batan – Indonesia

arbi_dimyati@hotmail.com

Abstract. Investigation on the relationship between the process conditions of milling time and the microstructure on the synthesis of the zirconia-ODS steel alloy has been performed. The elemental composition of the alloy was determined on 20 wt% Cr and zirconia dispersoid of 0.50 wt%. The synthesis was carried out by powder metallurgy method with milling time of 3, 5 and 7 hours, static compression of 20 Ton and sintering process for 4 minutes using the APS (Arc Plasma Sintering) equipment. SEM-EDX and XRD test was carried out to characterize the phase and morphology of the alloy and the effect to the mechanical properties was evaluated by the Vickers Hardness testing. The synthesis produced sample of ODS steel with good dense and very little porous with the Fe-Cr phase that clearly observed in the XRD peak pattern. In addition, milling time increased the homogeneously of Fe-Cr phase formulation, enhanced the grain refinement of the structure and increase the hardness of the alloy.

1. Introduction

The Oxide Dispersion Strengthen (ODS) alloy was recently developed for high temperature material application because of their good on mechanical strength and creep resistance [1]. This alloy also showed significantly improvement on the high temperature irradiation resistance and is suitable for use as a nuclear facility material. The use of zirconia as a dispersoid oxide to replace the Ytria was carried out to enhance the nuclear and radiation properties of the ODS alloy. Zirconium also shows a great affinity for C that expected to form carbides which act as a pinning dispersoid against grain boundary migration [2]. The ODS alloy are usually produced by powder metallurgy (PM) routes involving mechanical alloying of the oxide and pre-alloyed followed by consolidation technique, such as hot isostatic pressing and consolidation process by sintering. This PM method is still superior to disperse the oxide which plays an important role in the ODS alloy manufacturing that is very difficult to conduct by the casting process because of the agglomeration problem. However, the PM process requires a high cost due to the high price of powder element and high time and energy consumption in the sintering process [3].

Some research has been conducted to enhance the sintering process by developing the plasma sintering method using the SPS device. This SPS is able to enhance the consolidation of the powder, however the SPS still need high energy consumption. The new method of Arc Plasma Sintering (APS) developed in PSTBM-BATAN uses plasma as heat source can synthesize some alloy and reduce process time and energy consumption significantly [4]. The performance of the ODS steel alloy synthesized by APS was depend on the operating process of the milling, compressing and sintering time. Basically, milling process was conducted to mix the alloying element and refiner the grain size

IOP Conf. Series: Materials Science and Engineering **343** (2018) 012035 doi:10.1088/1757-899X/343/1/012035

of the powder. The longer time of milling process will produce the finer size of powder and predicted will refine the grain size and strengthened the alloy. However, the optimum of milling time is necessary to be investigated according to the highly cost for milling process.

Study and investigation for milling time optimation was performed by investigate the relation between the milling time and the microstructure. In this study, the relation between the milling time with the microstructure and phase resulted during the synthesis of ODS steel alloy and the change of the hardness was performed. The zirconia dispersion was used for the utilization of material in the nuclear facility to reduce the neutron absorption coefficient.

2. Experimental

The Fe-Cr ODS alloy was developed with the composition of 20 wt.% Cr and 0.5 wt.% of dispersion particles ZrO2. The samples were prepared by mixing and milling all the materials of Fe, Cr and ZrO2 powder with milling time of 3, 5 and 7 hours. After milling, the samples were pressed by isostatic pressing machine with the compression load of 20 Ton to form sample with coins formed. The samples were then consolidated by sintering process in the APS (Arc Plasma Sintering) for 4 minutes referred to the sintering time for processing the ODS steel with 12 Cr content [5]. APS is a new experimental sintering device developed at the PSTBM with schematic diagram as illustrated in Figure 1.

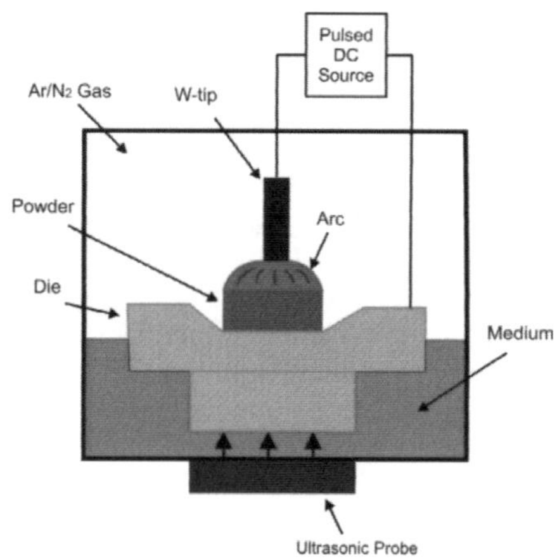

Figure 1. Schematic diagram of APS

The main component of the APS is arc plasma source that provide high temperature of space in a short time. This device allows fast process consolidation of powder metallurgy sample. The sintering process was carried out in fully argon atmosphere in order to protect against oxidation. The sample coin was placed on a copper cup and exposed by the plasma for 4 minutes with the power of 300 Watt with 25 A 12 V.

The morphology and composition of the alloys were studied by means of Optical Microscopy and Scanning Electron Microscopy (SEM) coupled to Energy Dispersive Spectroscopy (EDX). The EDX spectrums illustrate the chemical composition of the samples. The SEM- mapping for each element was performed to observe the homogeneity of alloying elements and the oxide dispersion in the matrix of the alloy. Analysis of the XRD-diffraction pattern was carried out to evaluate the phase of the alloy using the Match 3 program and was refined by Fullprof software. Mechanical properties of the consolidated samples and the effect of the processing and Cr composition were also studied using the Vickers micro hardness (HV) measurement.

3. Result and Discussion

The microstructure after sintering process using APS for 4 minutes observed by the optical microscopy showed good appearance with little porosity and good dense structure as showed in Figure

IOP Conf. Series: Materials Science and Engineering **343** (2018) 012035 doi:10.1088/1757-899X/343/1/012035

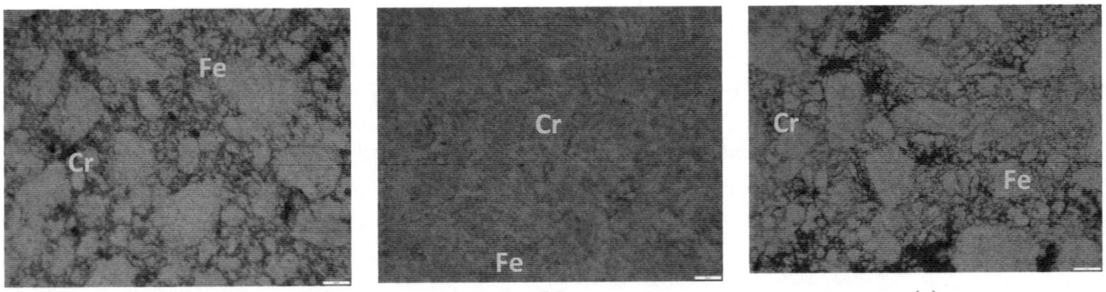

(a) (b) (c)

Figure 2. Morphology of ODS Steel Fe-20Cr-0.5ZrO_2 with Optical Microscopy P 500x with milling time of (a) 3 hours, (b)5 hours and (c) 7 hours

The SEM-EDX examination revealed the presence of particle with different size and composition, randomly dispersed in the ferritic matric. It is obviously clear that structure of all ODS alloy with some variation of Cr content and milling time were nearly same with general appearance of small and equiaxial grains and some elongated grains. The structure consists of two phases of the matrix phase mainly consists of iron phase and precipitate which mainly consists of Cr as given in mark (1) and mark 2 in Figure 3.

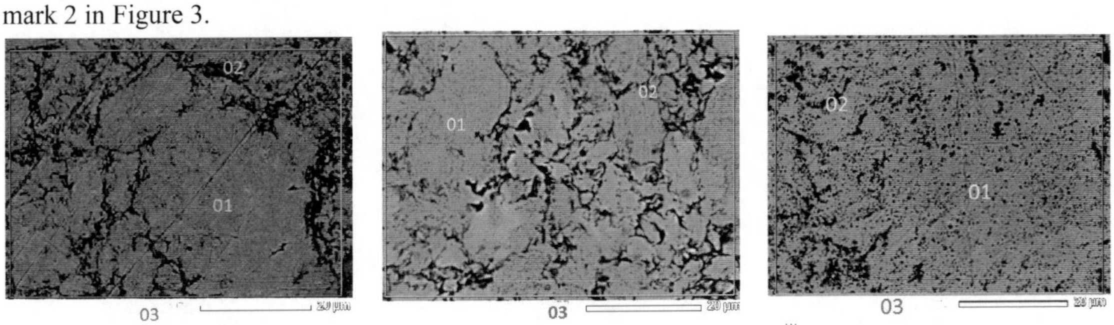

(a) 20 Cr-3H (b) 20Cr-5h (c) 20 Cr-7h

Figure 3. Micro structure of zirconia-ODS alloy Fe-20Cr-0.5 ZrO2 with of milling time process of (a) 3 hours, (b) 5 hours and (c) 7 hours.

The milling time will change the grain size of the structure where increasing of milling time produced finer grain size as obviously explained in Figure 3. The presence of zirconia in the alloy can be predicted by the wt% of Zr element which mainly observed in the diffusion zone of the matrix phase and the high Cr content phase. Observation on zirconia distribution in the alloy after consolidated by APS sintering process was carried out by SEM-area scanning test as resulted in the Zr distribution in Figure 4.

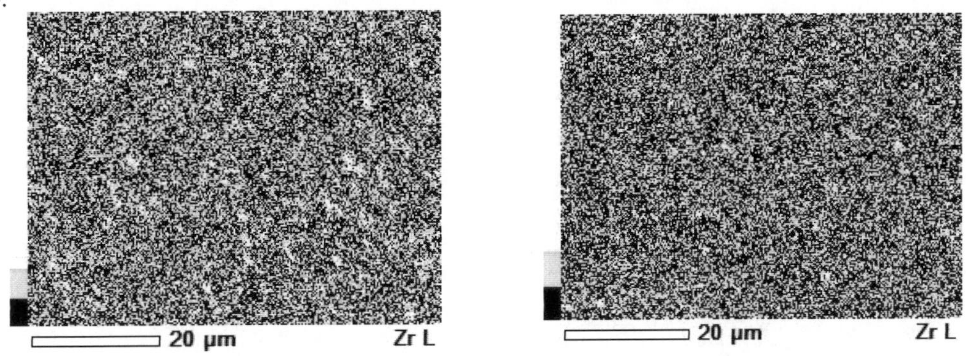

(a) 3 hours (b) 5 hours

Figure 4. Mapping area of Zr in the ODS alloy Fe-20Cr-0.5ZrO_2 with milling time of 3 hours (a) and 5 hours (b)

Based on the Zr content observed in Figure 4, it can be explained that the zirconia was distributed homogeneously in the alloy and predicted that increasing of the milling time will influence the homogenizing of the zirconia dispersion which is very important for enhancing the mechanical properties of the ODS steel alloy. The grain size and typically of oxide distribution was similar with the typically grain size distribution found in Spark Plasma Sintered that was reported as essential to prepare compact materials with strength and good ductility [6].

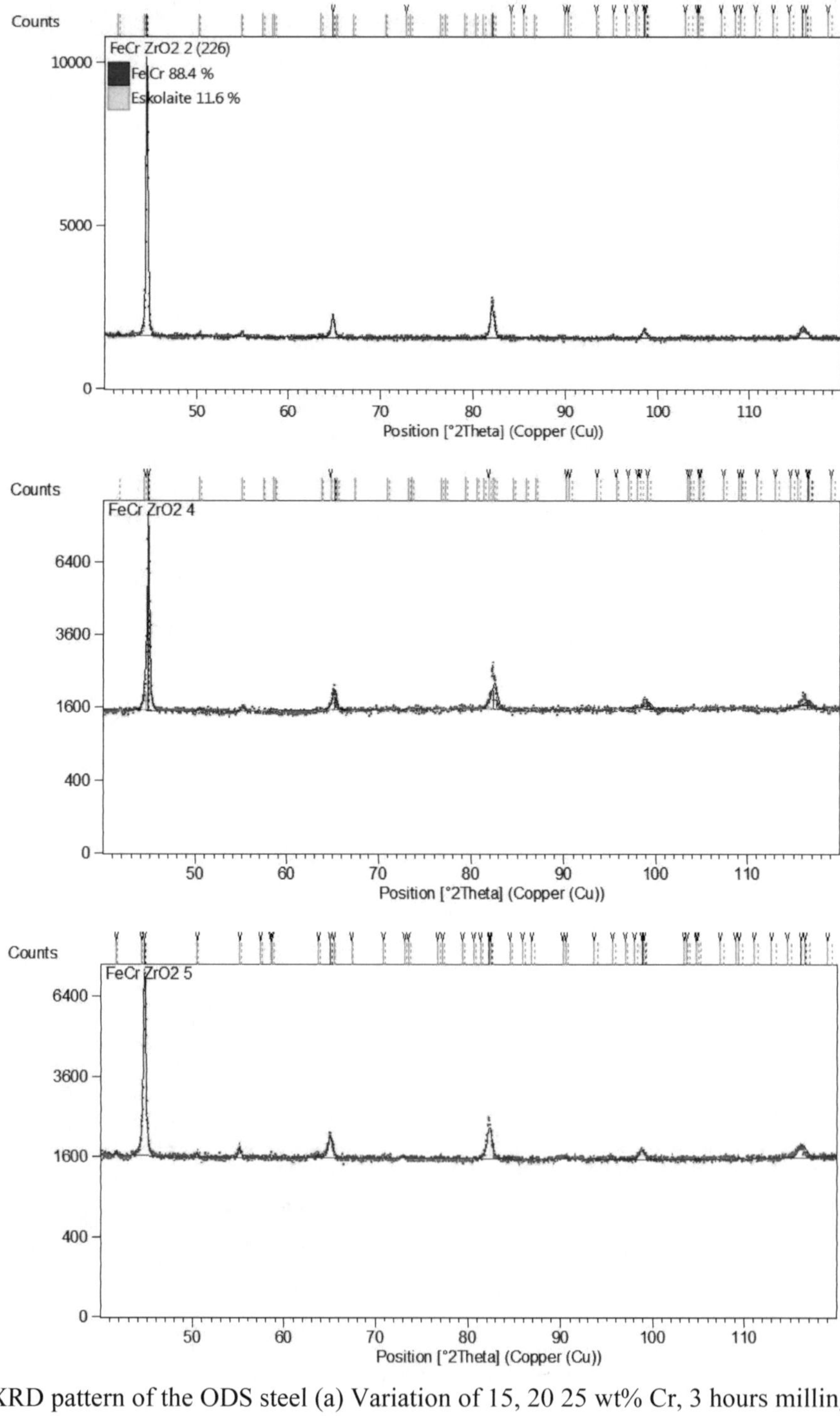

Figure 5. XRD pattern of the ODS steel (a) Variation of 15, 20 25 wt% Cr, 3 hours milling and (b) Fe-20Cr-0.5ZrO_2 with 3, 5 and 7 hours milling

The XRD test was performed to analyse the alloying process of the ODS steel by identified the phase formed on the alloy. Figure 5 showed the diffraction pattern of ODS steel alloy with variation of Cr content and milling time. The refine of the X-ray pattern was performed by Fullprof program to fit the pattern with the GOF result was less than 1.6. In general, the alloy was dominated with Fe-Cr phase confirmed the structure and composition of the element observed by the SEM-EDX. Analysis of the first peak of the XRD pattern was performed by more detail observation of the first peak (110) as illustrated in Figure 6.

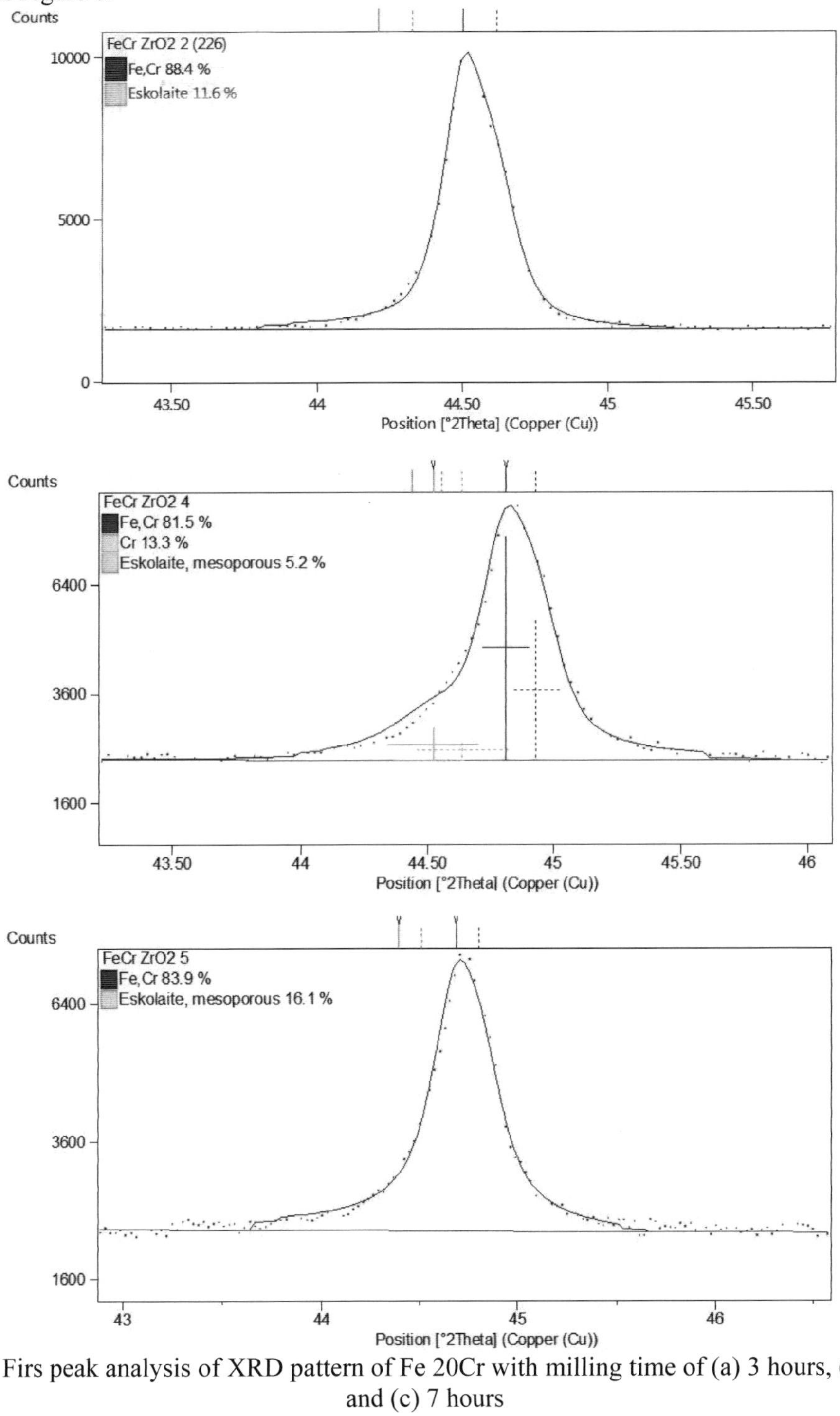

Figure 6. Firs peak analysis of XRD pattern of Fe 20Cr with milling time of (a) 3 hours, (b) 5 hours and (c) 7 hours

Addition of milling time increase the full-width at half-maximum (FWHM) of the pattern as observed in Figure 6 that revealed the grain size finer confirmed with the micro structure explained before. More detailed analysis in the first peak of all samples showed the contribution of Fe-Cr peak contains of Fe-Cr and Fe peak that revealed the alloying process for Fe-Cr phase formation is not yet completed during the very small time sintering as also explained in the SEM-EDX observation.

The crystal size was also determined on the XRD examined and the calculation result of crystal size for several milling times was illustrated in Figure 7. Increasing the milling time on the alloying process showed the smaller of crystal size.

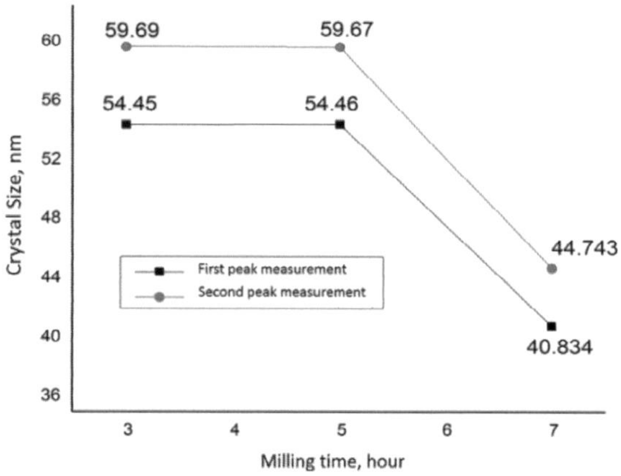

Figure 7. Crystal size of the ODS steel Fe20Cr0.5ZrO$_2$on some milling time

The hardness test was carried out to evaluate the mechanical characteristic of the alloy. The result of hardness test with variation Cr content and milling time of the ODS steel alloy was given in Figure 8.

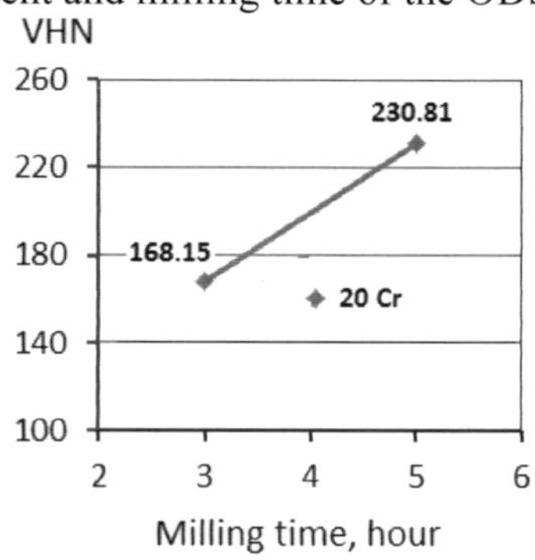

Figure 8. Hardness of ODS steel alloy for 3 and 5 hours milling time

After sintering process for 4 minutes the ODS steel alloys have significant hardness number from around 140 VHN to 210 VHN depend on the milling time. Increasing of milling time will increase the hardness that was caused by the grain refinement produced by the milling time close with the explanation of the microstructure discussed before. The hardness numbers obtained by the very low milling time of 4 minutes and low energy consumption of APS-sintering are good enough compare with the commercial ODS steel hardness of Fe-15 Cr-0.5Y_2O_3 around of 200 VHN.

4. Conclusion

Sintering method using the APS has successfully generated microstructure of the ODS steel alloy with dense structure and little porosity. The structure consists of mainly Fe-Cr phase with small equiaxed grain, homogeneous zirconia oxide distribution and significant hardness number. In addition, milling time will finer the grain and increase the hardness of the alloy. The longer of milling time process will also result finer grain size and increase the hardness.

5. References

[1] Andrea Garcia-Junceda et al, 2015 J. Am. Ceramic Society, 2015 pp 1-6.
[2] Klueh R L, J P R Shingledecker, R.W. Swindeman, D.T. Hoelzer 2005 *J. Nuclear Materials*
[3] Lavanya Raman, Karthick Gothandapai, B.S. Murty 2016 Defence Science J 2016 Vol 66, Juky2016 pp316 – 322, Vol 66, No 4, July 2016
[4] Dimyati, "APS, New Challenge for low emission bio-waste treatment," Proceeding of Asia-Pacific Microscopy Conference 11th, 2016
[5] Bandriyana B, et.all, 2017, *IOP Conf. Materials Sci. and Eng.* **176** 1-6.
[6] Taguchi M, Sumitomo, Ishibashi R, Aono Y, 2008, *Mat. Trans.* 49 6 1303 – 1310.

IOP Conf. Series: Materials Science and Engineering **343** (2018) 012036 doi:10.1088/1757-899X/343/1/012036

Temperature Control System for Mushroom Dryer

I A Wibowo[1], Nur Indah[1], D Sebayang[1] and N H Adam[2]

[1]Universitas Mercu Buana, Jakarta, Indonesia
[2]Universiti Tun Hussein Onn Malaysia, Batu Pahat, Johor, Malaysia

ignatius.agung@mercubuana.ac.id

Abstract. The main problem in mushroom cultivation is the handling after the harvest. Drying is one technique to preserve the mushrooms. Traditionally, mushrooms are dried by sunshine which depends on the weather. This affects the quality of the dried mushrooms. Therefore, this paper proposes a system to provide an artificial drying for mushrooms in order to maintain their quality. The objective of the system is to control the mushroom drying process to be faster compared to the natural drying at an accurate and right temperature. A model of the mushroom dryer has been designed, built, and tested. The system comprises a chamber, heater, blower, temperature sensor and electronic control circuit. A microcontroller is used as the controller which is programmed to implement a bang-bang control that regulates the temperature of the chamber. A desired temperature is inputted as a set point of the control system. Temperature of 45 °C is chosen as the operational drying temperature. Several tests have been carried out to examine the performance of the system including drying speed, the effects of ambient conditions, and the effects of mushroom size. The results show that the system can satisfy the objective.

1. Introduction

Mushrooms have high nutritional content. They are rich of minerals, proteins and fibres. In addition, they contain no cholesterol. Some types of mushroom are used for healthcare such as to increase human immune system, smooth blood flow, prevent high blood pressure, lower cholesterol in the blood, and neutralize toxic or poison in the body. Mushrooms are also widely known as a delicacy among food appreciators for centuries. [1,2]. In addition, today there is a tendency for people to reduce eating red meat, so many of them choose an alternative to switch to mushrooms to meet daily nutritional needs [3].

Mushrooms have the same properties as vegetables which are easily damaged cannot last long in natural conditions. Therefore, in the food processing industry, the mushrooms need to be preserved after harvest [4]. Mushrooms can be dried to preserve them. Drying is a process to reduce the water content of a material. This process requires energy to evaporate moisture content and transfer a certain amount of water vapour from the surface of the material to the surrounding environment [5].

Unlike vegetables which decrease their quality when dried, mushrooms can maintain their quality when dried properly. Wet mushrooms have 80% to 90% moisture content. They are chewy, smooth and flexible. Mushrooms become curved and hard when dried. Mushroom drying is aimed to reduce the water content of the mushroom which indirectly also kills microorganisms which cause damage to the mushroom. With the drying process it should reach 20% moisture content. Traditionally mushroom drying is performed naturally using sunlight. Although the cost is low, it is not effective because it depends on weather conditions which cannot be predicted and controlled. Mushroom can be easily damage if the drying process takes a long time or when the drying temperature is too low.

Another problem with natural drying is that the drying temperature is not accurate which deteriorate the quality of the dried mushrooms [6,7].

This paper presents the solution for the above mentioned problem. A prototype of a mushroom dryer has been developed. The dryer utilises a temperature control system to ensure the drying process is maintained at an operational temperature of 45°C which allow producing dried mushrooms with 20% moisture content [8].

2. System Development

2.1. *Drying Chamber*

The drying chamber is made of wood with the size of 32 cm × 24 cm × 28 cm as can be seen in figure 1. A system to control the temperature of the chamber is shown in figure 2. It consists of a lamp as a heater to warm up the chamber, a wire mesh rack to place the mushrooms to be dried, a blower to remove wet air from the chamber and a temperature sensor to detect chamber temperature.

Figure 1. Dryer chamber

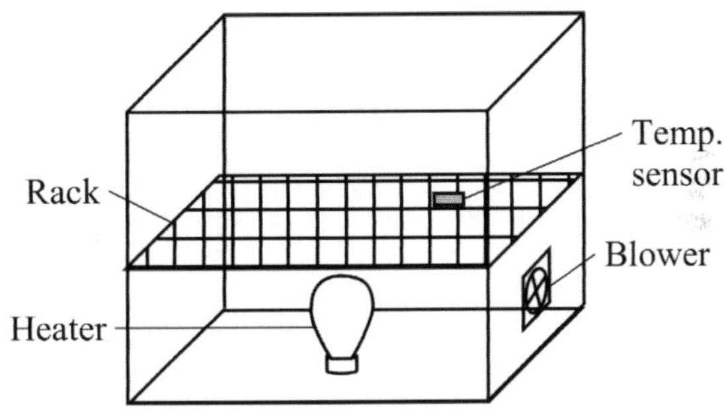

Figure 2. Drying system components

2.2. *Control and Monitoring System*

Figure 3 illustrates the diagram of electronic hardware to implement the temperature control and monitoring system. The heart of the system is microcontroller PIC16FF877A.

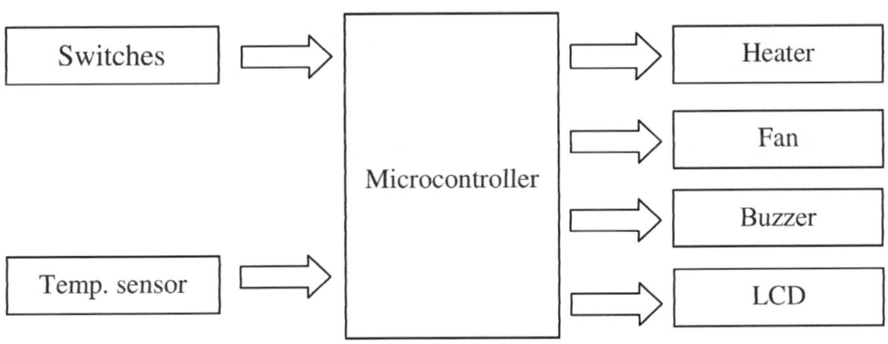

Figure 3. Components of control and monitoring system

Figure 4 shows the schematic of the microcontroller circuit. This type of microcontroller has a total of 40 pins where 32 of them are for input and output [9]. In this circuit, pins 2 and 3 are connected to humidity and temperature sensor outputs respectively. Pin 4 is for the reference voltage input. Pins 13 and 14 are connected to the timer crystal while pins 15, 16, and 17 are the inputs connected to the switches circuit. Pins 19 and 30 are connected to the heater and fan respectively, whereas pins 33 to 40 are used as outputs to the LCD.

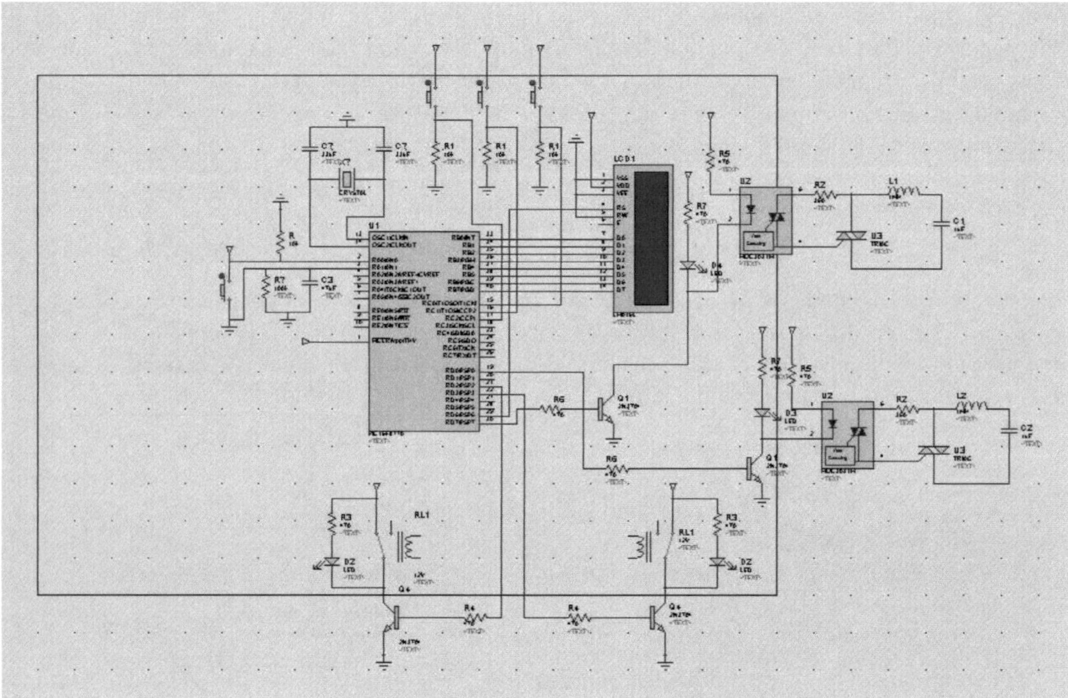

Figure 4. Schematic of the electronic circuit

Figure 5 shows the electronic circuit with microcontroller, switches to input the set point, and LCD to display temperature, humidity and time. A PWM circuit, sensor and lamp as the heater is shown in figure 6.

Figure 5. Controller and input-output interface

Figure 6. PWM circuit, sensor, and lamp

The block diagram of the control system for the mushroom dryer is shown in figure 7. It is a closed-loop control system with a bang-bang controller to control air temperature inside the drying chamber to be maintained at 45 °C ± 2 °C.

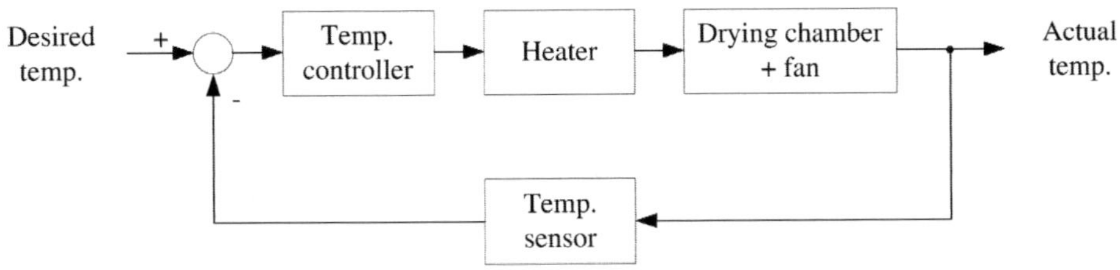

Figure 7. Control system block diagram

Figure 9 shows the Simulink model of the system to analyse the system behaviour and operation before the implementation.

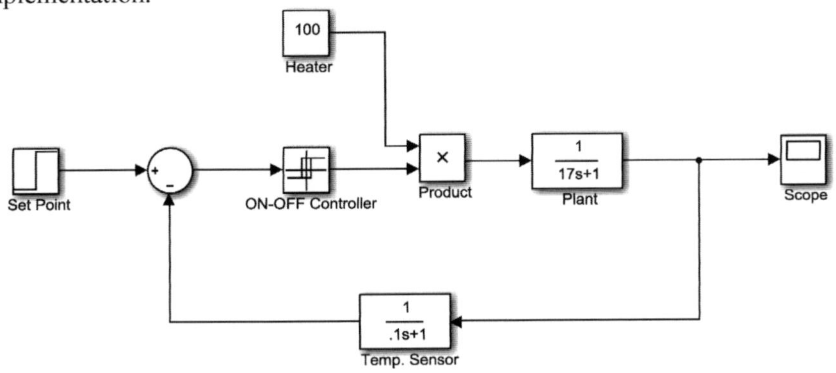

Figure 8. Simulink model

The operation of the mushroom dryer follows the flowchart in figure 8.

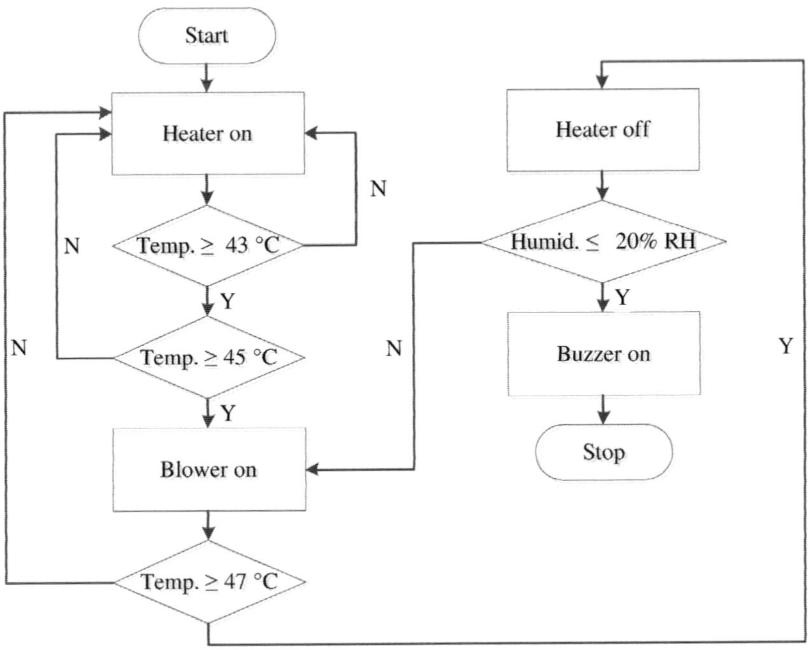

Figure 9. Flowchart of mushroom dryer operation

3. Results and Discussions

Figure 10 shows the results of experiments to investigate the required power of the heater. 25 W, 60 W, and 100 W lamps were tested as the heat source inside the dryer chamber. Time to reach the set point was found as 16 minutes, 11 minutes and 2 minutes for 25 W, 60 W, and 100 W lamps respectively. Since the 100 W lamp was the fastest and most effective for heating, it was chosen as the heater for this the dryer.

An experiment was done to test the respond of the system to the ambient temperature i.e. in dry and damp air conditions. Both curves in the graph show that the time to reach the set point is almost the same. Therefore it can be concluded that the performance of the system does not affected by the ambient conditions.

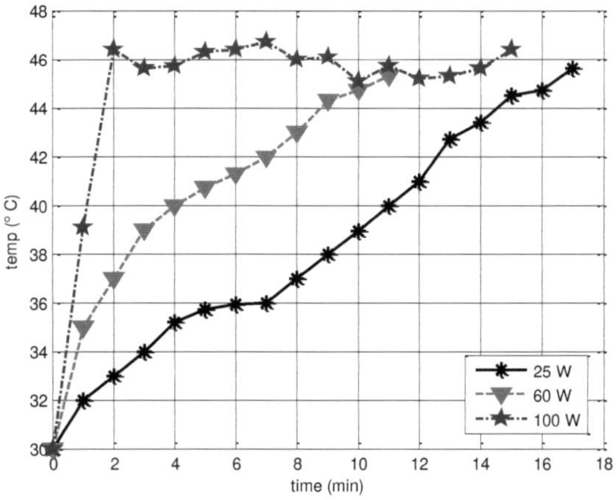

Figure 10. Results of heating power experiments

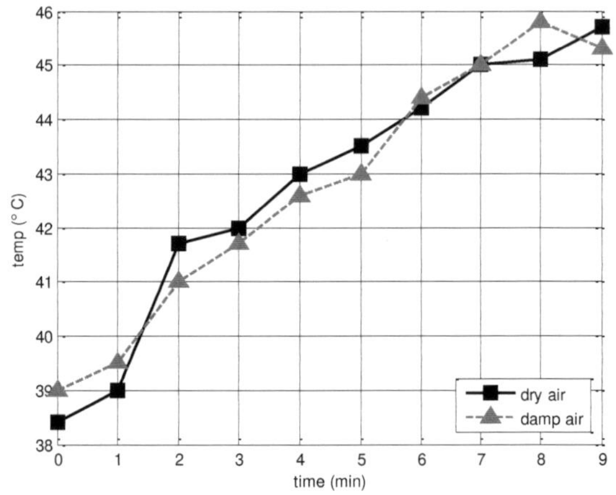

Figure 11. PWM circuit, sensor, and lamp

Another experiment investigates the performance of the dryer to dry thick or thin mushrooms. The results in figure 12 show that thicker mushrooms take longer to dry. However, the control system works for either thick or thin mushrooms.

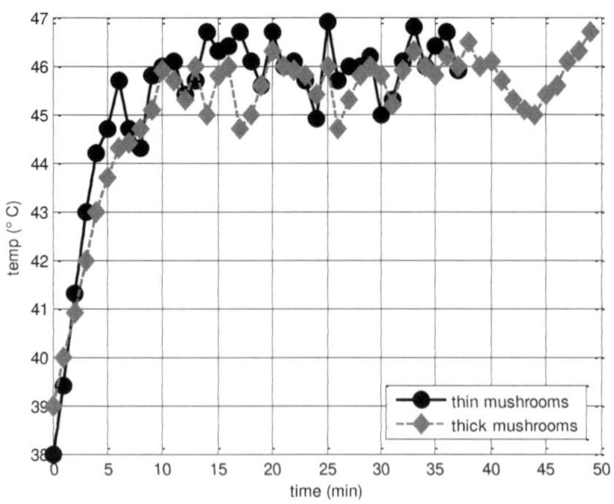

Figure 12. Control system block diagram

4. Conclusions
A prototype of a device to artificially dry mushrooms has been designed, built, and tested. The temperature control system applies bang-bang (ON-OFF) control to control air temperature inside the drying chamber to be maintained at 45 °C ± 2 °C. The device is able to work satisfactorily at various heater powers, ambient conditions and mushroom sizes.

References
[1] Doymaz E 2014 *J. Chem.* **2014**
[2] Du P, Cui B K and Dai Y C 2011 *J. Med. Plants Res.* **5(8)** pp 1374-1381
[3] Gunadi N G B and Putra I G C 2016 *J. Bak. Saras.* **5(1)** pp 77-84
[4] Suarnadwipa N and Hendra W 2008 *J. Ilm. Tek. Mes. Cakram* 2(1) pp 30-33
[5] Incopera F P and Dewit D P 1996 *Fundamentals of Heat and Mass Transfer* 4th *Ed.* (New York John Wiley & Sons)
[6] Alam N, Amin R, Khan A, Ara I, Shim M J, Lee M W and Lee T S 2008 *Mycobiology* **36(4)** pp 228–232
[7] Djatmiko B and Gautara 1983 *Praktek Pengolahan Hasil Pertanian* (Bogor:IPB) p 8
[8] Roger P 2006 *Mushrooms* (London: Macmillan) p 317
[9] Microchip 2013 *PIC16FF877A Datasheet* (Chandler: Microchip Technology Inc.)

IOP Conf. Series: Materials Science and Engineering **343** (2018) 012037 doi:10.1088/1757-899X/343/1/012037

The observation of resistivity change on the ultrasonic treated Fe-Cr ODS sinter alloy under magnetic field influence

Marzuki Silalahi[1], Setyo Purwanto[1], Mujamilah[1], Arbi Dimyati[1]

[1]Center for Science and Technology of Advanced Materials – National Nuclear Energy Agency

Kawasan PUSPIPTEK Serpong, Tangerang Selatan 15314, Indonesia

marzukisilalahi@yahoo.com

Abstract. About the observation of resistivity change on the ultrasonic treated Fe-Cr ODS sinter alloy under magnetic field influence. This paper reported about the observation of the resistivity change in the ultrasonic pre-treated Fe-Cr ODS sinter alloy under the influence of magnetic field at the Center for Science and Technology of Advanced Material, Nuclear Energy Agency of Indonesia. Fe-Cr ODS alloy were sinthesized by vacuum sintering of Fe- and Cr-powder dispersed Y_2O_3. However, before sintering the powder mixture was subjected to the irradiation process by ultrasonic for 50 hours at 20 kHz and then isostatic pressed up to 50.91 MPa to form a coin of 10 mm in diameter. LCR meassurement revealed the decreasing of resistivity about 3 times by increasing of applied magnetic field from 0 to 70 mT. In addition, VSM meassurement was performed on both as powder material and as sintered sample. The results showed increasing the magnetization with increasing magnetic field and the curve exhibits almost exact symmetry S-form with small hysterese indicating fast changing magnetization and demagnetization capability without energy loss. This opens strong speculations about the existence of magnetoresistant property of the material which is important for many application in field of sensors or electro magnetic valves.

1. Introduction

The ODS steel is an advanced material which is projected to be used for a system that is operated at high temperature such as for structural of advanced fission nuclear reactors and of fusion nuclear reactor [1]. The development of synthesis of ODS steel is in progress to find the adequate method and its characteristics. The ultrasonic method has been successefully to synthesize the Fe-Cr microalloy, following sintering in quartz capsule[2-4]. The Fe-Cr microalloy powders has bcc structure, homogenous, and no oxide as well as sintering product[2-4]. Besides, the ultrasonic method has the application for reducing the size particles that has effects on decreasing magnetoresistant value due to external magnetic field[5]. The ultrasonic method utilizes the mechanical efect of ultrasonic and cavitaion explosion that could has temperature of about 5000K and of thousand bar pressure[6]. In this experimetal work the ultrasonic method was used to synthesize the Fe-Cr ODS alloy following sintering in quartz capsule. The properties of product were characterized by using LCR and VSM.

2. Experiments

The material used to make an alloy is Fe powder with particle size about 3 μm, Cr with particle size about 23 μm, and Y_2O_3 with particle size < 50 nm and Toluene with 90% purity as media for ultrasonic irradiation treatment. The ratio composition in weight percent of powders mixture Fe:Cr: Y_2O_3 is 87: 12:1 respectively. The ultrasonically irradiation treatment of powders mixture was performed on the toluene solution in beaker glass at frequency of 20 kHz sonic vibracell amplitude for 50 hours. The characteristics of microstructure of treated ultrasonically powders was identified using Scanning Electron Microscopy (SEM) equipped with Energy Dispersive Spectroscopy (EDS) for micrograph and elemental composition, using X-Ray Diffraction (XRD) for the phase, and using Vibrating Sample Magnetometer (VSM) for measuring the magnetic properties. The compacting process of ultrasonically treated powders is done using isostatic pressed at at load of 50.91 MPa to form a coin of 10 mm in diameter. The coin was capsulated in quartz in vacumm condition. The sintering process was done in induction furnace at 1300 °C for 1 hour, and then left to cool down to room temperature in the furnace. Characterization of the sintered Fe-Cr has been carried out using X-ray diffraction, VSM and LCR meter for resistivity as function of applied magnetic field.

3. Result and Discussion

XRD diffraction pattern of ultrasonically treated powders and its sintered are shown in Figure 1.

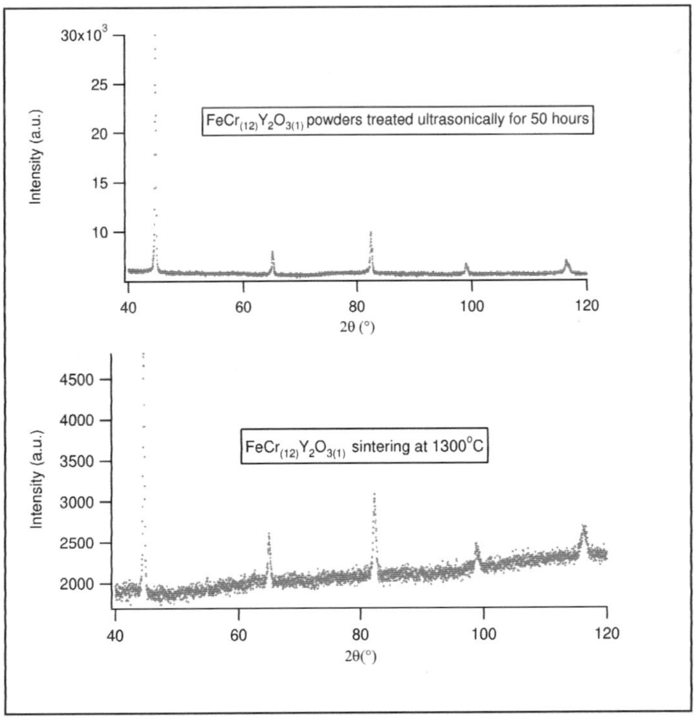

Figure 1. X-Ray diffraction intensity pattern of $FeCr_{(12)}Y_2O_{3(1)}$ powders treated ultrasonically for 50 h (a), and of sintered in quartz capsule at 1300 oC for 1 h. The both of peaks are consistent in similar position of bcc structure.

Figure 1 is a X-Ray diffraction intensity pattern of $FeCr_{(12)}Y_2O_{3(1)}$ powders treated ultrasonically for 50 h (a), and of sintered in quartz capsule at 1300 oC for 1 h. The both peak pattern of Figure 1 are similar with the bcc structure as found in earlier research[2,3]. Ultrasonic irradiations in toluene solution produce Fe-Cr microalloying as shown in Figure 2. As in Figure 2 the EDS of particles of area has composition of Fe = 98.26 % and Cr =1.74 %, and of spot has the composition of Fe = 3.44 % and Cr= 96.56 %. These particles are Fe dispersed Cr and Cr dispersed Fe.

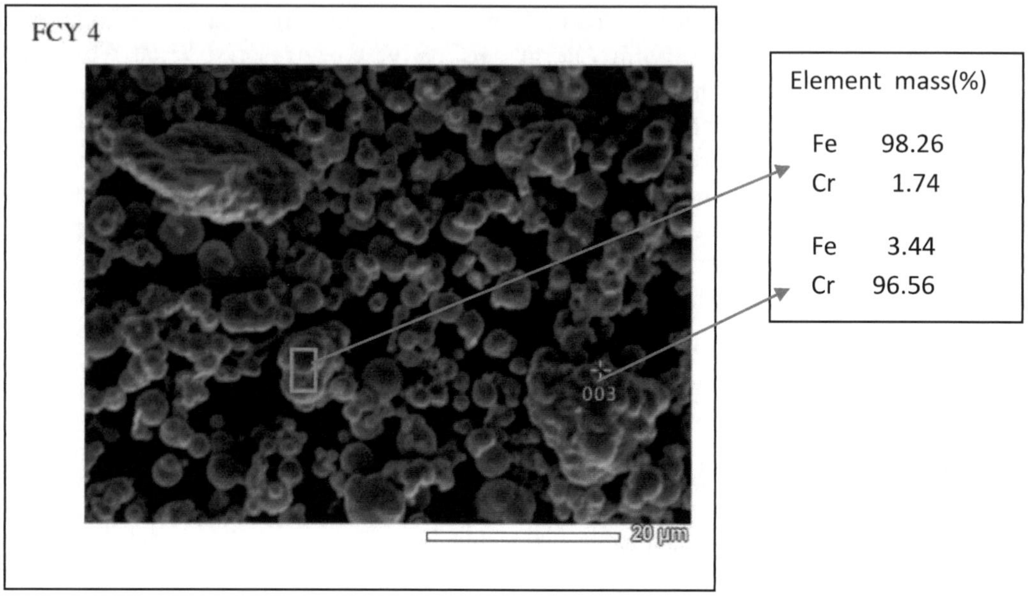

Figure 2. SEM Micrographs of the microstructure after irradiating for 50 hours at 20 khz of $FeCr_{(12)}Y_2O_{3(1)}$ powders. EDS result of powders of area were Fe dispersed Cr and point spot were Cr dispersed Fe.

The ultrasonic wave transmitted in toluene media can form stress and strain. These stress and strain can cause the particles to collide because the particles move to all directions[4] and also can cause cavitation nucleation[5,6]. The explotion of cavitation can produce high temperature and pressure that enabled Fe-Cr microalloying processes[2].

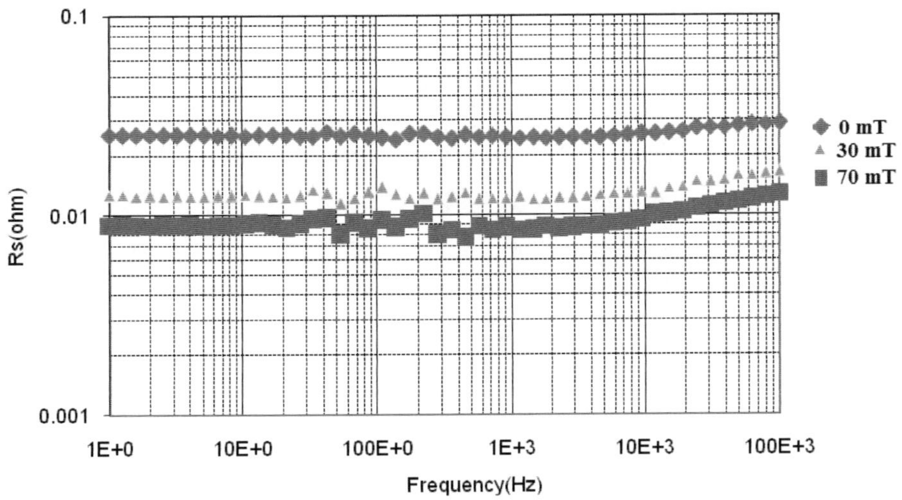

Figure 3. Resistivity data as function of applied magnetic field, blue (0mT), green (30 mT) and red (70 mT).

Figure 3 shown the Resistivity data of sintered $FeCr_{(12)}Y_2O_{3(1)}$ as function of applied magnetic field, blue (0mT), green (30 mT) and red (70 mT). The increasing of applied magntization up to 70mT decreased resistivity of 3 times. This phenomenon quite similar to the phenomenon of tunneling magnetoresistance in nanogranular system of Fe-Al_2O_3 as reported by S. Purwanto et al[7].

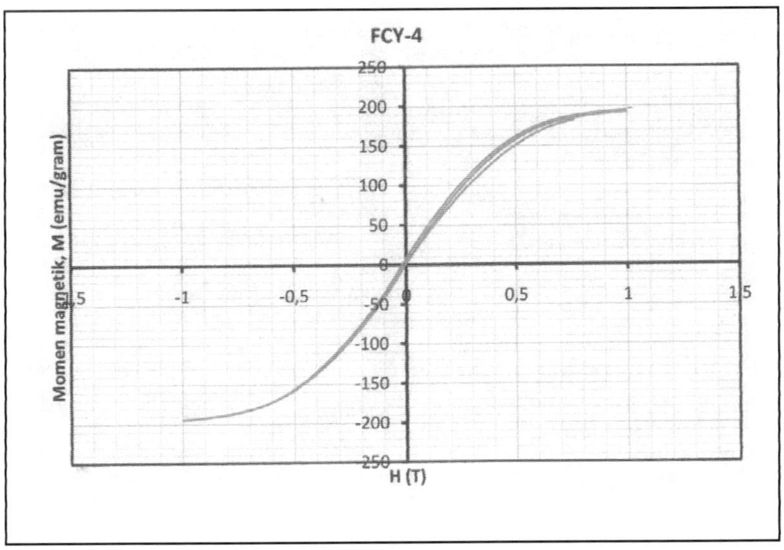

Figure 4. Magnetitation Diagram of $FeCr_{(12)}Y_2O_{3(1)}$ powder tretated ultrasonically for 50 hours.

In Figure 4 can be found that the $FeCr_{(12)}Y_2O_{3(1)}$ powder tretated ultrasonically for 50 hours showing high magnetic saturation and the absent of remanence and coercivity. The curve showed the almost exact symmetry S-form.

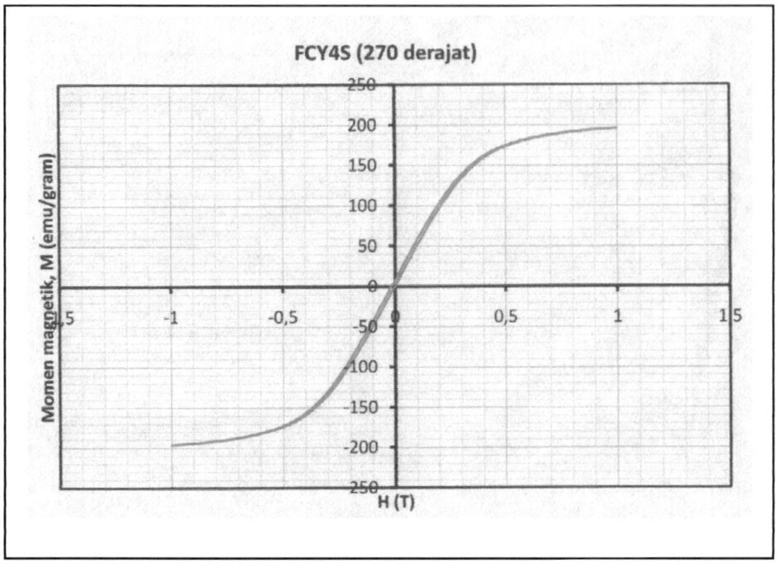

Figure 5. Graph magnetization of sintered $FeCr_{(12)}Y_2O_{3(1)}$ alloy taken in 270 °C angle.

Figure 5 shown the magnetization of sintered $FeCr_{(12)}Y_2O_{3(1)}$ alloy taken in 270 °C angle. For this magnetization angle, after sintered process the momen magnetic of Fe-Cr ODS irradiated ultrasonically was not changed.

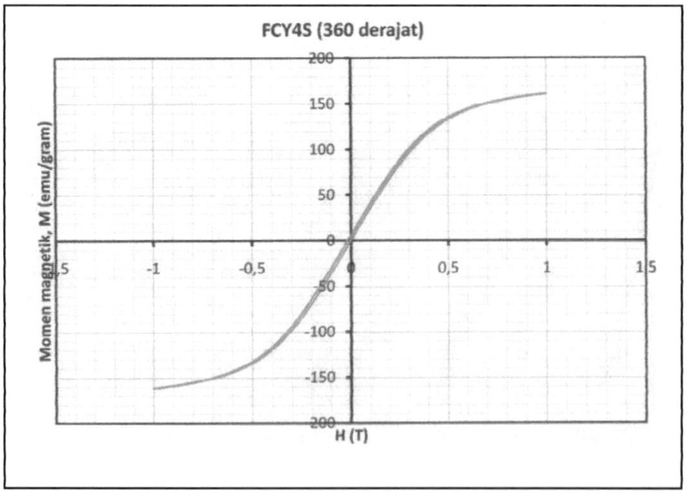

Figure 6. Graph magnetization of sintered $FeCr_{(12)}Y_2O_{3(1)}$ alloy taken in 360 °C angle.

In Figure 6 shown magnetization of sintered $FeCr_{(12)}Y_2O_{3(1)}$ alloy taken in 360 °C angle. The material is anisotropy as the momen magnetic is different in different angle. The structure and properties of Fe-based had been changed thorough adding Chromium, and also during milling the disordered magnetic phase, the reflectivity value of Fe-based powders (not adding Cr) was larger than FeSiAlCr powders[8]. Two kinds of atoms diffuse each other and solid solution reaction occurs on this interface[9]. A small magnetic moment or no moment alloy will be formed, and this results in reducing magnetic moment on magnetic layer. The existence of a few nanocrystalline phases results in a slight reduction of magnetization[10].

4. Conclusion

The Fe-Cr ODS alloy powders treated ultrasonically for 50 hours has the similar structure to the sintered in quartz capsule at 1300 ^{o}C. The ultrasonic irradiating effects on Fe-Cr ODS alloy for 50 hours has decreasing of resistivity about 3 times by increasing of applied magnetic field from 0 to 70 mT. The results showed increasing the magnetization with increasing magnetic field and the curve exhibits almost exact symmetry S-form with small hysterese indicating fast changing magnetization and demagnetization capability without energy loss. This opens strong speculations about the existence of magnetoresistant property of the material which is important for many application in field of sensors or electro magnetic valves .

5. References

[1]. Masaki Inoue, Takeji Kauto, Satoshi Ohtsuka, Research and Development of Oxide Dispersion Strengthened Ferritic Steel For Sodium Cooled Fast Breeder Reactor Fuells , Materials for Generation IV Nuclear Reactors, NATO Advanced Institut, Cargese, Corsica France, (2007).

[2]. Marzuki Silalahi, Arbi Dimyati, Sri Harjanto, Pudji Untoro, Bambang Suharno, Microalloying of Fe-Cr by using Ultrasonic Irradiation, International Journal of Technology (2014) 2: 169-182..

[3]. Marzuki Silalahi, Pudji Untoro, Bambang Suharno, Sri Harjanto, Ultrasonic treatment effect on the consolidation of Fe-Cr particle mixtures after compaction and sintering process, Majalah Metalurgi, V29.2.2014, ISSN: 0126-3188, 171-178.

[4]. Timothy J. Mason and Veronica Saez Bernal, An Introduction to Sonochemistry, John Wiley and Son, Ltd., 2012.

[5] M. Nad, "Ultrasonic horn design for ultrasonic machining technologies," *Appl. Comput. Mech.*, Vol.4, Issue.1, pp.79–88, 2010.

[6] T.Y Wu, N. Guo, C.Y. Teh, and J.X.W. Hay, *Advances in Ultrasound Technology for Environmental Remediation*, Springer Briefs in Green Chemistry for Sustainability, pp:5-11, 2013.

[7] S Purwanto. Tunneling Magnetoresistance (TMR) on Fe-Al2O3 Nanogranular film growth by Helicon Plasma Sputtering, Atom Indonesia vol. 34 no. 1 2008, ISSN: 0126-1568.

[8]. Tingdong ZHOU, Zhengyun WANG, Jiangkang TANG, and Haipeng LU, Structure and magnetic properties of Fe-based powders prepared by mechanical alloying, Acata Metall, Vol. 23, October 2010, pp 351-356.

[9] M. S. Lium E. Y. Jiang and Y. G. Liu, Vac Sci Techol (China) 14 (1994) 147

[10]. Qiang Luo, Fengxia Ye, Changjun Huang, Jin Jiao, Anisur Rahman, Peng Yu, Jie Li, and Jun Shen, "Size-dependent structure and magnetocaloric properties of Fe-based glass-forming alloy powders", AIP Advances 6, 045002 (2016); doi: 10.1063/1.4945754.

IOP Conf. Series: Materials Science and Engineering **343** (2018) 012037 doi:10.1088/1757-899X/343/1/012037

ACKNOWLEDGEMENTS

The authors would like to express their gratitude to Mr. Drs. Gunawan, MSc., for his valuable supports. The research was funded by DIPA 2015 Nuclear Reactor Materials Group PSTBM BATAN. The author would like to thank Mr. Drs. Bambang Sugeng to capture the spectrum of XRD, Mr. Agus Sujatno for image retrieval SEM / EDX, and Mr. Rohmad Salam for helping in sintering process.

Institute of Physics
Dirac House, Temple Back
Bristol BS1 6BE UK

ISSN: 1757-8981
ISBN 978-1-5108-6119-0

Final LIMTECH Colloquium and International Symposium on Liquid Metal Technologies 2017

IOP Conference Series: Materials Science and Engineering
Volume 228

Dresden, Germany
19 - 20 September 2017

IOP
Institute of Physics